Venomous Snakes

This is a volume in the series
Symposia of the Zoological Society of London

Number 70

From a symposium held at the Zoological Society of London on 27 and 28 April 1995

Venomous Snakes

Ecology, Evolution and Snakebite

Edited by R. S. THORPE

W. WÜSTER

and

ANITA MALHOTRA

School of Biological Sciences
University of Wales
Bangor

Published for THE ZOOLOGICAL SOCIETY OF LONDON

by CLARENDON PRESS · OXFORD

1997

Oxford University Press, Walton Street, Oxford OX2 6DP

Oxford New York
Athens Auckland Bangkok Bombay
Calcutta Cape Town Dar es Salaam Delhi
Florence Hong Kong Istanbul Karachi
Kuala Lumpur Madras Madrid Melbourne
Mexico City Nairobi Paris Singapore
Taipei Tokyo Toronto
and associated companies in
Berlin Ibadan

Oxford is a trade mark of Oxford University Press

Published in the United States
by Oxford University Press Inc., New York

A catalogue record for this book is available from the British Library

Library of Congress Cataloging in Publication Data
Venomous Snakes : ecology, evolution, and snakebite / edited by R.S. Thorpe, W. Wüster, and Anita Malhotra. – 1st ed.
(Symposium of the Zoological Society of London ; no.70)
"From a symposium held at the Zoological Society of London on 27 and 28 April 1995" – Ser. t.p.
Includes bibliographical references and index.
1. Poisonous snakes–Congresses. 2. Poisonous snakes–Venom–Congresses.
3. Snakebites–Congresses. I. Thorpe, R.S. (Roger S.)
II. Wüster, W. (Wolfgang) III. Malhotra, Anita.
IV. Zoological Society of London. V. Series.
QL1.Z733 no. 70
[QL666.06] 591 s–dc20 [597.96'0469] 96-20164

ISBN 0 19 854986 5

Typeset by Hewer Text Composition Services, Edinburgh
Printed in Great Britain by Bookcraft (Bath) Ltd
Midsomer Norton, Avon

To Garth Underwood
in recognition of his contribution to herpetology

Preface

For millennia, venomous snakes have aroused unique levels of fascination in humans. Today, venomous snakes are of great scientific interest to a uniquely broad range of researchers in biology: they are studied specifically to elucidate their taxonomy, ecology and evolutionary relationships, they have become popular as model organisms in behavioural and ecological studies, their venoms have contributed to important developments in physiology and medicine, and they are of course of considerable applied interest due to the public health problem of snakebite, while at the same time being of concern to conservationists.

At the same time, owing to the increasing specialization of scientists in general, much of the research into venomous snakes and their venoms has proceeded within the narrow confines of conventional disciplines. In particular, researchers in the toxinological/medical field have traditionally had little contact with evolutionary biologists and ecologists working with venomous snakes. This is likely to be to the detriment of both fields. It is no great secret that developments in our understanding of the systematics of many groups of venomous snakes have been slow to find their way into the toxinological/medical literature, and that this has caused considerable problems in its interpretation. At the same time, a greater understanding of the mode of action of venoms and toxins must be of interest to those attempting to elucidate the ecological and evolutionary function of venoms in snakes.

It is our belief that both the medical/toxinological as well as the ecological/evolutionary side of venomous snake research could benefit enormously from increased interaction between them. It is this belief that prompted us to organize the meeting 'Venomous Snakes: Ecology, Evolution and Snakebite' in London in April 1995. Our intention was to bring together leading researchers on all aspects of venomous snake biology to discuss a wide range of aspects of the biology of these organisms, and to forge new and stronger links between them. The range and quality of presentations at the meeting, the resulting interactions, and feedback from participants of all fields suggest that the meeting went some way towards achieving these aims.

The papers assembled in this volume do indeed span much of the spectrum of research carried out on venomous snakes. They include purely ecological papers such as those of Forsman and of Andrén *et al.* Systematics was a major focus of the meeting. The papers presented address a broad range of issues in venomous snake biology, including a general overview of venomous snake evolution (Underwood), analyses of higher taxonomic groups such as the sea snakes (Rasmussen), Old World vipers (Herrmann & Joger, Nilson & Andrén), New World pitvipers (Werman), the *Agkistrodon* complex (Parkinson *et al.*), and the genus *Bothrops* (Salomão *et al.*).

Two papers (Malhotra & Thorpe, Wüster *et al.*) discuss the systematics of complex species groups of considerable medical importance, the Asian green *Trimeresurus* and the South American *Bothrops atrox* complex. The tremendous impact of molecular methods in evolutionary research is reflected by the fact that six of the nine primarily systematic papers use various molecular methods for phylogeny reconstruction. It is particularly gratifying to see that a number of papers span the gap between the biomedical sciences and the evolutionary/ecological sciences. Daltry *et al.* and Williams & White discuss the evolutionary significance of geographic variation in venom composition, while Moura-da-Silva investigates the evolution of a group of venom toxins. Warrell discusses the interrelationships of taxonomy, geographic variation and the treatment of envenoming, Belt *et al.* provide a more detailed discussion of a particularly noteworthy example, Russell's viper, while Theakston and Williams & White review the treatment of envenoming. Finally Harris provides a timely review of the phospholipases A_2, a group of enzymes of great importance in snake venoms.

We hope that the papers contained in this volume will encourage a broad range of researchers on all aspects of venomous snakes and their venoms to interact to a greater extent, and to take steps towards a greater integration of their disciplines.

Finally we would like to express our warmest appreciation and gratitude to the Zoological Society of London, and in particular Unity McDonnell. Without their support, the meeting giving rise to this volume would not have taken place. Their professionalism and dedication ensured not only a very stimulating and pleasurable meeting for all participants, but also that its organization, and the ensuing editorial process for this volume, were a pleasure rather than a chore.

Bangor
29 November 1995

R. S. T.
W. W.
A. M.

Contents

11 Growth and survival of *Vipera berus* in a variable environment

ANDERS FORSMAN

12 The role of ecology in determining venom variation in the Malayan pitviper, *Calloselasma rhodostoma*

JENNIFER C. DALTRY, WOLFGANG WÜSTER & ROGER S. THORPE

13 Molecular evolution of phospholipase A_2s and metalloproteinase/disintegrins from venoms of vipers

ANA M. MOURA-DA-SILVA, R. D. G. THEAKSTON & JULIAN M. CRAMPTON

14 Geographical and intraspecies variation in the clinical manifestations of envenoming by snakes

DAVID A. WARRELL

15 Snake venom and snakebite in Australia

V. WILLIAMS & J. WHITE

16 Russell's viper in Indonesia: snakebite and systematics

PAUL J. BELT, ANITA MALHOTRA, ROGER S. THORPE, DAVID A. WARRELL & WOLFGANG WÜSTER

17 Toxic phospholipases in snake venom: an introductory review

JOHN B. HARRIS

Contributors

AHLQUIST, J. E., National Marine Fisheries Service, Southeast Fisheries Science Center, Charleston, SC 29422–0607, USA; *and* Graduate Program in Marine Biology, University of Charleston, SC; Department of Biochemistry & Molecular Biology, Marine Biomedical & Environmental Sciences Program, Medical University of South Carolina, Charleston, SC, USA.

ANDRÉN, C., Department of Zoology, Göteborg University, Medicinaregatan 18, S-413 90 Göteborg, Sweden.

BELT, P. J., Centre for Tropical Medicine, Nuffield Department of Clinical Medicine, University of Oxford, John Radcliffe Hospital, Headington, Oxford OX3 9DU, UK.

CRAMPTON, J. M., Wolfson Unit of Molecular Genetics, Liverpool School of Tropical Medicine, Pembroke Place, Liverpool L3 5QA, UK.

DALTRY, J. C., School of Biological Sciences, University of Wales, Bangor, Gwynedd LL57 2UW, UK.

FORSMAN, A., Department of Zoology, Uppsala University, Villavägen 9, S-752 36 Uppsala, Sweden.

FURTADO, M. F. D., Laboratório de Herpetologia, Instituto Butantan, Avenida Vital Brasil 1500, 05503–900 São Paulo-SP, Brazil.

HARRIS, J. B., School of Neurosciences, University of Newcastle-upon-Tyne, Newcastle-upon-Tyne NE2 4HH, UK; *and* Muscular Dystrophy Group Laboratories, Regional Neurosciences Centre, Newcastle General Hospital, Newcastle-upon-Tyne NE4 6BE, UK.

HERRMANN, H.-W., Kölner Aquarium am Zoo, Postfach 680369, 50706 Köln, Germany.

HOGE, S. A., Laboratório de Herpetologia, Instituto Butantan, Avenida Vital Brasil 1500, 05503–900 São Paulo-SP, Brazil.

HÖGGREN, M., Department of Genetics, Uppsala University, Box 7003, S-750 07 Uppsala, Sweden.

JOGER, U., Hessisches Landesmuseum, Friedensplatz 1, 64283 Darmstadt, Germany.

MALHOTRA, A., School of Biological Sciences, University of Wales, Bangor, Gwynedd LL57 2UW, Wales, UK.

MOODY, S. M., Department of Biological Sciences, Ohio University, Athens, OH 45701, USA.

MOURA-DA-SILVA, A. M., Wolfson Unit of Molecular Genetics, and Venom Research Unit, Liverpool School of Tropical Medicine, Pembroke Place, Liverpool L3 5QA, UK; *and* Laboratorío de Imunopatologia, Instituto Butantan, Avenida Vital Brasil 1500, São Paulo, Brazil.

NILSON, G., Department of Zoology, Göteborg University, Medicinaregatan 18, S-413 90 Göteborg, Sweden.

PARKINSON, C. L., Department of Biology, University of Louisville, Louisville, KY 40292, USA.

PUORTO, G., Laboratório de Herpetologia, Instituto Butantan, Avenida Vital Brasil 1500, 05503–900 São Paulo-SP, Brazil.

RASMUSSEN, A. R., Zoological Museum, University of Copenhagen, Universitetsparken 15, DK-2100 Ø, Denmark.

SALOMÃO, M. G., School of Biological Sciences, University of Wales, Bangor, Gwynedd LL57 2UW, UK; *and* Laboratório de Herpetologia, Instituto Butantan, Avenida Vital Brasil 1500, 05503–900 São Paulo-SP, Brazil.

TEGELSTRÖM, H., Department of Genetics, Uppsala University, Box 7003, S-750 07 Uppsala, Sweden.

THEAKSTON, R. D. G., Venom Research Unit, Liverpool School of Tropical Medicine, Pembroke Place, Liverpool L3 5QA, UK.

THORPE, R. S., School of Biological Sciences, University of Wales, Bangor, Gwynedd LL57 2UW, Wales, UK.

TOUZET, J.-M., Fundacíon Herpetológica Gustavo Orcés, Casilla 17 03 448, Quito, Ecuador.

UNDERWOOD, G., Department of Zoology, The Natural History Museum, Cromwell Road, London SW7 5BD, UK.

WARRELL, D. A., Centre for Tropical Medicine, University of Oxford, Nuffield Department of Clinical Medicine, John Radcliffe Hospital, Headington, Oxford OX3 9DU, UK.

WERMAN, S. D., Department of Biology, Mesa State College, Grand Junction, CO 81502, USA.

WHITE, J., Haematology Department, Adelaide Children's Hospital, North Adelaide, SA 5006, Australia.

WILLIAMS, V., Haematology Department, Adelaide Children's Hospital, North Adelaide, SA 5006, Australia.

WÜSTER, W., School of Biological Sciences, University of Wales, Bangor, Gwynedd LL57 2UW, UK.

Organizers of symposium

PROFESSOR R. S. THORPE, School of Biological Sciences, University of Wales, Bangor, Gwynedd LL57 2UW, UK.

DR ANITA MALHOTRA, School of Biological Sciences, University of Wales, Bangor, Gwynedd LL57 2UW, UK.

DR W. WÜSTER, School of Biological Sciences, University of Wales, Bangor, Gwynedd LL57 2UW, UK.

Chairmen of sessions

PROFESSOR H. GREENE, Museum of Vertebrate Zoology, 3101 Valley Life Sciences Building, Berkeley, California 94720, USA.

PROFESSOR R. S. THORPE, School of Biological Sciences, University of Wales, Bangor, Gwynedd LL57 2UW, UK.

DR G. UNDERWOOD, Department of Zoology, The Natural History Museum, Cromwell Road, London SW7 5BD, UK.

PROFESSOR D. A. WARRELL, Centre for Tropical Medicine, University of Oxford, Nuffield Department of Clinical Medicine, John Radcliffe Hospital, Headington, Oxford OX3 9DU, UK.

DR V. WILLIAMS, Haematology Department, Adelaide Children's Hospital, North Adelaide, SA 5006, Australia.

DR W. WÜSTER, School of Biological Sciences, University of Wales, Bangor, Gwynedd LL57 2UW, UK.

1 An overview of venomous snake evolution

GARTH UNDERWOOD

Synopsis

Scolecophidians have serous as well as mucous supralabial glands, but do not otherwise foreshadow later developments.

Remaining snakes are primarily of secretive or burrowing nocturnal habits with a simple duplex retina (with rods and small single cones), mucous labial glands and serous superior and inferior rictal glands, opening into the corner of the mouth. They have limited mobility of the jaws and use constriction to subdue large prey of small diameter. Phisalix reports superior rictal glands to be toxic; perhaps they have a defensive function. Amongst erycines, boines, pythonines, tropidophines and bolyerines the rictal glands regress; they are able to swallow bulky prey, subdued by constriction.

Like *Anilius*, early caenophidians retained superior and inferior rictal glands and similar habits. The foundation of the caenophidian radiation was the acquisition of dental glands and the development of mobile jaws facilitating the swallowing of large prey subdued by the venom. Elapomorphines and atractaspids are at this grade. Phospholipases appear to be universally present, other venom components are variously secreted by different lineages. Teeth and muscles were variously modified for the injection of the venom.

With large single cones, boigines precede the forms at the next level, which also have double cones in the retina. Crotalines retain primitive nocturnality in association with the pit organ. Viperines and other caenophidians are of less secretive and more diurnal habits. Above the level of the elapids the inferior rictal gland appears to be lost. Within the 'colubroid' radiation arose the colubrines *Dispholidus* and *Thelotornis* and the natricine *Rhabdophis*. Some colubroid lineages were able to overcome prey simply by swallowing, some revived constriction; these lineages show regression of the venom apparatus. With the appearance of the thanatophidian lineages some animals acquired immunity and some became mimics.

Introduction

Sphenodon has many small tubular glands across the roof of the mouth. The Iguania have well-developed superior and inferior labial and dental glands. Few other lizards have superior glands. The Anguimorpha have particularly well-developed infralabial

Symp. zool. Soc. Lond. (1997) No. 70: 1–13
ISBN 0 19 854986 5

glands; the varanoids also have a mandibular gland, including the venom gland of *Heloderma* (Kochva 1978).

The Scolecophidia have mucous infralabial glands and serous as well as mucous supralabial glands (Gabe & Saint Girons 1969) but we have no evidence of toxicity at this level.

The remaining snakes, the Alethinophidia, include the crown group, the Caenophidia, and a number of lower-grade lineages to which I refer as henophidians. Cundall, Wallach & Rossman (1993: fig.2) published an analysis of these lineages based on 65 characters. They have, in sequence, Scolecophidia, *Anomochilus*, uropeltines, *Cylindrophis*, *Anilius*, *Xenopeltis*, *Loxocemus* and a trichotomy of pythonines, boines plus erycines and then, in sequence, bolyerines, tropidophines, *Acrochordus* and 'Colubroidea'.

Jaws and dentition

Anomochilus, *Cylindrophis*, uropeltines and *Xenopeltis* have processes of the palatine bones which descend between the posterior processes of the vomers (Haas 1931; Bellairs 1949; Cundall & Rossman 1993). They also have short quadrate bones and contact between the premaxilla and the maxilla. Comparison with lizards suggests that the latter is a primitive condition; its retention may be functionally related to burrowing. The mobility of the upper jaws is evidently limited. *Anilius* differs in that the choanal processes of the palatines descend behind the vomers, with a ligamentous link, as in all other snakes. These snakes are able to consume prey of small diameter but of relatively great weight (Greene 1983). Cundall (1987) suggests that the hinge in the lower jaw is functionally important in the swallowing of large prey. Gans (1961) reports that *Anilius* shows inertial feeding, like *Varanus*.

Snakes above this level show much greater mobility of the upper jaws. Anthony (1955) points out that in pythonines and boines the maxilla is relatively long both before and behind the prefrontal, as compared with most caenophidians. Similarly the squamosal is proportionately longer in relation to the quadrate and the dentary is longer in relation to the mandible than in most caenophidians. With mobility of the maxilla there may be hinges between it and the prefrontal and ectopterygoid. Nearly all other snakes have sufficient mobility of the jaws to permit some rotation of the maxilla (Kardong 1982).

In most snakes the palatine bone has a mesial choanal process, linked to the vomer, and a lateral process, linked to the maxilla. In *Atractaspis* there is a gap between the pterygoid and a palatine which retains its usual processes (Underwood & Kochva 1993). This permits the pterygoid to move forwards with the ectopterygoid when the fang is erected, without disturbing the palatine bone. In vipers the palatine, without processes, is reduced to a forward extension of the pterygoid, again permitting free forward movement.

Cundall (1983) finds that in several caenophidians the palatine and pterygoid play the principal role in the swallowing of prey. He suggests that the maxillary teeth function mainly in the capture of prey and that the mandibles function mainly to hold the prey against the palato-pterygoid teeth.

Anthony (1955) recognizes a variety of conditions of the maxillary teeth of most snakes, apart from elapids and vipers, without any apparent relationship with the state of the venom apparatus. Where the posterior teeth are modified there is usually a preceding diastema. This presumably allows room for the palatal sheath which surrounds the fangs.

Jackson & Fritts (in press) find, by scanning EM, a difference between anterior maxillary teeth and modified posterior teeth. The anterior teeth each bear a labial and a lingual ridge. By contrast the posterior teeth have anterior and posterior ridges, whether they be solid, grooved or canaliculate; they appear to have rotated by 90°. This difference does not hold where the anterior and posterior teeth are not otherwise differentiated. They find that the anterior fangs of elapids have the morphology of the posterior teeth of opisthodont snakes. Grooves are usually confined to posterior fangs. However, in elapids, teeth behind the fangs, where they occur, are usually grooved. In *Ophiophagus hannah*, for example, there are also grooves on palatine and dentary teeth. Some opisthoglyphs such as species of *Ahaetulla* may have grooves on anterior maxillary teeth and on the dentary teeth. Canaliculate teeth are found only in *Atractaspis*, vipers and elapids. Recent scanning EM studies have shown that some fangs, for example *Azemiops*, have cutting edges (Mebs, Kuch & Meier 1994).

Glands

Kardong (1982) makes a useful distinction between a toxic secretion, demonstrated by its effect on test animals, and a venom, judged by its role in subduing prey. He suggests some possible roles for those toxins which do not appear to function as venoms in the life of the snake.

Saint Girons (1987) distinguishes between labial glands, with multiple ducts opening along the margins of the lips, and dental glands which open into a pocket around the base of a tooth. Taub (1967) introduced the term Duvernoy's gland, to replace 'parotid gland', for dental glands other than the venom glands of vipers and elapids. This makes a distinction between gland types which is being eroded by the observation that all Duvernoy's glands that have been investigated secrete toxins, some of which have been shown to function as venoms. I prefer to call all of them dental glands.

Gabe & Saint Girons (1969) and Kochva (1978) distinguish between mucous and two or three types of serous cells. Kochva points out that there is still uncertainty about the relationship between the cell types recognized by histologists and their secretions. It is, however, fairly clear that mucous cells secrete mucins and that all venom glands contain serous cells of some type. I distinguish only between mucous and serous.

Phisalix (1922) described the 'glande temporale antérieure'. As the position of this gland is not always 'anterior temporal' I prefer McDowell's (1987) term rictal. Smith & Bellairs (1947) and Gabe & Saint Girons (1969) give an account of the microscopic anatomy and histology. These glands may rise mesial or lateral to the quadrato-maxillary ligament. They are serous, with some mucous cells in the ducts, which open into the corner of the mouth.

I report here some new observations, mostly made by dissection. It should be stressed that, as pointed out by Smith & Bellairs (1947), it may be necessary to cut sections to be certain that a small body is a gland. In particular, ducts are rarely seen in dissection and mucous and serous glands are uncertainly distinguishable.

In Alethinophidia the supralabial glands are mucous, with some serous cells in some forms (Taub 1967). Taub reports that the xenodermatines, *Xenodermus* and *Fimbrios*, have cords of serous cells alternating with cords of mucous cells 'along the entire supralabial region'; he does not comment on the relationship of the serous cords with the teeth.

The simple branched tubular dental glands are usually without a reservoir, with the notable exception of *Dispholidus*. They open by a single duct into the pocket surrounding the base of the fangs. We have, as yet, only a preliminary account of the venom glands of elapomorphines (Salomão & Ferrarezzi 1993). The venom gland of *Atractaspis* has radially disposed tubules, with little branching, around a central lumen; there are mucous cells towards the base of each tubule. Elapids have a venom gland with compound branched tubules which pass into an accessory gland with tubules of mucous cells. The venom gland of viperids has branched tubules opening into a large lumen. The lumen narrows to pass through the accessory gland, the anterior part of which has mucous cells (Kochva 1978).

Haas (1930) reported a lateral rictal gland in *Atractaspis corpulenta*. Wollberg, Kochva & Underwood (in prep.) confirmed this observation and discovered a hitherto unreported *inferior* rictal gland. Further examinations were therefore extended to the lower lips.

Anilius has a large lateral superior rictal gland, with a large duct opening into the corner of the mouth, and a small duct draining the posterior part of the gland. There is also a large serous inferior rictal gland, with a duct to the corner of the mouth (Fig. 1).

Superior rictal glands are found in *Anomochilus*, uropeltines, *Cylindrophis*, *Calabaria* and several species of *Eryx* (Phisalix 1922; Smith & Bellairs 1947; Gabe & Saint Girons 1969; Koçhva 1978; Cundall & Rossman 1993), followed in the latter two genera by a pocket. Tropidophines have a rictal gland on the anterior and dorsal wall of a shallow horizontal rictal pocket. *Xenopeltis* has a deep rictal pocket with a thick wall. Pythonines, boines, *Ungaliophis* and *Exiliboa* have a horizontal rictal pocket with a thin wall (Fig. 1). All of these rictal structures lie mesial to the quadrato-maxillary ligament. *Loxocemus*, *Bolyeria* and *Casarea* have a simple rictal fold. Of these only *Cylindrophis* is known so far to have an inferior rictal gland.

Acrochordus has no more than a simple rictal groove. *Pareas stanleyi* is reported by Taub (1967), on the basis of sections, to have a 'Duvernoy's' gland. By dissection of several species (including *stanleyi*) I find a strip of gland dorsal to the long rictal groove without any extension forward to the level of the posterior maxillary teeth and suggest that it is a rictal gland. I find no quadrato-maxillary ligament. In xenodermatines the superior rictal gland is relatively large, in a dorsal temporal and mesial position (Fig. 1); *Achalinus* also has a narrow inferior rictal gland.

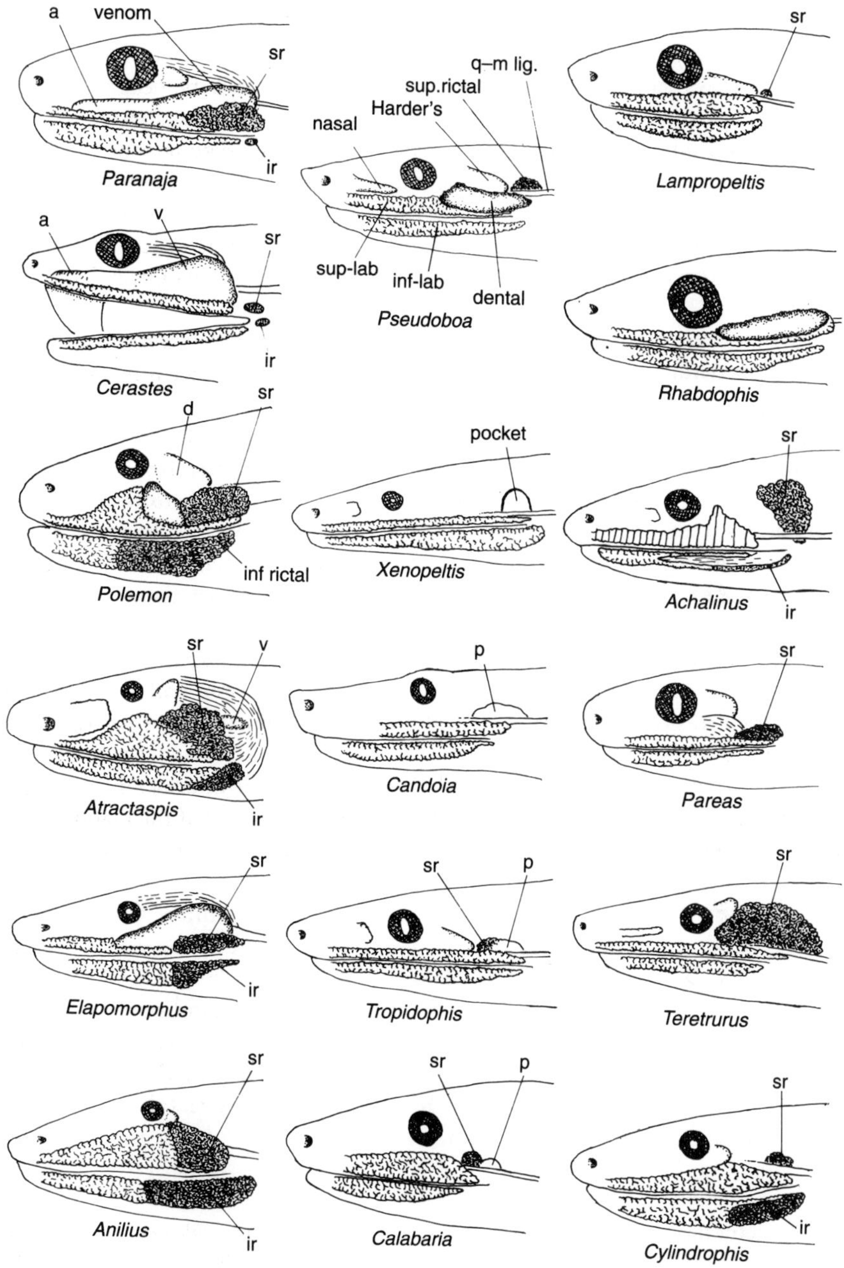

Fig. 1. Semidiagrammatic sketches showing: accessory (a), dental (d), venom (v), Harder's, nasal, superior and inferior rictal (sr & ir) and supra- and infralabial (sup-lab & inf-lab) glands; quadrato-maxillary ligament (q-m lig) and rictal pocket (p) of: *Achalinus rufescens* (xenodermatine, supralabial: alternate serous & mucous cords in *Fimbrios*), *Anilius scytale*, *Atractaspis corpulenta*, *Calabaria reinhardti* (erycine), *Candoia asper* (boine), *Cerastes*, *Cylindrophis rufus*, *Elapomorphus nasutus*, *Lampropeltis getulus*, *Paranaja multifasciata*, *Pareas monticola*, *Polemon bocourti*, *Pseudoboa neuwiedii*, *Rhabdophis tigrinus*, *Teretrurus rhodogaster* (uropeltine), *Tropidophis haetianus*, and *Xenopeltis unicolor*.

Elapomorphus and *Atractaspis* have superficial superior and inferior rictal glands; in the latter the superior gland has a large duct and two small ducts draining the posterior lobes, rather like *Anilius*. *Aparallactus*, *Polemon* and *Chilorhinophis* are similar but the compact dental and superior rictal glands follow the supralabial gland in sequence with little overlap (Underwood & Kochva 1993; Wollberg *et al.* in prep.). A small, superficial, superior rictal gland has been reported for some *Vipera* and *Crotalus* (Kochva 1978). Saint Girons & Ineich (1993) describe a unique condition in *Echis*: the gland opens to the exterior within the border of the last supralabial scale.

By dissection I find small, lateral superior and, sometimes, inferior rictal glands in some *Atheris*, *Bitis*, *Cerastes* and *Causus*. *Paranaja* has a lateral superior rictal and a small inferior rictal gland (Fig. 1). Superior rictal glands are found in *Aspidelaps* and *Elapsoidea*.

Martin (1899) describes the development of the venom apparatus of *Vipera aspis*. A longitudinal ridge of ectoderm in the roof of the mouth is divided into a lateral glandular ridge and a mesial maxillary dental lamina. At a head width of 5 mm the dental lamina bears a series of 12 dental rudiments. Towards the posterior end the glandular ridge bears a large bud, precursor of the venom gland. Anteriorly the glandular ridge bears six buds, each of which nearly corresponds with one of the dental buds. Martin calls these 'bourgeons venimeux accessoires'. The venom gland develops in association with a posterior dental rudiment. In Martin's figure there is one dental bud behind the fang precursor (Fig. 2). This recalls Rasmussen's (1979) report that some species of *Crotaphopeltis* and *Dipsadoboa* have, behind the fangs, a single small tooth, which may be solid or grooved. At the next stage all accessory glandular and dental buds disappear. Martin interprets this as suggesting an ancestral condition in which a series of venom glands was associated with a series of maxillary teeth.

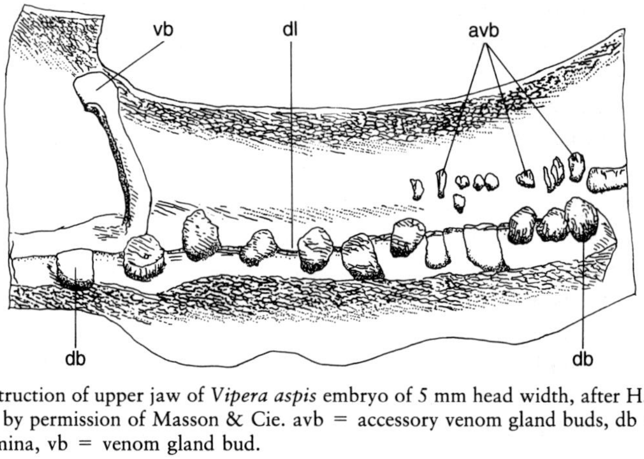

Fig. 2. Reconstruction of upper jaw of *Vipera aspis* embryo of 5 mm head width, after H. Martin; from Phisalix (1922), by permission of Masson & Cie. avb = accessory venom gland buds, db = dental bud, dl = dental lamina, vb = venom gland bud.

The development of the dental glands of *Telescopus fallax* and *Vipera palaestinae* is similar (Kochva 1963). Kochva (1978) also notes that a similar pattern has been observed in the proteroglyphous *Walterinnesia aegyptia* and *Naja naja*. Gygax (1971) describes the development of *Natrix tessellata*. The dental gland appears as a lateral bud at the posterior end of the dental lamina, some 20 days before the appearance of the supralabial gland primordia. Lateral to the posterior end of the inferior dental lamina there is, at the same stage of development, a small bud which looks like the rudiment of an inferior dental gland. Holding the view that the isodont condition is primitive he wonders whether it has phylogenetic implications.

The adductor externus superficialis, adductor externus profundus and pterygoideus muscles and a compressor glandulae, of uncertain homology, variously insert on the capsule of the dental gland in a few opisthoglyphous snakes such as *Dispholidus*, *Brachyophis* and *Elapomorphus* and in *Atractaspis*, viperids and elapids (Rosenberg 1967; Kochva 1978).

In 'higher' Caenophidia the superior rictal gland is mesial, as reported by Smith & Bellairs (1947), Gabe & Saint Girons (1969) and McDowell (1987). Absence of the gland is presumably secondary.

Other characters

Can we place the thanatophidian lineages in an evolutionary sequence? I briefly review some characters which give indications, albeit sometimes uncertain. Henophidians serve as outgroup. Outside the thanatophidian lineages there are some more or less clearly discernible groups; I make particular reference only to the South American xenodontines, *sensu* Cadle (1984) and, amongst the 'highest' caenophidians, the nearly cosmopolitan colubrines and natricines. The many genera outside these groups undoubtedly include many forms of toxicological interest, as yet unrecognized or little investigated; Weinstein & Kardong (1994) review some of these.

Henophidians have small, type C, cones and type D rods in the retina. The outer, visual cell nuclei outnumber the inner nuclei, of bipolar, amacrine and horizontal cells. This suggests a duplex retina with a high level of summation in association with nocturnal habits. *Pareas* and atractaspids show this pattern. Boigines represent the next level with large single cones, type A, in addition to C and D cells (Rasmussen 1985).

The remaining investigated caenophidians have the peculiarly ophidian double cones, type B. Some which retain rods still have more outer than inner nuclei. These include the nocturnal crotalines of which *Crotalus* is known to have a typical vertebrate rhodopsin (Crescitelli 1958) and *Agkistrodon* to have typical rods with few synapses (Hirosawa 1978). In this respect these crotalines resemble birds and mammals which have oligosynaptic rods, with rhodopsin, and polysynaptic cones. This appears to be an inheritance from duplex amniote ancestors. No known lizard has a duplex retina.

Diurnal caenophidians have a high ratio of inner to outer nuclei (inner nuclear quotient: Rasmussen 1990) suggesting that the rods, where present, have acquired cone-type polysynaptic footpieces; such caenophidians include viperines and elapids, colubrines and natricines. It looks like a high-resolution retina. It is remarkable that crotalines retain what appears to be a primitive nocturnal condition, in association with the development of the pit organ. By contrast the viperines have undergone a nocturnal–diurnal transformation in parallel with other lineages. It is also remarkable that secretive and nocturnal elapids have what looks like a 'diurnal' retina (Underwood 1970).

Table 1. Tabulation of some of the evidence bearing on the phylogeny of snakes of toxicological interest. Where there is variation in a taxonomic unit the inferred primitive condition is given

	Character nos														
	1	2	3	4	5	6	7	8	9	10	11	12	13	14	15
Lampropeltis	1	-	-	m	-	-	-	-	-	s	sl	-CBA	>	+	-
Dispholidus	1	+	-	-	-	-	+	-	-	s	sl	?	?	+	-
Thelotornis	1	+	-	-	-	-	+	-	-	s	sl	?	?	+	-
Colubrinae	1	+	-/+	m/-	-	-	+	-	-	b	sl	DCBA	>	+	-
Rhabdophis	1	+	+	-	-	-	+	-	-	b	s2	DCBA	>	+	-
Natricinae	1	+	+	m/-	-	-	+	-	-	b	s2	DCBA	>	+	-
Philodryas	1	+	-	m	-	-	+	-	-	b	f	(2)	(2)	+	-
Paranaja	1	+	+	1	-	+	+	-	-	b	f	?	?	-	+
Homoroselaps	0	+	-	-	-	-	+	-	-	s	s	-CBA	>	-	+
Viperinae	1	+	+	1	-	+	+	(1)	-	b	f	DCBA	>	+	-
Causus	1	+	+	1	-	+	+	-	-	b	f	DCBA	>	+	-
Crotalinae	1	+	+	1	-	+	+	-	-	b	f	DCBA	<	+	-
Crotaphopeltis	0	+	-	-	-	+?	+	-	-	s	s	DCB-	<	+	-
Pseudoboa	0	+	-	m	-	-	+	-	-	b	f	DC--	<	+	+
Atractaspis	0	+	-	1	-	+	+	-	-	b	f	DC--	<	-	+
Elapomorphus	0	+	-	1	-	+	+	-	+	b	f	?	?	-	+
Xenodermatinae	0	-	-	m	-	+	?	-	-	b	f	?	?	-	-
Pareas	0	-	-	1	-	-	-	-	+	b	f	DC--	<	-	-
Acrochordus	0	-	+	-	-	-	-	+	+	b	f	DC--	<	-	-
Bolyerines	0	-	+	-	-	-	-	+	+	b	f	?	?	-	-
Tropidophines	0	-	+	m	+	-	-	+	+	b	f	DC--	<	-	-
Boines	0	-	-	-	+	-	-	+	+	b	f	DC--	<	-	+
Erycines	0	-	-	m	+	-	-	+	+	b	f	DC--	<	-	+
Xenopeltis	0	-	-	-	+	-	-	+	+	b	f	DC--	<	-	+
Anilius	0	-	-	1	-	+	-	+	+	b	f	?	?	-	+
Cylindrophis	0	-	-	m	-	+	-	+	+	b	f	?	?	-	+
Scolecophidia	0	-	-	-	-	-	-	+	+	s	s	D---	<	-	+

(1) Small right common carotid in some
(2) *Philodryas* not known but other diurnal xenodontines DCBA.

Character numbers:

1. Palatine nerve emerges on sphenoid-parietal suture or on underside of sphenoid (0/1)
2. Posterior maxillary teeth simple/modified (-/+).
3. Posterior hypapophyses: absent/present (-/+)
4. Superior rictal gland: lateral/mesial/absent (1/m/-)
5. Rictal pocket: absent/present & mesial (-/+)
6. Inferior rictal gland: present/absent (+/-)
7. Dental gland: absent/present (-/+)
8. Right common carotid artery: present/absent (+/-)
9. Rectal caecum: present/absent (+/-)
10. Hemipenis: bilobed/simple (b/s)
11. Hemipenis: sulcus spermaticus: forked or simple (f/s); s1 & s2 differently asymmetrical
12. Retina: visual cell types (A, B, C, D)
13. Retina, inner nuclei exceed outer nuclei: yes/no (>/<)
14. Scale-pits on trunk: absent/present (-/+)
15. Tubercles on trunk scales: present/absent (+/-)

In most henophidian lineages the palatine nerve enters a short Vidian canal to emerge into the braincase and leave by a foramen in the sphenoid–parietal suture;

this also holds for *Pareas*, xenodermatines, elapomorphines and atractaspids. In a more derived condition the Vidian canal is fully roofed inside the braincase, but this is not externally distinguishable from the previous condition. Further derived is a fully roofed Vidian canal which opens anteriorly on the underside of the sphenoid bone; this is seen in most other snakes, including viperids, elapids, colubrines and natricines (Rieppel 1978).

Most lizards and henophidians have a rectal caecum. The only caenophidians known to retain it are *Pareas*, elapomorphines and the uncertain atractaspids, *Micrelaps* and *Brachyophis*.

The hemipenes of most henophidians are bilobed with a forked sulcus spermaticus. These conditions are assumed to be primitive for caenophidians. The sulcus is simple in colubrines and differently simple in most natricines (Dowling & Savage 1960).

Amongst henophidians, tubercles are generally present on the scales of the head and trunk, pits are known only in a few pythons. Presence of pits and absence of tubercles on the trunk are assumed to be derived in caenophidians. Epigeal henophidians show the scale-rows falling to a minimum on the neck, rising to a maximum towards midbody and falling on the posterior trunk. Amongst burrowing henophidians with a cylindrical body, *Anilius*, *Cylindrophis* and some uropeltids show a similar pattern. This is also found in cylindrical atractaspids.

Discussion

The study of the clinical, pharmacological, physiological and biochemical aspects of snake venom is highly developed. The unravelling of the evolutionary history of snake venom still has far to go. The following scenario is based in part on a number of unconfirmed observations; it is hoped that these will be tested. In that the Scolecophidia eat relatively small prey (Greene 1983) they do not foreshadow developments in other snakes. They do, however, have serous, as well as mucous, supralabial glands, of unknown function.

Phisalix & Caius (1918) injected aqueous extracts of rictal glands into birds. The snakes were *Eryx johni* and seven species of uropeltid (four *Uropeltis*, two *Platyplectrurus* and one *Teretrurus*). The effects variously included: irregular breathing, bronchial rattle, narcosis, alternating periods of stupefaction and alertness, muscular weakness, convulsions, local haemorrhage, pulmonary congestion and lachrymal and nasal hypersecretion. They note that lizards and frogs are less susceptible to the secretion of *Teretrurus madurensis* than birds are. It is astonishing that, in 75 years, these very interesting observations do not appear to have been followed up by later investigators. They cannot too easily be dismissed as reaction to foreign protein. Phisalix & Caius state explicitly, in the case of the 'parotid' gland, 'La sécrétion, telle qu'elle sort de la glande, possède exactement la même action toxique que les macérations de celle-ci'.

As the gland has no association with the teeth it seems unlikely that it plays a role in subduing prey. This leads to the speculative hypothesis that perhaps the original function of toxins in snakes was defensive. Prey were subdued by constriction

(Greene & Burghardt 1978). In some henophidians the gland has given way to a pocket, in others it is lost. Some, possessing the gland, such as *Anilius*, *Cylindrophis* and some uropeltids, have striking coloration which could serve a warning function.

Many caenophidians retain a superior rictal gland; the morphological similarity of the rictal glands of *Anilius* and *Atractaspis* is close. Early caenophidians acquired a venom-secreting dental gland. In henophidians and in those caenophidians without a dental gland the posterior maxillary teeth are not enlarged; this suggests that they are primarily an aid to penetration of dental gland secretion. Did the first dental gland evolve by the activation of the rictal gland gene complex in a new situation? In many caenophidians the dental gland secretion appears to have a digestive function as well as serving to subdue prey. With the capacity to subdue larger prey went enhancement of the capacity to swallow them.

Table 1 summarizes some data which suggest a sequence of early thanatophidian lineages. The secretive nocturnal or burrowing life style of the South American elapomorphines and African atractaspids, and CD-pattern retina of the latter, may be retained from early snakes (Walls 1942; Underwood 1970). This retinal pattern is also shown by *Pareas* and the nocturnal opisthoglyphous *Pseudoboa*. The nocturnal boigines, *Boiga*, *Crotaphopeltis* and *Telescopus semiannulatus*, have C and D plus type A visual cells (Rasmussen 1985). Viperids, elapids, some xenodontines and other caenophidians also have type B double cells in the retina. This ABCD visual cell pattern is associated with primarily diurnal habits; the high proportion of inner nuclei suggests high visual resolution.

Cadle (1984), on immunological evidence, places *Apostolepis* (an elapomorphine) and *Pseudoboa* in the xenodontines. Taken at face value and placed in the present context this implies that the South American xenodontines represent the stem lineage of Caenophidia above the level of the xenodermatines, with the African atractaspids and Afro-Asian boigines as derivative branches! Cadle (1994) found *Atractaspis* to be immunologically distant from other 'atractaspids'.

Some snake venom components evidently originated by the activation in the dental gland of genes which code for proteins which have a function elsewhere in the body. Phospholipase A_2 is so widely distributed that it was presumably a component of 'ancestral' venom. It should be possible to compile a list of these primary gene products with the hope of discovering whether a sequence of 'adoption' is recognizable. Such a sequence could be obscured by gene losses. Kochva (1978) gives an account of the problems of distinguishing the protein families in venom. Some lineages are characterized by unique gene products, such as sarafotoxin only in the *Atractaspis* lineage and deriving from an endothelin progenitor. Reduplication of such 'adopted' genes complicates the use of these data for phylogenetic analysis by introducing the problem of orthologous and paralogous sequences. It should not, however, obscure a sequence of gene families, if there be one.

Conversion to a high-resolution retina of 'diurnal' type appears to be completed in the elapids. Nevertheless they have tubercles but lack scale-pits on head and trunk, features usually associated with secretive or burrowing habits.

Most of the unanalysed 'colubroids' are primarily diurnal. Some of them have lost rods from the retina and superficially resemble lizards in having three cone types.

Walls (1942: *Hypsiglena*) documents a return to nocturnality by development of low-threshold, rod-like cones, reminiscent of geckos. In various lineages toxic secretions of the dental gland were evidently maintained. A venomous function has been acquired in the colubrines *Dispholidus* and *Thelotornis* and the natricine *Rhadophis*.

The early acquisition of venom appears to have laid the foundation of the radiation of Caenophidia. The evolution of dangerously venomous snakes affected other members of the contemporary fauna. For example, Perales *et al.* (1994) found in the blood of each of three species of South American marsupials an anti-bothropic complex which they partially sequenced. These complexes had no recognizable homology with any known protein. This suggests that this adaptive feature is of some antiquity. Cerdas & Lomonte (1982) report that *Clelia clelia*, which preys on pitvipers, is unaffected by the venom of several species of pitvipers, coral snakes and even cobras.

Venomous snakes variously developed warning colour patterns and warning behaviours. This gave rise to a variety of mimics. The most complex of these is coral snake mimicry, reviewed by Mertens (1956). Brodie (1993) found evidence to show that it influences the behaviour of bird predators.

Acknowledgements

I am indebted to Jean Claude Rage for a copy of the paper of H. Martin. Carl Gans, Elazar Kochva, Findlay Russell, H. Saint Girons and David Warrell kindly commented on Phisalix's report of toxins in the 'glande temporale antérieure' and on other matters. K. Jackson and T. Fritts allowed me to cite their paper in advance of publication.

References

Anthony, J. (1955). Essai sur l'évolution anatomique de l'appareil venimeux des ophidiens. *Annls Sci. nat. (Zool.)* (11) **17**: 7–53.

Bellairs, A. d'A. (1949). Observations on the snout of *Varanus*, and a comparison with that of other lizards and snakes. *J. Anat.* **83**: 116–146.

Brodie, E. D. (1993). Differential avoidance of coral snake banded patterns by free-ranging avian predators in Costa Rica. *Evolution* **47**: 227–235.

Cadle, J. E. (1984). Molecular systematics of neotropical xenodontine snakes. I. South American xenodontines. *Herpetologica* **40**: 8–20.

Cadle, J. E. (1994). The colubrid radiation in Africa (Serpentes: Colubridae): phylogenetic relationships and evolutionary patterns based on immunological data. *Zool. J. Linn. Soc.* **110**: 103–140.

Cerdas, L. & Lomonte, B. (1982). Estudio de la capacidad ofiófaga y la resistencia de la zopilota (*Clelia clelia*, Colubridae) de Costa Rica a los venenos de serpiente. *Toxicon* **20**: 936–939.

Crescitelli, F. (1958). The natural history of visual pigments. *Ann. N.Y. Acad. Sci.* **74**: 230–255.

Cundall, D. (1983). Activity of head muscles during feeding by snakes: a comparative study. *Am. Zool.* **23**: 383–396.

Cundall, D. (1987). Functional morphology. In *Snakes: ecology and evolutionary biology*: 106–140. (Eds Seigel, R. A., Collins, J. T. & Novak, S. S.). Macmillan, New York.

Cundall, D. & Rossman, D. (1993). Cephalic anatomy of the rare Indonesian snake *Anomochilus weberi*. *Zool. J. Linn. Soc.* **109**: 235–273.
Cundall, D., Wallach, V. & Rossman, D. (1993). The systematic relationships of the snake genus *Anomochilus*. *Zool. J. Linn. Soc.* **109**: 275–299.
Dowling, H. G. & Savage, J. M. (1960). A guide to the snake hemipenis: a survey of basic structure and systematic characteristics. *Zoologica, N.Y.* **45**: 17–28.
Gabe, M. & Saint Girons, H. (1969). Données histologiques sur les glandes salivaires des lépidosauriens. *Mém. Mus. natn. Hist. nat. Paris (A) (Zool.)* **58**: 1–112.
Gans, C. (1961). The feeding mechanism of snakes and its possible evolution. *Am. Zool.* **1**: 217–227.
Greene, H. W. (1983). Dietary correlates of the origin and radiation of snakes. *Am. Zool.* **23**: 431–441.
Greene, H. W. & Burghardt, G. M. (1978). Behavior and phylogeny: constriction in ancient and modern snakes. *Science* **200**: 74–77.
Gygax, P. (1971). Entwicklung, Bau und Funktion der Giftdrüse (Duvernoy's gland) von *Natrix tessellata*. *Acta trop.* **28**: 226–274.
Haas, G. (1930). Über die Kaumuskulatur und die Schädelmechanik einiger Wühlschlangen. *Zool. Jb. (Anat.)* **52**: 95–218.
Haas, G. (1931). Über die Morphologie der Kiefermuskulatur und die Schädelmechanik einiger Schlangen. *Zool. Jb. (Anat.)* **54**: 333–416.
Hirosawa, K. (1978). A microscopic observation on the retina of habu, *Trimeresurus flavoviridis*. *Snake* **10**: 20–23.
Jackson, K. & Fritts, T. (In press). Evidence from tooth surface morphology for a posterior origin of the proteroglyph fang. *Amphibia-Reptilia*.
Kardong, K. V. (1982). The evolution of the venom apparatus in snakes from colubrids to viperids and elapids. *Mems Inst. Butantan* **46**: 105–118.
Kochva, E. (1963). Development of the venom gland and trigeminal muscles in *Vipera palaestinae*. *Acta anat.* **52**: 49–89.
Kochva, E. (1978). Oral glands of the Reptilia. In *Biology of the Reptilia* **8**: 43–161. (Eds Gans, C. & Gans, K. A.). Academic Press, London & New York.
Martin, H. (1899). Recherches sur le développement de l'appareil venimeux de la *Vipera aspis*. *C. r. Ass. Anat.* **1**: 56–66.
McDowell, S. B. (1987). The architecture of the corner of the mouth of colubroid snakes. *J. Herpet.* **20**: 353–407.
Mebs, D., Kuch, U. & Meier, J. (1994). Studies on venom and venom apparatus of Fea's viper, *Azemiops feae*. *Toxicon* **32**: 1275–1278.
Mertens, R. (1956). Das Problem der Mimikry bei Korallenschlangen. *Zool. Jb.* (*Syst.*) **84**: 541–576.
Perales, J., Moussatché, H., Marangoui, S., Oliveira, B. & Dumont, G. B. (1994). Isolation and partial characterisation of an anti-Bothropic complex from the serum of South American Didelphidae. *Toxicon* **32**: 1237–1249.
Phisalix, M. (1922). *Animaux venimeux et venins*. Masson et Cie., Paris.
Phisalix, M. & Caius, R. P. F. (1918). V. L'extension de la fonction venimeuse dans l'ordre entier des ophidiens et son existence chez des familles ou elle n'avait pas été soupçonnée jusqu'ici. *J. Physiol. Path. gén.* **17**: 923–964.
Rasmussen, J. B. (1979). An intergeneric analysis of some boigine snakes — Bogert's (1940) group XIII and XIV (Boiginae, Serpentes). *Vidensk. Meddr dansk naturh. Foren.* **141**: 97–155.
Rasmussen, J. B. (1985). A re-evaluation of the systematics of the African rear-fanged snakes of Bogert's groups XIII–XVI, including a discussion of some evolutionary trends within

Caenophidia. In *Proceedings of the international symposium on African vertebrates: systematics, phylogeny and evolutionary ecology*: 531–538. (Ed. Schuchmann, K.-L.). Zoologisches Forschungsinstitut und Museum Alexander Koenig, Bonn.

Rasmussen, J. B. (1990). The retina of *Psammodynastes pulverulentus* (Boie, 1827) and *Telescopus fallax* (Fleischmann, 1831) with a discussion of their phylogenetic significance (Colubroidea, Serpentes). *Z. zool. Syst. EvolForsch.* **28**: 269–276.

Rieppel, O. (1978). The evolution of the basicranium in the Henophidia (Reptilia: Serpentes). *Zool. J. Linn. Soc.* **66**: 411–431.

Rosenberg, H. I. (1967). Histology, histochemistry, and emptying mechanism of the venom glands of some elapid snakes. *J. Morph.* **123**: 133–155.

Saint Girons, H. (1987). Évolution de la fonction venimeuse chez les reptiles. In *Compte-rendu du colloque organisé à la Faculté Catholique des Sciences*: 9–22. Societé Herpétologique de France & Fondation Marcel Merieuse, Lyon.

Saint Girons, H. & Ineich, I. (1993). Données histologiques sur la glande labiale supérieure externe des Viperidae du genre *Echis*. *Amphibia-Reptilia* **14**: 315–319.

Salomão, M. G. & Ferrarezzi, H. (1993). A morphological, histochemical and ultrastructural analysis of the Duvernoy's glands of elapomorphine snakes: the evolution of their venom apparatus and phylogenetic implications. In *III congresso Latino Americano de herpetologia*: 43. Instituto de Biologia da Universidade Estadual de Campinas, Campinas.

Smith, M. A. & Bellairs, A. d'A. (1947). The head glands of snakes, with remarks on the evolution of the parotid gland and teeth of the Opisthoglypha. *J. Linn. Soc. Lond.* (*Zool.*) **41**: 351–368.

Taub, A. M. (1967). Comparative histological studies on Duvernoy's gland of colubrid snakes. *Bull. Am. Mus. nat. Hist.* **138**: 1–50.

Underwood, G. (1970). The eye. In *Biology of the Reptilia* **2**: 1–97. (Eds Gans, C. & Parsons, T. S.) Academic Press, London & New York.

Underwood, G. & Kochva, E. (1993). On the affinities of the burrowing asps *Atractaspis* (Serpentes, Atractaspididae). *Zool. J. Linn. Soc.* **107**: 3–64.

Walls, G. L. (1942). The vertebrate eye and its adaptive radiation. *Bull. Cranbrook Inst. Sci.* **19**: 1–785.

Weinstein, S. A. & Kardong, K. V. (1994). Properties of Duvernoy's secretions from opisthoglyphous and aglyphous colubrid snakes. *Toxicon* **32**: 1161–1185.

Wollberg, M., Kochva, E. & Underwood, G. (In preparation). *On the rictal glands of some atractaspid snakes.*

2 Systematics of sea snakes: a critical review

ARNE REDSTED RASMUSSEN

Synopsis

The taxonomy of elapid sea snakes is reviewed. There is no general agreement on generic and suprageneric relationships. The problematic genera are reviewed on the basis of published data and personal observations. The author recognizes the following genera in three groups of sea snakes: the *Aipysurus* group — *Aipysurus* and *Emydocephalus*; the *Hydrophis* group — *Acalyptophis, Astrotia, Enhydrina, Ephalophis, Hydrelaps, Hydrophis, Kerilia, Kolpophis, Lapemis, Parahydrophis, Pelamis, Thalassophina* and *Thalassophis*; the *Laticauda* group — *Laticauda*. Some of the most obvious aquatic adaptations are discussed in relation to the classification of the sea snakes.

The classification of sea snakes since Wall (1909)

Sea snakes form a group of aquatic snakes with a vertically flat paddle-like tail, which is not found in any other terrestrial or aquatic snakes. Sea snakes belong to the same group as terrestrial elapids; they are all characterized by the anterior position of the poison-fangs on the maxillary bone (proteroglyphous) (McCarthy 1985).

The first monograph on sea snakes was published by Wall (1909) and included 11 genera. The status of the group, as the subfamily Hydrophiinae, was as in Boulenger's (1896) classification. M. A. Smith (1926) published the second monograph on sea snakes in two decades. This detailed classification still remains the standard work on this very difficult group of aquatic elapids. Smith (1926) regarded the sea snakes as one family, Hydrophiidae, and separated it into two subfamilies; Laticaudinae with *Laticauda* as the most primitive genus, from which *Emydocephalus* and *Aipysurus* branched off, and Hydrophiinae, which included all other sea snakes (14 genera). Later, M. A. Smith (1931) described a new genus, *Ephalophis*, which differs from all other known Hydrophiinae in having broad ventrals, a different shape of the rostral shield and the tail more rounded than usual in Hydrophiinae. Smith concluded then that *Ephalophis* in other respects resembled the subfamily Hydrophiinae too closely to be separated from them; however, in 1943 Smith concluded that the two subfamilies were united through *Ephalophis*.

In 1957, in a paper about the morphology and phylogeny of the Squamata, Underwood wondered if the Australian elapids could be descended from primitive laticaudine sea snakes, indicating that Hydrophiidae is not a good monophyletic group.

Symp. zool. Soc. Lond. (1997) No. 70: 15–30
ISBN 0 19 854986 5

Two years later Dowling (1959) included all sea snakes in the subfamily Hydrophiinae, concluding that Smith's (1926) three genera in Laticaudinae were not obviously related to each other, and that the sea snakes only differed from terrestrial elapids in their aquatic adaptation. Later, Dowling (1967) reviewed the classification of the colubrids, dividing the family Elapidae into four subfamilies. Two of these were sea snakes: Laticaudinae (recent sea snakes) and Hydrophinae (advanced sea snakes). Underwood (1967) followed the usual practice of placing all sea snakes together in the Hydrophiinae. However, Underwood also mentioned that all sea snakes share the absence of choanal and maxillary processes of the palatine, giving some support to the idea that the Australasian elapines may have been derived from early sea snakes.

The same year, McDowell (1967) postulated that the genus *Laticauda* was distinct from all other sea snakes in having well-defined cup-shaped calyces on the hemipenis, a sulcus spermaticus which is forked within this calyculate region, a venom gland which has its posterior end bent downwards behind the corner of the mouth and a maxilla and ectopterygoid which meet in an end-to-end joint. McDowell also mentioned that all sea snakes other than *Laticauda* have the neural spines of the caudal vertebrae strongly elongated (thus giving support to the caudal 'fin'), nostrils rotated upward and backward on the snout to lie dorsally and posterior to the conchal pad, and a premaxilla which is loosely attached to the skull, allowing the tip of the upper lip to close down as a valve. On the basis of the above-mentioned characters, McDowell (1967) stated that 'Since there is no detailed resemblance in marine specializations between *Laticauda* and the Hydrophiidae it seems likely that the specializations have been independently acquired. *Emydocephalus* and *Aipysurus* are like other true Hydrophiidae in all these features, and the much-cited resemblance between these two genera and *Laticauda*, the failure of the palatine to equal the maxilla in forward extent, is not detailed'.

McDowell (1967) also discussed the musculus adductor externus superficialis, and divided it into the following four different types: the *Glyphodon* type, the *Oxyuranus* type, the *Pseudechis* type and the *Demansia* type. All four types are found in sea snakes. Later, McDowell (1969) made a note about *Ephalophis greyi* and the origin and classification of sea snakes. In this paper he separated the subfamily Hydrophiinae (including *Aipysurus* and *Emydocephalus*) into the following three groups based on scalation, vertebral anatomy, venom gland musculature and skull morphology: the *Aipysurus* group (including the genera *Aipysurus*, *Emydocephalus* and *Ephalophis*), the *Hydrelaps* group (including the genus *Hydrelaps*) and the *Hydrophis* group (including all other sea snakes in the subfamily Hydrophiinae). The adductor externus superficialis was discussed again and McDowell concluded that *Ephalophis greyi* and *Hydrophis mertoni* have a *Demansia* type of muscle, and other members of that subfamily have secondarily developed a *Glyphodon* type of musculature.

In a paper from 1972, McDowell reviewed the *Hydrophis* group again, using the venom gland muscle together with the occipital head of the depressor mandibulae, skull characters, dentition, position of the heart, hemipenis and scale characters. The result was a resurrection of the genus *Disteira*, including three species from

Hydrophis, together with the genera *Enhydrina* and *Astrotia*, and an expansion of the genus *Lapemis* to include members of the genera *Kolpophis* and *Thalassophina* (former *Praescutata* — McDowell 1972: 244). The genus *Hydrophis* was reduced by the removal of the three species to *Disteira* and expanded by inclusion of members of the genus *Microcephalophis*. McDowell also divided *Hydrophis* into three subgenera: *Hydrophis*, *Leioselasma* and *Chitulia* (formerly *Aturia*, see Williams & Wallach 1989). The following genera were unchanged: *Acalyptophis*, *Kerilia*, *Thalassophis* and *Pelamis*. Concerning the origin of the Hydrophiinae, McDowell (1969, 1972, 1974) regarded *Ephalophis* as the most primitive genus, linking Hydrophiinae to the *Demansia* group of terrestrial Australian elapids, particularly to the genera *Rhinoplocephalus* and *Drepanodontis*.

Burger & Natsuno (1974) suggested the formation of a new subfamily, Ephalophiinae, based on the *Aipysurus* group and the *Hydrelaps* group of McDowell (1969). Three new groups were suggested in the subfamily Ephalophiinae: the *Aipysurus* group (including *Aipysurus* and *Emydocephalus*), the *Ephalophis* group (including *Ephalophis* and *Parahydrophis* (new genus for *Ephalophis mertoni* — McDowell 1969) and the *Hydrelaps* group (including *Hydrelaps*). The new subfamily was based on the following characters (Burger & Natsuno 1974): ventrals corresponding to vertebrae, ventrals one-third or more body width and exceeding three widths of adjacent dorsals, anal divided in two, dorsal scales strongly imbricate, caudal vertebrae with paired haemapophyses that enclose blood vessels, heart in anterior third of body and gap between heart and liver exceeding length of heart. Burger & Natsuno (1974) included all other hydrophiine sea snakes (*sensu* McDowell 1972) in the subfamily Hydrophiinae with the following changes: the genus *Hydrophis* (*sensu* McDowell 1972) was expanded with an additional subgenus, *Disteira*; Burger & Natsuno stated that the different shape of musculus adductor externus superficialis was not enough to separate *Disteira* from *Hydrophis*; the genus *Kolpophis* was separated from *Lapemis* (*sensu* McDowell 1972) owing to the great difference between the two groups in the number of ventrals and dorsal scales. All other genera were kept as in McDowell (1972).

Using serological data, Minton & da Costa's (1975) results indicated that the genera *Aipysurus* and *Emydocephalus* may represent an older marine radiation compared to seven species from the *Hydrophis* group and that the genus *Laticauda* shows no closer serological relationship to *Aipysurus* than to the *Hydrophis* group.

Cogger (1975) did not include *Disteira* in the genus *Hydrophis*, nor did he accept *Disteira* as redefined by McDowell (1972); he re-created the genera *Enhydrina* and *Astrotia*.

Based on 43 morphological characters, Voris (1977) recognized three major groups of sea snakes: (1) the genus *Laticauda*; (2) the genera *Aipysurus* and *Emydocephalus*; and (3) the *Hydrophis* group including all other genera of sea snakes. He also suggested that these groups may have independent origins among the elapids or a single origin with a very early separation.

H. M. Smith, Smith & Sawin (1977) suggested four tribes in the subfamily Hydrophiinae (*sensu* McDowell 1972), without mentioning the genera belonging to the tribes. The genus *Laticauda* was classed with the American and South Asiatic coral snakes in the subfamily Elapinae.

Underwood (1979) followed Burger & Natsuno (1974) except for *Enhydrina*, where he followed Cogger (1975). Mao & Chen (1980) followed Cogger (1975) in a paper including drawings of the skull, muscles of the head and hemipenis of the species from Taiwan.

In 1981, Kharin suggested splitting the genus *Aipysurus* into the two following subgenera: *Smithohydrophis* for *Aipysurus foliosquama* and *A. apraefrontalis* and *Aipysurus* for all other species in the genus *Aipysurus*. Cadle & Gorman (1981) recognized the same three major groups as Voris (1977) by microcomplement fixation analysis using antisera to albumins from different species. Despite the fact that Cadle & Gorman's data suggested that the *Aipysurus* group (*sensu* the genera *Aipysurus* and *Emydocephalus*) was clearly distinct from both *Laticauda* and the *Hydrophis* group, they assigned the *Aipysurus* group to be cladistically closer to the *Hydrophis* group. They also concluded that the diversification of the *Hydrophis* group is relatively recent, and that the genus *Hydrophis* is paraphyletic. The question of the relationships of *Laticauda* to other sea snakes was not resolved, but no special relationship between the genus *Micrurus* and *Laticauda* was found (Cadle & Gorman 1981; see also Cadle & Sarich 1981). Using chromosome data, Gorman (1981) agreed with McDowell (1972) in including the genus *Microcephalophis* in the genus *Hydrophis*, but concluded that *Laticauda* is most similar to the Australian *Notechis* and not *Micrurus*.

Mao *et al.* (1983), using the same methods as Cadle & Gorman, reached the same conclusions concerning the relationship of *Laticauda* with *Micrurus*. Mao *et al.* (1983) also found that the Hydrophiinae are much closer immunologically to the Australasian elapids than the genus *Laticauda* is.

In 1983, Mengden published a review of the taxonomy of the Australian elapid snakes including aquatic elapids. Mengden concluded that three key questions remained to be resolved at the family level. One of them concerned aquatic elapids and was as follows: 'What is the precise relationship between the laticaudid and hydrophiid sea snakes and the Australian elapids, and is there evidence to demonstrate an independent origin of hydrophiid and laticaudid sea snakes?'

In 1984 Kharin published one paper about the *Hydrophis* group, in which he raised McDowell's (1972) subgenera in *Hydrophis* to generic level (Kharin 1984a), and another paper in which he suggested that *Laticauda semifasciata* and *L. schistorhynchus* are sufficiently different from other species of *Laticauda* to be placed in a separate genus: *Pseudolaticauda* (Kharin 1984b).

Using immunological evidence, Schwaner *et al.* (1985) suggested that the hydrophiines (including the *Aipysurus* group) diverged from the *Notechis* lineage very recently; perhaps only five million years ago (based on the 'molecular clock').

In 1985 McCarthy preferred Underwood's (1967, 1979) classification in an analysis of the monophyly of elapid snakes, refuting H. M. Smith *et al.*'s (1977) classification mainly because it implies that the relationships between proteroglyphous groups are better understood than is currently the case (McCarthy 1985).

Later, McCarthy (1986) analysed both the relationships between species and the wider relationships of *Laticauda*, using 65 morphological characters and both phenetic and phylogenetic methods. At the species level McCarthy refuted Kharin's

(1984b) proposal and mentioned that it appears possible that *L. colubrina* is phylogenetically closer to *L. semifasciata* than it is to *L. laticaudata*, a proximity which would be obscured if they were placed in different genera; he therefore retained the genus *Laticauda* in its former broad sense. McCarthy's conclusion on the wider relationships was as follows:

> it may be concluded that laticaudines and hydrophiines do indeed seem to be comparatively closely related. There is however still a residue of conflicting data which tend to support McDowell's contrasting proposition that *Laticauda* is more closely related to particular groups of terrestrial elapines than to hydrophiines (a topology suggested both by phenetic analysis and by some parsimony runs).

Toriba & Yosida (1986) remarked that Kharin's (1984b) characters seem to be insufficient for the generic division of *Laticauda* but, referring to Mao & Chen (1980), they also noted that there are several other characteristics supporting this division, which were not mentioned by Kharin (1984b).

In 1986, McDowell published a very comprehensive paper about the architecture of the corner of the mouth of colubrid snakes, in which he reviewed the 'euproteroglyph' venom gland musculature from the 'analytical morphological' view (in contrast to his paper of 1967, where he used the 'analytical taxonomic' viewpoint (McDowell 1986: 385)). Using this view, McDowell separated the sea snake genera (*sensu* McDowell 1972 except *Ephalophis* and *Parahydrophis*, which were *sensu* Burger & Natsuno 1974) into the following four types: the *Glyphodon* type, including *Acalyptophis, Aipysurus, Emydocephalus, Hydrophis, Kerilia, Lapemis, Pelamis* and *Thalassophis;* the *Oxyuranus* type, including *Hydrelaps* and *Disteira (*except *D. stokesi*); the *Pseudechis* type, including *Disteira stokesi, Laticauda colubrina* and *L. laticaudata*; and the *Demansia* type, including *Ephalophis, Parahydrophis* and *L. schistorhynchus*. He also mentioned that genera such as *Ephalophis, Hydrelaps* and *Laticauda* did not show adaptations that would prevent them (or, more likely, their ancestors) from invading land, indicating that the original habitat of Colubroidea was the sea coast and mangrove thickets, with several independent terrestrial incursions giving rise to the colubrid and proteroglyphous lineages (McDowell 1986).

In 1987, Toriba & Nakamoto followed McCarthy's (1986) revision of laticaudine sea snakes. Cogger, Heatwole *et al.* (1987) adopted the classification of Burger & Natsuno (1974) in their treatment of the genus *Laticauda*; they also refuted Kharin's (1984b) proposal of separating *Laticauda* into two genera, on the basis of Voris's results from 1977. Using fingerprint comparisons of albumin, Guo, Mao & Yin (1987) concluded that *Laticauda* and *Hydrophis* belong to a natural group, with a relationship far closer than that indicated by morphological criteria. They did not include any terrestrial Australian elapids. McDowell's classification was also questioned by Young (1987), on the grounds that functional morphological criteria had not been utilized.

Slowinski (1989) also refuted Kharin's (1984b) new genus in laticaudine sea snakes in a parsimony analysis of the amino acid sequences of the short-chain neurotoxins. Kharin (1989) followed Cogger (1975) and Cogger, Cameron & Cogger (1983) concerning the genus *Disteira* and added a new species to the genus.

Gopalakrishnakone & Kochva (1990) analysed the venom glands and some associated muscles in sea snakes, and concluded that the general morphology of the venom glands in hydrophiine snakes is indeed quite uniform and most closely resembles that of the *Notechis* group. In some sea snakes, the compressor muscle shows an uninterrupted portion running continuously from the skull to the lower jaw, possibly suggesting an earlier divergence of this group (Gopalakrishnakone & Kochva 1990).

In a checklist, Golay *et al.* (1993) followed Burger & Natsuno (1974), Cogger (1975) and McCarthy (1986). Testing Burger & Natsuno's (1974) groups in the subfamily Ephalophiinae, both Guinea (1993) and A. R. Rasmussen (in prep.) found them to be paraphyletic.

A. R. Rasmussen (1994) presented a cladistic analysis of the *Hydrophis* subgenus *Chitulia* (*sensu* McDowell 1972), based on 22 morphological characters. The results indicated that the subgenus *Chitulia* was paraphyletic and that the group has been based solely on plesiomorphic character states. Furthermore, Kharin's (1984a) proposal to raise the three subgenera created by McDowell to generic status was rejected (A. R. Rasmussen 1994).

The current status of the sea snakes

It is clear that there is at present no general agreement on the status of the sea snakes at the suprageneric level. However, at the generic level there seems to be a broader consensus. In the following I will discuss all the problematic genera, and give my opinion on their current status.

The following genera have generally been accepted since M. A. Smith (1926, 1931): *Acalyptophis*, *Aipysurus*, *Emydocephalus*, *Ephalophis*, *Hydrelaps*, *Hydrophis*, *Kerilia*, *Lapemis*, *Laticauda*, *Pelamis* and *Thalassophis* (M. A. Smith 1943; McDowell 1969, 1972, 1974; Burger & Natsuno 1974; Cogger 1975, 1992; Cogger, Cameron & Cogger 1983; Mengden 1983; McCarthy 1986; Bussarawit, Rasmussen & Andersen 1989; A. R. Rasmussen 1989, 1992, 1993, 1994, in prep.; McCarthy & Warrell 1991; Golay *et al.* 1993). The genus *Parahydrophis* Burger & Natsuno, 1974 has also been accepted (Cogger, Cameron & Cogger 1983; Cogger 1992; McDowell 1986; Golay *et al.* 1993).

The *Laticauda* group

The genus *Laticauda* was traditionally treated as one genus until Kharin (1984b) made a new genus for two of the species (see above). Against Kharin's proposal speak the results of Voris (1977) and McCarthy (1986), using 43 and 65 morphological characters, respectively, and Slowinski (1989), using amino acid sequences of shortchain neurotoxins. They all polarized their characters and both McCarthy and Slowinski used parsimony methods in analysing them. All three analyses showed that the species of the *Laticauda* group form a tight group of closely related species, which leaves me in no doubt in rejecting Kharin's (1984b) proposal and recognizing only one genus in the *Laticauda* group. At the suprageneric level most authors recognize *Laticauda* as constituting a group on its own, but whether this group is most closely related to the other sea snakes, or to some group of terrestrial elapids in the Australian region, is still unclear (Voris 1977; Cadle & Gorman 1981; McCarthy 1986).

The genera *Disteira* and *Lapemis*

The resurrection of the genus *Disteira*, including *kingi*, *major*, *nigrocinctus*, *schistosa* and *stokesi*, and the expansion of the genus *Lapemis* to include *annandalei*, *curtus* (syn: *hardwickii* (Gritis & Voris 1990)), and *viperina* by McDowell (1972) have in my opinion not reduced the taxonomic problems with these species and in their relationships with the genus *Hydrophis*. The only character keeping the genus *Disteira* together is the broad dorsal portion of the venom gland muscle (levator anguli oris of adductor externus superficialis (McDowell 1986)) also called the *Oxyuranus* type (McDowell 1972). The *Oxyuranus* type is also found in the genus *Hydrelaps*, which McDowell (1972) regarded as a particularly primitive form, and in some terrestrial genera of elapids such as *Acanthophis*, *Dendroaspis*, *Echiopsis*, *Micrurus*, *Notechis*, *Salomonelaps* and *Tropidechis*. This indicates that the features are widely distributed and do not represent a good apomorphic condition for McDowell's (1972) concept of *Disteira*. Later, McDowell (1986) separated *stokesi* from the other species in *Disteira* for the following reason: 'However, in *Disteira* ["*Astrotia*"] *stokesi*, the *Oxyuranus* pattern seen in other members of *Disteira* has been modified into the *Pseudechis* pattern by having the rear border of the levator anguli oris attach to the quadrate just ventrolateral to the attachment of the muscle of the nuchal crest of the braincase . . .'

Therefore *stokesi* lacks the only character by which McDowell (1972) characterizes the genus *Disteira*. Furthermore, it can be separated from the other species in *Disteira* by the number of body vertebrae (Table 1), in having the most anterior position of the heart (Table 1), and by a unique division of the ventrals throughout the body. These are good reasons to resurrect the genus *Astrotia* for *stokesi*.

The species *schistosa* is also unique in its number of body vertebrae (Table 1), and *schistosa* can be separated from *kingi*, *major* and *nigrocinctus* in having ventrals which are very small (but distinct) throughout, a higher number of scale rows on the neck and body (Table 1) and a curved palatine. The species *schistosa* is also unique in having the mental narrow, elongated and partly hidden in a groove in the symphysis. These unique characters are in my opinion important enough to give *schistosa* separate generic status, resurrecting the genus *Enhydrina*. This is in accordance with Cogger (1975).

I also find it hard to believe that *kingi* and *major* (both from the Australian region) should be more closely related to the small-headed *nigrocinctus* (from the Bay of Bengal) than to, for example, *H. obscurus*, *H. stricticollis*, *H. mamillaris* and *H. klossi*, all of which are small-headed species of the genus *Hydrophis* from South-East Asia. Many sea snake species in Australian waters (including *major* and *kingi*) have spots or blotches on their body, which are not found on the small-headed *Hydrophis* species from South-East Asia, including *nigrocinctus*. The small-headed *Hydrophis* also have few teeth and nearly the same number of vertebrae as *nigrocinctus* (180–238), whereas *kingi* has in excess of 60 vertebrae more than *nigrocinctus*. A high number of vertebrae is also found in the species *H. elegans* from Australian waters, possibly providing a good apomorphic state for *H. elegans* and *kingi*.

Table 1. Variation of external and internal characters of five species of *Disteira* and three species of *Lapemis* (both *sensu* McDowell 1972) and the genera *Aipysurus* and *Hydrophis* (*sensu* M.A. Smith 1926)

Species	Ventrals no.	Ventrals distinct throughout	%VS–heart[a]	VB–body[b]	Scale rows neck	Scale rows body	Maxillary teeth	Palatine curved
Disteira								
kingi	311–342	Yes	*c.* 46–52	263–271	27–29	36–40	2–4	No
major	210–270	Yes	*c.* 38–44	166–179	31–35	37–45	6–7	Feebly
nigrocinctus	296–330	Yes	*c.* 50–54	180–197	27–33	39–45	0–3	No
schistosa	239–322	Very small	*c.* 37–41	153–163	40–55	49–66	3–4	Yes
stokesi	226–286	Divided	*c.* 24–28	142–146	37–47	47–63	6–8	Yes
Lapemis								
annandalei	320–368	Very small	??	130–156	62–82	74–93	6–7	No
curtus	114–274	No	*c.* 21–25	128–141	23–35	25–43	3–6	No
viperina	226–274	Yes, enlarge anteriorly	*c.* 23–27	150–161	27–34	37–50	5	Yes
Aipysurus 7 species	130–175	Yes	*c.* 24–37	130–175	15–21	17–25	5–11	No
Hydrophis 30 species	220–514	Yes	*c.* 33–53	152–249	18–45	28–63	4–18	No/feebly

[a]%VS-heart = relative position of tip of the heart in number of ventrals
[b]VB-body = number of body vertebrae.

The species *kingi*, *major* and *nigrocinctus* are all in agreement with M. A. Smith's (1926) diagnosis of the genus *Hydrophis*. For the reasons given above, I suggest that they be reassigned to the genus *Hydrophis*, but not as a subgenus, as suggested by Burger & Natsuno (1974), because I do not recognize the subgeneric division of *Hydrophis* (A. R. Rasmussen 1994).

The genus *Lapemis* (*sensu* McDowell 1972) is also kept together by one character only, the lack of a supraoccipital head on the depressor mandibulae. This condition is also found in the genera *Aipysurus* and *Emydocephalus* (McDowell 1972). Generally, the diagnostic characters mentioned by McDowell (1972) for the genus *Lapemis* are also found in the genus *Hydrophis*. However, there are some unique characters found in *viperina* (the shape of the parietal and the ventrals which are broad anteriorly and smaller posteriorly, but distinct throughout), and in *annandalei* (a much higher count of scale rows on the neck and body compared with other sea snakes, see Table 1).

McDowell also mentioned that the three *Lapemis* species differ from each other in the following characters (besides those mentioned in Table 1): palatine exceeding maxilla in forward extent (*annandalei* and *viperina*), or nearly straight (*curtus*); sphenoid entering border of anterior orifice of cavum epiptericum (*annandalei* and *viperina*), or not (*curtus*); with a moderately developed median keel (*viperina*), or quite unkeeled (*annandalei* and *curtus*); head shields unfragmented (*viperina* and *curtus*), or parietal fragmented (*curtus*), or all except frontal fragmented (*annandalei*). Furthermore, the number of teeth on the palatine, ectopterygoid and dentary is different in the three species. The skulls of *curtus* and *viperina* (Fig. 1) differ in the structure of the transverse occipital crest, *viperina* having a very small, or nearly non-existent crest, whereas *curtus* has a very heavy crest running from the occipital bones to the midline and further forward on the parietal (Fig. 1). This indicates that the loss of the depressor mandibulae has taken place by convergence, as the supporting bone structure is totally different in the two species. On the basis of the above-mentioned characters, I find no support for keeping the genus *Lapemis* together as suggested by McDowell, but prefer to reinstate the genus *Thalassophina* for *viperina* following M. A. Smith (1926) and the genus *Kolpophis* for *annandalei* following M. A. Smith (1926) and Burger & Natsuno (1974), retaining *curtus* in the genus *Lapemis*.

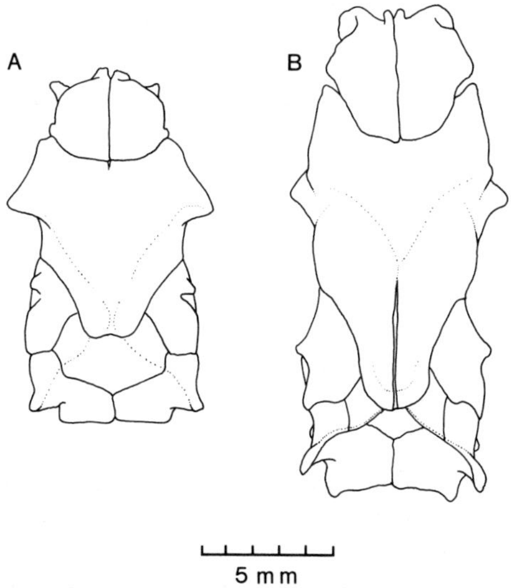

Fig. 1. Dorsal aspect of parietal: (A) *viperina*; (B) *curtus*.

The subfamily Ephalophiinae

Burger & Natsuno's (1974) subfamily Ephalophiinae has been treated as natural by some authors (e.g., Underwood 1979). However, all the character states used to define the group are in my opinion based on plesiomorphic states found in most terrestrial snakes and in many Australian elapids: ventrals corresponding to vertebrae, ventrals one-third or more of body width and exceeding three widths of adjacent dorsals, anal divided in two, dorsal scales strongly imbricated, caudal vertebrae with paired haemapophyses that enclose the blood vessel, heart in anterior third of body and gap between heart and liver exceeding length of heart. Also, Voris's (1977), and A. R. Rasmussen's (in prep.) results show that the genera *Ephalophis*, *Parahydrophis* and *Hydrelaps* are closer to the *Hydrophis* group than to *Aipysurus* and *Emydocephalus*.

Aquatic adaptations or good apomorphic states?

A preliminary investigation of sea snakes and seven terrestrial elapids from New Guinea and Australia used parsimony methods on 52 different morphological characters (A. R. Rasmussen in prep.). The results indicated that the sea snakes could be separated into three independent groups of aquatic elapids. It was indicated that what on the face of it look like good apomorphic characters (e.g., salt gland, flat tail, lungs, eyes), could be convergent adaptations to the environment. In particular, the flat tail has been thought of as an obvious apomorphic state. However, McDowell (1967) pointed out that the tail in the genus *Laticauda* is different from that in other species of sea snakes. Notably, the caudal 'fin' is not strongly supported by neural spines. Later, McCarthy (1987) mentioned that *Ephalophis* and *Aipysurus* also lack the extensive bony support for the paddle found in the more derived members of the group. McDowell (1969, 1972) and McCarthy (1987) also mentioned the paired ventral processes (haemapophyses).

Having studied the osteological modifications in the tail of sea snakes by X-ray technology (Fig. 2), I found that the structures in the tail fall into four distinct conditions (observed on midtail vertebrae): (1) pleurapophysis and haemapophysis nearly the same length, but haemapophysis very broad at the distal end, both nearly as long as the width of the vertebrae (Fig. 2) — these conditions are found in the genera *Parahydrophis* and *Ephalophis*; (2) pleurapophysis and haemapophysis both elongated and more than 1.5 times longer than the width of the vertebrae, but haemapophysis longest and not broader distally than proximally (Fig. 2) – these conditions are found in *Aipysurus* and *Emydocephalus*; (3) only pleurapophysis elongated and longer than the width of the vertebrae, haemapophysis very small, or nearly reduced (Fig. 2) — these conditions are found in *Hydrelaps*, and the *Hydrophis* group (*sensu* McDowell 1972); (4) both pleurapophysis and haemapophysis evolved, but pleurapophysis longest and nearly the same size as the width of the vertebrae (Fig. 2) — these conditions are found in some terrestrial Australian elapids and in *Laticauda*.

I found that all sea snakes except *Laticauda* have elongated neural spines (Fig. 2). However, in *Ephalophis*, *Hydrelaps* and *Parahydrophis* the neural spines are shorter than in the other sea snake species examined (Fig. 2). The sublingual salt gland has also been investigated in sea snakes (see Dunson & Dunson 1974; Dunson 1975). The gland is found in *Acrochordus granulatus*, as well as in sea snakes (including *Laticauda*), indicating that at least two lines have evolved a sublingual salt gland independently. Measuring the fluid concentration and secretion rate, Dunson (1975) indicated that the secretory fluid concentration is constant within genera such as *Aipysurus* and *Hydrophis*, unlike the secretory rate. The comparison of fluid concentration in Dunson's table 1 (1975: 347) showed that, of all the sea snakes that were investigated, the species examined in the genus *Aipysurus* have the highest fluid concentrations of chloride (Cl) sodium (Na) and potassium (K) ions. This is in spite of the fact that the species examined in *Aipysurus* eat totally different food items (e.g., *A. eydouxi* eats fish eggs and *A. laevis* eats fish).

Heatwole & Seymour (1975) have also found some differences between *Aipysurus* and *Emydocephalus* on the one hand and *Hydrophis* and *Lapemis* on the other, in respect to the volumes of the two components of the vascular lung. The tracheal lung in *Aipysurus* and *Emydocephalus* is smaller in percentage of total volume of the vascular lung than in *Hydrophis* and *Lapemis*, and the bronchial lung in *Hydrophis* and *Lapemis* is smaller in percentage of the total volume of the vascular lung than in *Aipysurus* and *Emydocephalus*.

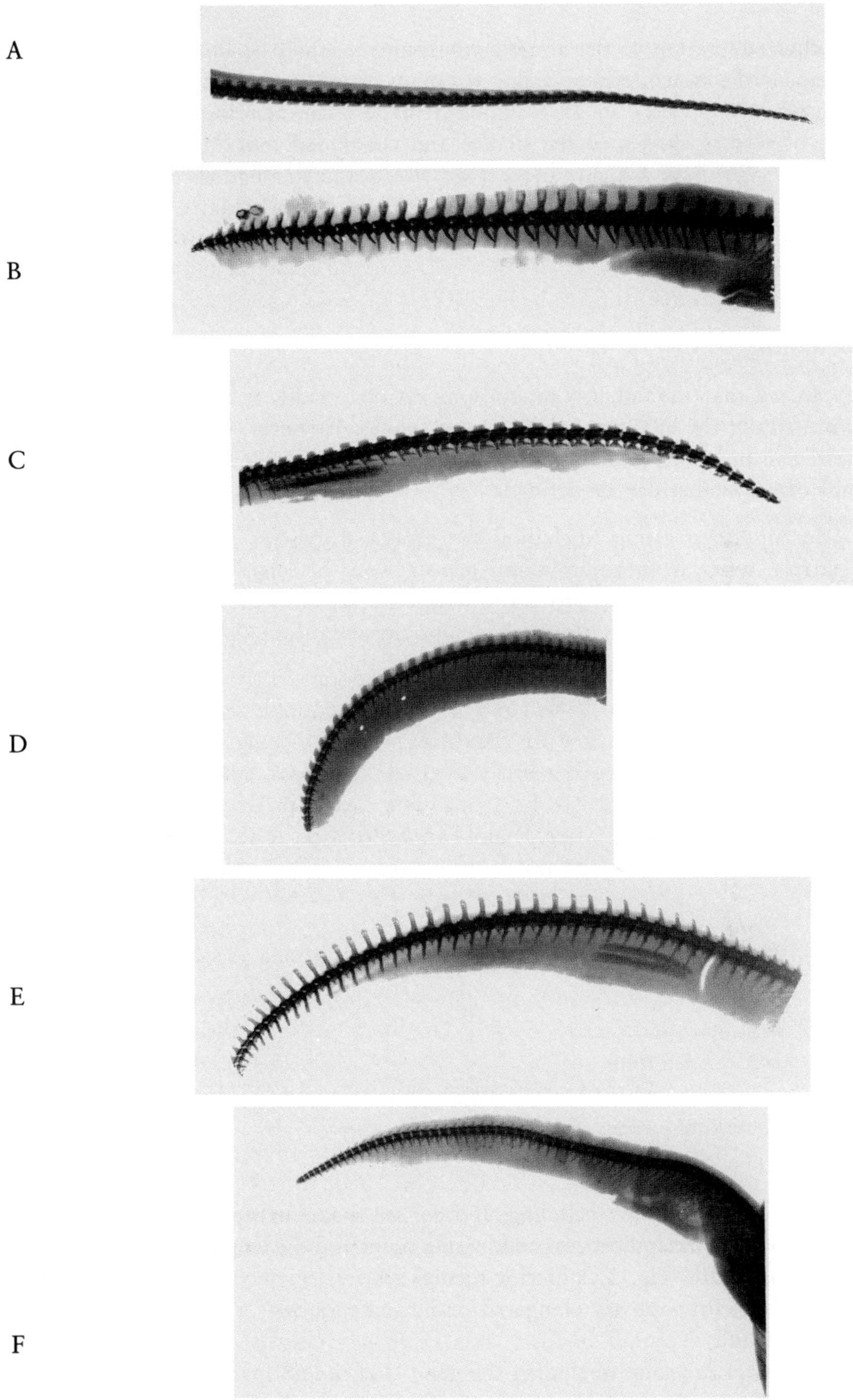

Fig. 2. Showing the X-ray of the tail (top to bottom): *Hemiaspis signata* (a terrestrial Australian elapid), *Aipysurus duboisi*, *Ephalophis greyi*, *Hydrelaps darwiniensis*, *Hydrophis major* and *Laticauda colubrina*.

Another adaptation to the aquatic environment could be the retina, which has been used in the systematics of snakes for many years (Underwood 1967, 1979; J. B. Rasmussen 1985, 1990). In 1975 Hibbard made some histological studies of the retinas of several species of sea snakes and concluded that *Aipysurus* (including *Emydocephalus*) have a higher percentage of rod and small cone receptors, a more distinct stratification of the inner plexiform layer of the retina and better development of the optic tectum and probably have generally better vision than other sea snake species studied.

Conclusion

One can see that the aquatic adaptations do not readily resolve the relationship between any of the different groups of sea snakes. However, by using the shape of parietal and maxillary bone, together with the conditions mentioned above, three groups of sea snakes can be defined.

1. The *Aipysurus* group (including the genera *Aipysurus* and *Emydocephalus*): parietal without an anterolateral process and the length less than the width (Fig. 3), anterior medial process of maxilla absent or distinctly behind the fang (Fig. 4) and with an elongated haemapophysis in the tail.

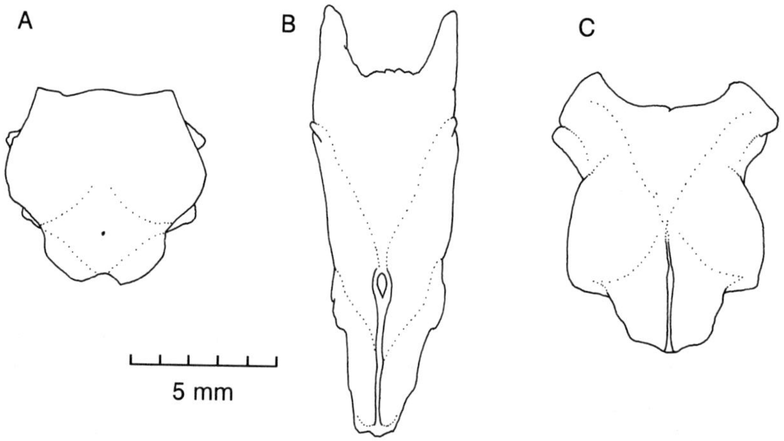

Fig. 3. Dorsal aspect of parietal: (A) *Aipysurus eydouxi*; (B) *Hydrophis cyanocinctus*; (C) *Laticauda laticaudata*.

2. The *Hydrophis* group (including all other sea snakes except *Laticauda*): parietal with an anterolateral process ending in a point and the length equal to or longer than the width (Fig. 3), anterior medial process of maxilla close to the fang (Fig. 4), pleurapophysis elongated and haemapophysis reduced or small and broad distally.
3. The *Laticauda* group (including the genus *Laticauda*): parietal with an anterolateral process not ending in a point (Fig. 3) and a very short maxillary bone (Fig. 4) (for further apomorphic states, see McCarthy 1986).

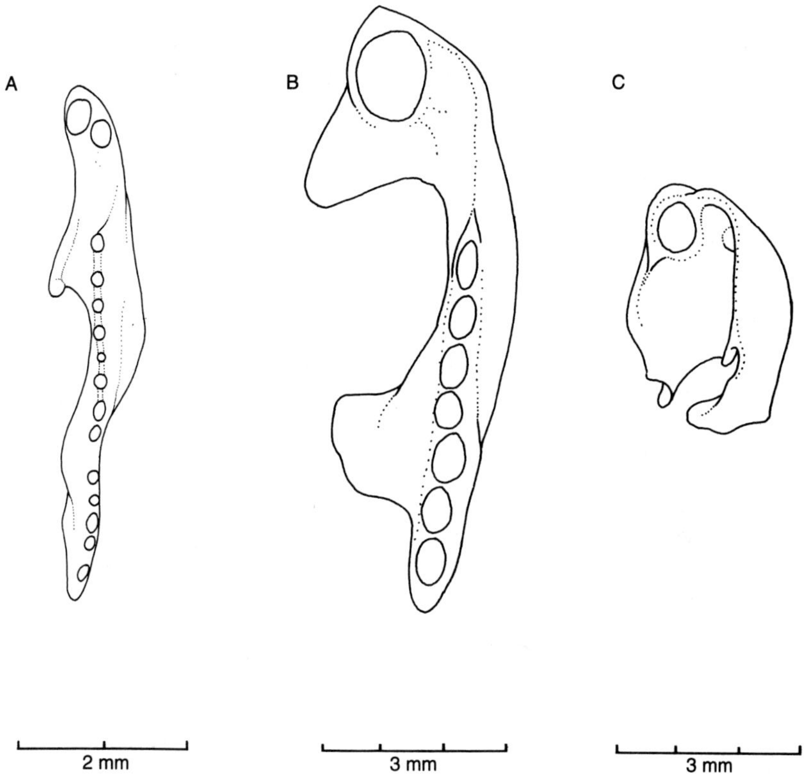

Fig. 4. Ventral aspect of maxillary bone: (A) *Aipysurus eydouxi*; (B) *Astrotia stokesi*; (C) *Laticauda laticaudata*.

The character transformations of the above-mentioned conditions still have to be tested in a more thorough cladistic analysis. Obviously, many questions remain to be answered before we can say anything about the broader relationships of the sea snakes. In particular, the polarizations of many characters in both aquatic and terrestrial elapids may be changed if some of the aquatic elapids are the most primitive members among proteroglyphous snakes, giving rise to one or more terrestrial elapids as indicated by Underwood (1957, 1967) and McDowell (1985, 1986).

Acknowledgements

I thank M. Andersen, D. Eibye-Jacobsen, P. Gravlund, A. Greer, A. S. Henriksen, J. B. Rasmussen and R. Sadlier for their many helpful discussions and advice; The Australian Museum, Sydney, for laboratory facilities in 1993. The study was supported by Dansk Naturhistorisk Forening, Hemmingsens legat and The Danish National Research Council. I wish to thank G. Brovad for photographs and M. Andersen for drawings.

References

Boulenger, G. A. (1896). Catalogue of the snakes in the British Museum (Natural History). Volume III, containing the Colubridae (Opisthoglyphae and Proteroglyphae), Amblycephalidae, and Viperidae. British Museum of Natural History, London.

Burger, W. L. & Natsuno, T. (1974). A new genus for the Arafura smooth seasnake and redefinitions of other seasnake genera. *Snake* **6**: 61–75.

Bussarawit, S., Rasmussen, A. R. & Andersen, M. (1989). A preliminary study on sea snakes (Hydrophiidae) from Phuket harbour, Phuket Island, Thailand. *Nat. Hist. Bull. Siam Soc.* **37**: 209–225.

Cadle, J. E. & Gorman, G. C. (1981). Albumin immunological evidence and the relationships of sea snakes. *J. Herpet.* **15**: 329–334.

Cadle, J. E. & Sarich, V. M. (1981). An immunological assessment of the phylogenetic position of the New World coral snakes. *J. Zool., Lond.* **195**: 157–167.

Cogger, H. G. (1975). Sea snakes of Australia and New Guinea. In *The biology of sea snakes*: 59–139. (Ed. Dunson, W. A.). University Park Press, Baltimore, London & Tokyo.

Cogger, H. G. (1992). *Reptiles and amphibians of Australia.* (5th edn). Reed International Books, Chatswood, New South Wales and Cornell University Press, New York.

Cogger, H. G., Cameron, E. E. & Cogger, H. M. (1983). Hydrophiidae. In *Zoological catalogue of Australia. Amphibia and Reptilia* **1**: 241–256. (Ed. Walton, D. W.). Canberra: Australian Government Publishing Service.

Cogger, H., Heatwole, H., Ishikawa, Y., McCoy, M., Tamiya, N. & Teruuchi, T. (1987). The status and natural history of the Rennell Island sea krait, *Laticauda crockeri* (Serpentes: Laticaudidae). *J. Herpet.* **21**: 255–266.

Dowling, H. G. (1959). Classification of the Serpentes: a critical review. *Copeia* **1959**: 38–52.

Dowling, H. G. (1967). Hemipenes and other characters in colubrid classification. *Herpetologica* **23**: 138–142.

Dunson, W. A. (1975). Salt and water balance in sea snakes. In *The biology of sea snakes*: 329–354. (Ed. Dunson, W. A.). University Park Press, Baltimore, London & Tokyo.

Dunson, W. A. & Dunson, M. K. (1974). Interspecific differences in fluid concentration and secretion rate of sea snake salt glands. *Am. J. Physiol.* **227**: 430–438.

Golay, P., Smith, H. M., Broadley, D. G., Dixon, J. R., McCarthy, C., Rage, J. C., Schätti, B. & Toriba, M. (1993). *Endoglyphs and other major venomous snakes of the world. A checklist.* Azemiops Herpetological Data Centre, Geneva.

Gopalakrishnakone, P. & Kochva, E. (1990). Venom glands and some associated muscles in sea snakes. *J. Morph.* **205**: 85–96.

Gorman, C. G. (1981). The chromosomes of *Laticauda* and a review of karyotypic evolution in the Elapidae. *J. Herpet.* **15**: 225–233.

Gritis, P. A. & Voris, H. K. (1990). Variability and significance of parietal and ventral scales in the marine snakes of the genus *Lapemis* (Serpentes: Hydrophiidae), with comments on the occurrence of spiny scales in the genus. *Fieldiana (Zool.)* **56**: 1–13.

Guinea, M. L. (1993). A phylogenetic analysis of the thick sea snakes (Hydrophiidae: Ephalophiinae). In *Second world congress of herpetology: abstracts*: 108. (Eds Davies, M. & Norris, R. M.). Department of Zoology, University of Adelaide, Adelaide, South Australia.

Guo, Y. -W., Mao, S. -H. & Yin, F. -Y. (1987). Comparison of cobra plasma albumins with those of banded krait and sea snakes. *Comp. Biochem. Physiol. (B)* **87**: 559–566.

Heatwole, H. & Seymour, R. (1975). Diving physiology. In *The biology of sea snakes*: 289–328. (Ed. Dunson, W. A.). University Park Press, Baltimore, London & Tokyo.

Hibbard, E. (1975). Eyes and other sense organs of sea snakes. In *The biology of sea snakes*: 355–384. (Ed. Dunson, W. A.). University Park Press, Baltimore, London & Tokyo.

Kharin, V. E. (1981). A review of sea snakes of the genus *Aipysurus* (Serpentes: Hydrophiidae). *Zool. Zh.* **60**: 257–264.

Kharin, V. E. (1984a). A review of sea snakes of the group *Hydrophis* sensu lato (Serpentes, Hydrophiidae). 3. The genus *Leioselasma. Zool. Zh.* **63**: 1535–1546.

Kharin, V. E. (1984b). Revision of sea snakes of subfamily Laticaudinae Cope, 1879, sensu lato (Serpentes, Hydrophiidae). *Proc. zool. Inst., Leningrad* **124**: 128–139.

Kharin, V. E. (1989). A new sea snake species of the genus *Disteira* (Serpentes, Hydrophiidae) from the waters of the Malay Archipelago. *Vêst. Zool.* **1989**: 29–32.

Mao, S. -H. & Chen, B. -Y. (1980). Sea snakes of Taiwan. A natural history of sea snakes. *NSC spec. Publ.* No. 4: 1–64.

Mao, S. -H., Chen, B. -Y., Yin, F. -Y. & Guo, Y. -W. (1983). Immunotaxonomic relationships of sea snakes to terrestrial elapids. *Comp. Biochem. Physiol. (A)* **74**: 869–872.

McCarthy, C. J. (1985). Monophyly of elapid snakes (Serpentes: Elapidae). An assessment of the evidence. *Zool. J. Linn. Soc.* **83**: 79–93.

McCarthy, C. J. (1986). Relationships of the laticaudine sea snakes (Serpentes: Elapidae: Laticaudinae). *Bull. Br. Mus. nat. Hist. (Zool.)* **50**: 127–161.

McCarthy, C. J. (1987). Sea snake puzzles. In *Proceedings of the 4th ordinary general meeting of the Societas Europaea Herpetologica*: 279–284. (Eds van Gelder, J. J., Strijbosch, H. & Bergers, P. J. M.). Faculty of Sciences, Nijmegen.

McCarthy, C. J. & Warrell, D. A. (1991). A collection of sea snakes from Thailand with new records of *Hydrophis belcheri* Gray. *Bull. Br. Mus. nat. Hist. (Zool.)* **57**: 161–166.

McDowell, S. B. (1967). *Aspidomorphus*, a genus of New Guinea snakes of the Family Elapidae, with notes on related genera. *J. Zool., Lond.* **151**: 497–543.

McDowell, S. B. (1969). Notes on the Australian sea-snake *Ephalophis greyi* M. Smith (Serpentes: Elapidae, Hydrophiinae) and the origin and classification of sea snakes. *Zool. J. Linn. Soc.* **48**: 333–349.

McDowell, S. B. (1972). The genera of sea-snakes of the *Hydrophis* group (Serpentes: Elapidae). *Trans. zool. Soc. Lond.* **32**: 189–247.

McDowell, S. B. (1974). Additional notes on the rare and primitive sea-snake, *Ephalophis greyi. J. Herpet.* **8**: 123–128.

McDowell, S. B. (1985). The terrestrial Australian elapids: general summary. In *Biology of Australian frogs and reptiles*: 261–264. (Eds Grigg, G., Shine, R. & Ehmann, H.). Surrey Beatty & Sons Pty & The Royal Zoological Society of New South Wales, Chipping Norton, NSW.

McDowell, S. B. (1986). The architecture of the corner of the mouth of colubroid snakes. *J. Herpet.* **20**: 353–407.

Mengden, G. A. (1983). The taxonomy of Australian elapid snakes: a review. *Rec. Aust. Mus.* **35**: 195–222.

Minton, S. A. & da Costa, M. S. (1975). Serological relationships of sea snakes and their evolutionary implication. In *The biology of sea snakes*: 21–31. (Ed. Dunson, W. A.). University Park Press, Baltimore, London & Tokyo.

Rasmussen, A. R. (1989). An analysis of *Hydrophis ornatus* (Gray), *H. lamberti* Smith, and *H. inornatus* (Gray) (Hydrophiidae, Serpentes) based on samples from various localities, with remarks on feeding and breeding biology of *H. ornatus. Amphibia-Reptilia* **10**: 397–417.

Rasmussen, A. R. (1992). Rediscovery and redescription of *Hydrophis bituberculatus* Peters, 1872 (Serpentes, Hydrophiidae). *Herpetologica* **48**: 85–97.

Rasmussen, A. R. (1993). The status of the Persian Gulf sea snakes *Hydrophis lapemoides* (Gray, 1849) (Serpentes, Hydrophiidae). *Bull. Br. Mus. nat. Hist. (Zool.)* **59**: 97–105.

Rasmussen, A. R. (1994). A cladistic analysis of the *Hydrophis* subgenus *Chitulia* (Serpentes, Hydrophiidae). *Zool. J. Linn. Soc.* **111**: 161–178.

Rasmussen, J. B. (1985). A re-evaluation of the systematics of the African rear-fanged snakes of Bogert's groups XIII–XVI, including a discussion of some evolutionary trends within Caenophidia. *In Proceedings of the international symposium on African vertebrates: systematics, phylogeny and evolutionary ecology*: 531–548. (Ed. Schuchmann, K. -L.). Museum Alexander Koenig, Bonn.

Rasmussen, J. B. (1990). The retina of *Psammodynastes pulverulentus* (Boie, 1827) and *Telescopus fallax* (Fleischmann, 1831) with a discussion of their phylogenetic significance (Colubroidea, Serpentes). *Z. zool. Syst. EvolForsch.* **28**: 269–276.

Schwaner, T. D., Baverstock, P. R., Dessauer, H. C. & Mengden, G. A. (1985). Immunological evidence for the phylogenetic relationships of Australian elapid snakes. In *Biology of Australian frogs and reptiles*: 177–184. (Eds Grigg, G., Shine, R. & Ehmann, H.). Surrey Beatty & Sons Pty & The Royal Zoological Society of New South Wales, Chipping Norton, NSW.

Slowinski, J. B. (1989). The interrelationships of laticaudine sea snakes based on the amino acid sequences of short-chain neurotoxins. *Copeia* **1989**: 783–788.

Smith, H. M., Smith, R. B. & Sawin, H. L. (1977). A summary of snake classification (Reptilia, Serpentes). *J. Herpet.* **11**: 115–121.

Smith, M. A. (1926). *Monograph of the sea-snakes (Hydrophiidae).* British Museum (Natural History), London.

Smith, M. A. (1931). Description of a new genus of sea-snake from the coast of Australia, with a note on the structures providing for complete closure of the mouth in aquatic snakes. *Proc. zool. Soc. Lond.* **1931**: 397–398.

Smith, M. A. (1943). *The fauna of British India Ceylon and Burma, including the whole of the Indo-Chinese sub-region. Reptilia and Amphibia* **3**. Taylor and Francis, London.

Toriba, M. & Nakamoto, E. (1987). Reproductive biology of erabu sea snake, *Laticauda semifasciata* (Reinwardt). *Snake* **19**: 101–106.

Toriba, M. & Yosida, T. H. (1986). Karyotypes of the laticaudine sea snakes from Ryukyu Archipelago. *Snake* **18**: 70–76.

Underwood, G. (1957). On lizards of the family Pygopodidae. A contribution to the morphology and phylogeny of the Squamata. *J. Morph.* **100**: 207–268.

Underwood, G. (1967). *A contribution to the classification of snakes.* British Museum (Natural History), London.

Underwood, G. (1979). Classification and distribution of venomous snakes in the world. In *Handbook of experimental pharmacology* **52**. *Snake venoms*: 15–40. (Ed. Lee, C.). Springer-Verlag, Berlin and Heidelberg.

Voris, H. K. (1977). A phylogeny of the sea snakes (Hydrophiidae). *Fieldiana (Zool.)* **70**: 79–169.

Wall, F. (1909). A monograph of the sea snakes. *Mem. Asiat. Soc. Beng.* **2**: 169–251.

Williams, K. L. & Wallach, V. (1989). *Snakes of the world* **1**. *Synopsis of snake generic names.* Robert E. Krieger Publishing Co., Malabar, Florida.

Young, B. A. (1987). The comparative cephalic osteology and myology of three species of sea snakes. *Zool. Jb. Abt. Anat. Ontog. Tiere* **116**: 421–452.

3 Evolution, systematics and biogeography of Palaearctic vipers

GÖRAN NILSON and CLAES ANDRÉN

Synopsis

Very few synapomorphies have been proposed for the 'Eurasian group' (*Vipera s. l.*, *Eristicophis* and *Pseudocerastes*) within the Viperinae. Several studies have demonstrated difficulties in distinguishing the group in cladistic terms. However, different species groups show monophyletic patterns. The different spectra of habitat choice, altitude and latitude distributions, reproductive strategies etc. are analysed, and various phylogenies as well as the evolutionary history of lineages within the group are re-examined.

Species concepts are focused on interbreeding and on common descent. Thus the breeding systems are of interest in analysing systematics. Reproduction is characterized by spermiogenesis during autumn (the '*aspis*' type) or during spring (the '*berus*' type). In the latter, mating is triggered by the spring moult. The different strategies can be connected to different habitat types and ecology. Oviparous species with the '*berus*' type of reproduction inhabit southern steppe/semidesert areas. Viviparous species with the '*berus*' type of reproduction inhabit environments at high altitude or latitude, i.e., climatically more severe environments, such as alpine habitats or boreal forests. Viviparous species with the '*aspis*' type of reproduction inhabit lowland temperate areas. Is this pattern reflecting a broad adaptive flexibility amongst vipers, or is it a consequence of the phylogeny? The splitting of lineages is a Paleogene or early Miocene event. All extant taxa within a lineage seem to share the same reproductive strategy, and thus reproduction seems to follow the evolutionary history of the different lineages. Taxa with similar reproductive systems are further united by similar climatic and habitat adaptations. The phylogenetic splitting in lineages represents a major ecological shift and lineages express their own ecological adaptations. These facts should be reflected in the taxonomy of the group.

Introduction

Most venomous snakes of the Palaearctic region belong to the genus *Vipera* and related groups within the subfamily Viperinae, the 'Eurasian vipers'. This assemblage of snakes contains up to 74 extant and extinct taxa, including subspecies and Neogene and Quaternary species. The actual number can vary, depending on the classification used.

Symp. zool. Soc. Lond. (1997) No. 70: 31–42
ISBN 0 19 854986 5

Very few apomorphies have been proposed for the 'Eurasian group' (*Vipera s. l.*, *Eristicophis* and *Pseudocerastes*). According to Groombridge (1980, 1986), the single apomorphy proposed is the presence of a nasorostral scale. He points out the difficulties of distinguishing the group in cladistic terms, which is also reflected in the results of Marx & Rabb (1965) and Ashe & Marx (1988).

Materials and methods

Most taxa have been studied in the field. Almost all habitats have been visited. A large number of taxa have been bred in captive breeding programmes at the Department of Zoology, University of Göteborg (ZIG) over a number of years. Tissues and osteological material from ZIG have been used in co-operation or by other herpetologists, and results published elsewhere during these years are used in this review.

Nomenclature and species complexes

The Eurasian vipers have earlier been divided into subgenera or genera. Genus *Vipera* was introduced by Laurenti in 1768 in his description of *Vipera Francisci Redi*—the Italian asp viper, a species already named as *Coluber aspis* by Linnaeus 1758 in his 10th edition of *Systema Naturae*. The Italian asp viper is presently known as *Vipera aspis francisciredi*. Merrem 1820 introduced *Pelias* based on *Vipera berus* for the smaller vipers. This was followed by Bonaparte 1835 in his description of *Pelias ursinii* in Italy. In 1896 Boulenger and Alcock & Finn respectively described *Pseudocerastes* and *Eristicophis* as monotypic genera. *Daboia* was originally introduced by Gray in 1842 for *Daboia russelli*. In 1927 Reuss introduced the name *Macrovipera* for the *Vipera lebetina* group.

The genus *Vipera* (*s.l.*) has been divided by various authors in recent times. Saint Girons (1978), followed by Szyndlar (1991) for Neogene species, used the division 'European' and 'Oriental' vipers: 'European' for the mainly European *berus* and *aspis* complexes, and 'Oriental' for the larger, mainly Asian *lebetina*, *xanthina* and *russelli* complexes. *Daboia* was extended to all the 'Oriental vipers' by Obst (1983) including *Pseudocerastes* as one of its subgenera. Obst further used *Vipera* and *Rhinaspis* as subgenera in his *Vipera s.str.*, representing the *berus* and *aspis* groups respectively. However, *Rhinaspis* Reuss, 1933 is synonymous with *Vipera* Laurenti, 1768 (Nilson 1989).

Zerova (1992) used the names 'shieldheaded' vipers for the *berus* complex and 'scaleheaded' for the rest in genus *Vipera s.l.* She reintroduced the genus *Pelias* for the *berus* complex, her 'shieldheaded' vipers, and divided her 'scaleheaded vipers' into the two subgenera *Vipera* — 'small scaleheaded vipers' (= *aspis* group) and *Daboia* — 'large scaleheaded vipers' (= Oriental vipers, *sensu* Saint Girons).

Herrmann, Joger & Nilson (1992a) restricted *Daboia* to *V. russelli*, and re-established the name *Macrovipera* from synonymy for the *Vipera lebetina* group.

Thus during the last 20 years four different generic names have been used for *Vipera* (*s.l.*): *Vipera s.str.*, *Pelias*, *Macrovipera* and *Daboia*. Moreover, the two species groups in *Vipera s.str.* have been used in different combinations together with the other groups. In the following discussion they are treated as two separate groups: *Vipera* 1 (= *Vipera s.str.*, the *aspis* complex) and *Vipera* 2 (the *xanthina* complex).

Climatic preferences and biogeography

Moist and wet habitats are preferred by the smaller vipers of the *berus* group (*berus*, *kaznakovi*, *dinniki*, *barani*, *darevskii* etc.) (e.g. Brodmann 1987; Orlov & Tuniyev 1990; own obs.). The *ursinii* group prefers drier habitats in lowland or alpine regions (Kramer 1961; Nilson & Andrén in prep.). These are all taxa referred to *Pelias*.

Dry habitats on rocky or sandy ground in southern Europe and the Mediterranean Basin are inhabited by species like *aspis*, *ammodytes*, *latastei* (*Vipera* 1) (e.g. Brodmann 1987 and references therein), and in the Near East by *xanthina*, *raddei* and related forms (*Vipera* 2) (e.g. Nilson & Andrén 1986 and references therein). Thus the two *Vipera* groups inhabit drier habitats from lowland to mountain regions. Dry steppes and semideserts, as well as dry agricultural land, are inhabited by the large members of the *lebetina* group, the blunt-nosed vipers (*Macrovipera*) (e.g. Brodmann 1987 and references therein). The monotypic *Daboia* is the single tropical member of '*Vipera*' *s.l.* (e.g. Wüster *et al.* 1992 and references therein).

Rocky and sandy deserts of central Asia are inhabited by the monotypic genera *Pseudocerastes* and *Eristicophis*, respectively (e.g. Joger 1984; Latifi 1991).

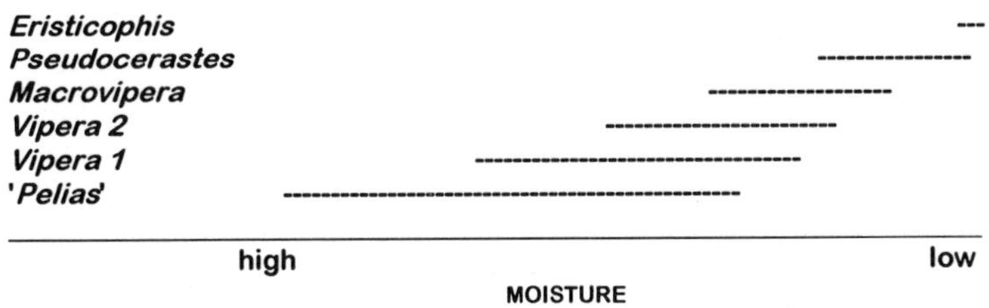

Fig. 1. Relative distribution of the Palaearctic 'Eurasian' vipers in a moisture continuum.

In a continuum from wet to dry the *Pelias* group is found in the wetter section, except for *Vipera ursinii* which also occurs in drier regions. The other Palaearctic groups have a clinal distribution towards drier habitats in a series: *Vipera* 1 — *Vipera* 2 — *Macrovipera* — *Pseudocerastes* — *Eristicophis*. *Eristicophis* is the most arid-adapted genus, occurring in sand-dune deserts (Fig. 1). *Daboia* is divergent, and represents a taxon adapted to the tropical zone.

A gradient can be seen when it comes to latitudinal distribution of the different groups, with *Pelias* as the most northern in its distribution. *Vipera berus* crosses the Polar circle, with its northern limit on the Kola Peninsula and northernmost Scandinavia. The remaining groups are distributed in a series towards the south (Fig. 2), with *Daboia* crossing the equator in south-east Asia. A similar gradient, although converted, can be seen in the altitudinal distribution. All groups occur at or near sea level, but the *Pelias* group as a group reaches the highest altitudes, and the remainder display a decreasing maximum altitudinal occurrence. However, isolated populations within several of the groups can have a higher distribution. *Pseudocerastes* and *Macrovipera* rarely climb above 1000 m altitude, but exceptions are known from India and Kashmir respectively. *Vipera* 2 reaches almost 3000 m in the Elburz mountains in northern Iran (*Vipera latifii*) as does *Vipera* 1 in the Atlas mountains in Morocco (*Vipera monticola*) (Fig. 2). When altitude and latitude are plotted together, a pattern emerges where *Pelias* is a lowland group towards the north and 'climbs' up in the mountains towards the south, and is replaced by the other groups in similar series. *Vipera* 1 and *Vipera* 2 inhabit higher altitudes in the southernmost part of their ranges in a similar way to *Pelias*, although in a less pronounced manner, and are replaced by *Macrovipera* at lower elevations (Fig. 2). Going from west to east, this can be exemplified by the alpine *Vipera monticola* (*Vipera* 1) and *Macrovipera mauritanica* in Morocco; alpine *Vipera ursinii ursinii* (*Pelias*) in France and Italy and sympatric lowland inhabitants *Vipera aspis aspis* and *V. a. francisciredi* (*Vipera* 1); alpine *Vipera ursinii macrops*/*V. u. graeca* (*Pelias*) in former Yugoslavia and Greece respectively and sympatric lowland inhabitants *Vipera a. ammodytes*/*V. a. meridionalis* (*Vipera* 1); *Vipera ursinii anatolica* (*Pelias*) in southern Turkey and sympatric *Vipera xanthina* (*Vipera* 2) and *Macrovipera lebetina obtusa* in coastal lowlands; *Vipera ursinii eriwanensis* (*Pelias*), *Vipera raddei raddei* (*Vipera* 2) and *Macrovipera lebetina obtusa* in the Aras valley, eastern Turkey and Armenia; and *Vipera ursinii ebneri* (*Pelias*), *Vipera latifii* (*Vipera* 2) and *Macrovipera lebetina obtusa* in the Elburz mountains, northern Iran (own obs.). Is this pattern a secondary ecological adaptation, reflecting a broad adaptive flexibility, or is this a result of their phylogenetic history?

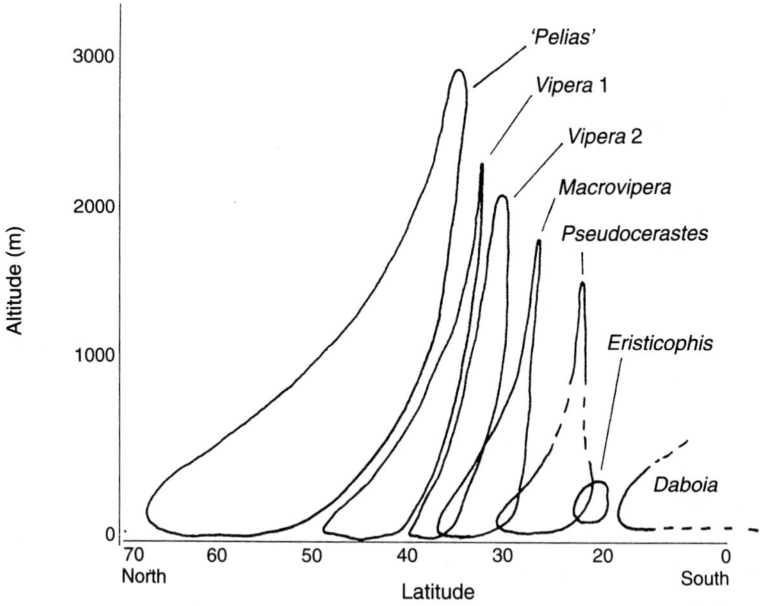

Fig. 2. Combined altitude and latitude distribution of the Palaearctic 'Eurasian' vipers. See text for discussion.

Reproductive strategies

Species concepts are focused either on interbreeding or on common descent. Thus the breeding systems are of interest in analysing systematics.

The reproductive behaviour for vipers includes combat and mating (e.g. Nilson 1980, 1981; Nilson & Andrén 1988). Birth of developed young or deposition of eggs take place during the later part of the active season. Both oviparity and viviparity occur but are connected to certain species groups (Table 1). The large species of *Macrovipera*, *Pseudocerastes*, *Eristicophis* and *Vipera palaestinae* are oviparous. The remainder, i.e. *Pelias*, *Vipera* 1, *Vipera* 2 and *Daboia* are all viviparous.

Table 1. Taxa adapted to the different reproductive strategies

Oviparity	Viviparity		
	Mixed — *berus*		Mixed — *aspis*
lebetina	*berus*		*seoanei*
mauritanica	*ursinii*		*latastei*
schweizeri	*renardi*?	⟷	*renardi*?
palaestinae	*kaznakovi*?	⟷	*kaznakovi*?
Pseudocerastes persicus	*xanthina* complex		*ammodytes*
	ammodytes?		*a. montandoni*
			a. meridionalis
			aspis
			a. aspis
			a. zinnikeri
Macrovipera	'*Pelias*'		*Vipera* 1
Pseudocerastes	*Vipera* 2		*Vipera seoanei*
Vipera palaestinae	(some *ammodytes*?)		(*V. kaznakovi*?)

Two kinds of reproductive cycles, the '*aspis*' and the '*berus*' type, have been described. Saint Girons (1976, 1982) termed these types 'mixed', as the spermatogenesis is extended over more than one season. The autumn is the most important season for spermatogenetic activities in the '*aspis*' type, which is characterized by a summer–autumn spermatocyto- and spermiogenesis, with the production of ripe sperm starting in the autumn, and continuing during the spring. As a consequence both autumn and spring mating occurs. Referred to as mixed, this type of spermatogenetic cycle is approaching the aestival kind, or postnuptial in the sense of Schuett (1992).

The '*berus*' type is characterized by autumn spermatocytogenesis and spring spermiogenesis, with the sperm not ready until just prior to the start of the spring mating, which in turn is triggered by the spring moult (Nilson 1980). The mating activities of the '*aspis*' type are not triggered by a moult. The '*berus*' type belongs to category II with mixed spermatogenetic cycle (*sensu* Schuett 1992).

In females a type II vitellogenesis (Aldridge 1979) occurs in all taxa examined, with a secondary vitellogenesis beginning during autumn. The young are born in the following autumn. Females of the '*aspis*' type can mate both during the autumn and during the spring while females of the '*berus*' type only mate during the spring.

Vipera berus and related taxa are '*berus*' breeders. The European alpine *ursinii* populations also seem to have the '*berus*' reproductive strategy (Table 1). The alternative pattern, the '*aspis*' mode, has been described from most *aspis* subspecies, from *latastei* and *monticola*, and also from *seoanei*, which traditionally is thought to be related to *berus*. Thus *seoanei* and *Vipera aspis* subspecies all seem to be typical '*aspis*' breeders. Moreover, in our material, the lowland *kaznakovi* as well as *montandoni* and *meridionalis* of the *ammodytes* group show spring and autumn mating activities in captivity, indicating the '*aspis*' strategy. However, the situation is not perfectly clear. According to Saint Girons (e.g. 1980, 1982) besides *berus* and *ursinii*, *ammodytes*, *montandoni* and *kaznakovi* also use the '*berus*' strategy. The Asian populations of the *renardi* group, which show late autumn mating behaviour in captivity, are also of uncertain status.

The members of the *xanthina* complex are all '*berus*' strategists, which also seems to be the case for all the oviparous members of the *lebetina* group.

Thus taxa have evolved different breeding strategies, and exhibit different biogeography. The oviparous taxa are all restricted to southern hot steppe and semidesert areas. They are all characterized by the '*berus*' strategy. Species belonging to this group are *lebetina*, *mauritanica* and *schweizeri*, i.e. *Macrovipera*, and *Pseudocerastes* and *Eristicophis*. *Vipera palaestinae* also has this pattern. The viviparous taxa that have the '*aspis*' reproduction, with autumn and spring mating, are all lowland temperate taxa, like *Vipera aspis*, *V. latastei* and some or perhaps all *V. ammodytes*. Furthermore, *V. seoanei* expresses this pattern. *Vipera renardi* and *kaznakovi* remain questionable. Thus the '*aspis*' category is restricted mostly to *Vipera* 1.

The viviparous taxa that have the '*berus*' reproduction and a spring mating triggered by moulting are all the taxa that are found in climatically more severe environments, such as in alpine habitats or at high latitudes. Taxa of this group are *Vipera berus*, alpine *V. ursinii* of Europe, (some *ammodytes*?) and the *xanthina* complex, i.e. mostly *Pelias* and *Vipera* 2 (Table 1).

Table 2. Ecological and reproductive characteristics for the different taxa

Macrovipera *Pseudocerastes* (*Eristicophis*) *V. palaestinae*	*Pelias* *Vipera* 2	*Vipera* 1
Low latitude Low altitude Hot climate Dry habitat Oviparous '*berus*' reprod.	High latitude or high altitude Temperate Wet or dry Viviparous '*berus*' reprod.	Low latitude and middle altitude Warm to temperate Dry Viviparous '*aspis*' reprod.

So there seems to be a pattern where *Macrovipera*, *Pseudocerastes* and *Eristicophis*, and *Vipera palaestinae* display oviparity and a '*berus*' type of reproductive strategy, and inhabit dry semidesert areas at lower altitudes and latitudes. *Vipera* 1 display viviparity and the '*aspis*' reproductive strategy and inhabit temperate lowlands, while *Pelias* and *Vipera* 2 exhibit viviparity and a '*berus*' reproductive cycle, and inhabit mountainous environments or high latitudes (Table 2). Again, does this pattern reflect adaptive flexibility, or phylogenetic history?

Phylogenetic reconstructions

Attempts have been made to reconstruct the evolutionary history of the Palaearctic vipers. Focusing on the branching events, a reanalysis of these reconstructions may help in understanding the phylogeny of the groups discussed here.

Saint Girons (1980) proposed a phylogeny within the 'European vipers' based on a phenetic approach. This represents a traditional view of the relationships within this group (i.e. *Pelias* and *Vipera* 1), with a trichotomy whereby the *ursinii* subspecies, the *berus* group including *kaznakovi* and *seoanei*, and the *aspis* group (*Vipera* 1) are sister lineages. The latter is in turn divided into a trichotomy with the *aspis*, *latastei* and *ammodytes* groups as sister clades.

A similar result was calculated by Herrmann, Joger, Nilson & Sibley (1987), based on immunology, with a trichotomy of *Vipera* 1, the *berus* group and the *ursinii* group, which together form a sisterclade to *Vipera* 2. However, *Vipera kaznakovi*, traditionally referred to *Pelias*, is clustered in the *Vipera* 1 clade. *Vipera palaestinae* and *Macrovipera* are clustered together at the base of the tree.

In a later study based on molecular methods (Herrmann, Joger & Nilson 1992b), clear differences within this 'European' group can be observed. On the basis of immunological distances (hierarchical cluster analysis) the *ursinii* group is an earlier sidebranch to the rest of *Pelias* and *Vipera* 1. Again *kaznakovi* (together with its sister species *dinniki*) is in the *Vipera* 1 group by clustering with *ammodytes*, as is *seoanei* which clusters with *aspis*. With the distance Wagner procedure, *seoanei* clusters with *berus* while *kaznakovi* is at the base of the *Vipera* 1 stem. *Vipera seoanei* (originally described as a subspecies of *berus*) and *kaznakovi* are traditionally referred to the *berus* complex (*Pelias*). Thus, the positions of these taxa are not properly resolved, but the actual (*seoanei*) and possible (*kaznakovi*) '*aspis*'-type reproductive strategy suggests a real connection to *Vipera* 1. Together *Pelias* and *Vipera* 1 form a sister clade to *Vipera* 2 (through *xanthina*) and in turn a sister group to *Macrovipera* (through *lebetina*).

Phylogenetic relationships within the *Vipera ursinii* complex based on immunological distances (Fitch–Margoliash, Wagner and UPGMA trees) all gave the same branching patterns (Joger, Herrmann & Nilson 1992; Nilson, Andrén & Joger 1993). The *ursinii* taxa, which constitute a part of *Pelias*, are divided in obviously well-distinguished lineages, actually demonstrating a series of sibling species with immunological distances well within the range of those separating good species.

A cladistic analysis based on morphology and molecular methods (Nilson, Höggren *et al.* 1994) shows that the *ursinii* group is well separated from the *kaznakovi* group. Together they form a sister clade to *berus*. *Vipera aspis* is connected to the base of this group. The connection of *kaznakovi* to *aspis* was not supported in this study.

For the 'Oriental' vipers, Groombridge (1980) reconstructed the phylogeny on the basis of a cladistic analysis of morphology. *Pseudocerastes* and *Eristicophis* form a clade which is a sister group to *lebetina* (*Macrovipera*) (Groombridge 1986). All these taxa together form a sister group to another clade containing *Daboia* and *V. palaestinae*. In turn *Vipera* 2 is a sister group to the whole of this cluster, and together they form a sister lineage to *Vipera* 1. *Pelias* is a sister group to all. The different genera and species groups can be distinguished in phylogenetic terms.

Herrmann, Joger & Nilson (1992a) reconstructed the phylogeny on the basis of immunological distances. Again *Pseudocerastes* is a sister group to *Macrovipera* (*lebetina* and *mauritanica*), but different analyses gave slightly different results for other taxa. *Vipera palaestinae* clustered either to the base of this clade (FITCH program in PHYLIP, version 3.1 for Fitch–Margoliash tree), or to the base of the sister clade represented by *xanthina* and *raddei* (= *Vipera* 2) (KITCH program from PHYLIP for reciprocal data matrix).

In the same study, an UPGMA tree based on immunological albumin comparison revealed *Macrovipera*, *Daboia* and a clade consisting of *Eristicophis* and *Pseudocerastes* as a trichotomy which forms a sister group to *Vipera* 2 together with *V. palaestinae*. By applying time calibration to the data (1 ID = 0.5 million years), and by dating the separation of *Macrovipera schweizeri* on the western Cyclades at approximately five million years ago ('Messinian event'), a picture emerges that the splitting of *Vipera* 2 from the branch leading to *Macrovipera*, *Daboia* and the desert vipers took place roughly 15 million years ago, during the middle Miocene, and that the splitting of the latter took place about 13 million years ago.

The position of *V. palaestinae* remains uncertain. It has been clustered together with *Vipera* 2, with *Macrovipera* and with *Daboia* in different phylogenetic reconstructions.

A hypothetical phylogenetic relationship of 'Eurasian vipers' could be as shown in Fig. 3. *Pseudocerastes* and *Eristicophis* cluster together in all cases. *Pelias* and *Vipera* 1 share some species (*kaznakovi*, *seoanei*) with unresolved positions. Together *Pelias*, *Vipera* 1 and *Vipera* 2 are viviparous, and more or less mountainous. *Macrovipera* is an oviparous steppe inhabitant while *Daboia* is tropical and viviparous.

The traditional question of cladistic classification is, should large monophyletic groups be retained or should major monophyletic clades within them be recognized as discrete taxa? The traditional genus *Vipera s. l.* must be considered as polyphyletic. If this reflects the true evolutionary history, it suggests a recognition of taxonomic units on a more fragmented level. On the basis of the reconstructed phylogenies, the seven taxonomic lineages discussed here may each have a long evolutionary history. The fossil record can give us an answer.

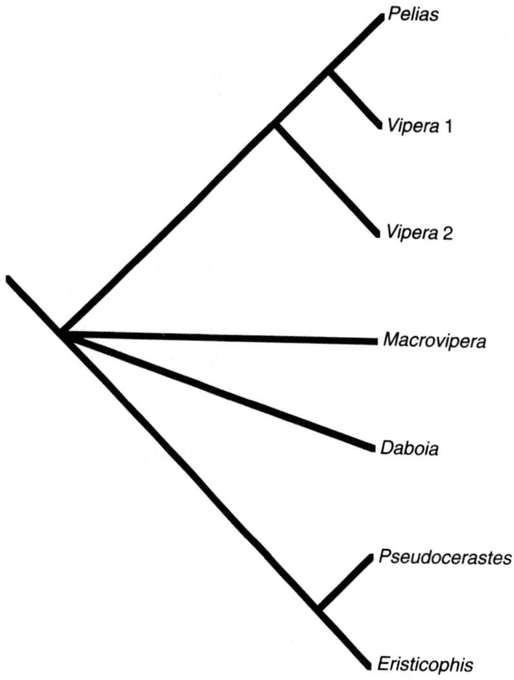

Fig. 3. Hypothetical phylogenetic relationships of 'Eurasian vipers'.

The fossil record

The fossil record has been thoroughly analysed during the last decade.

The *Pelias* lineage is traced back to the lower Miocene with a large series of well-documented fossils (Szyndlar 1984) (Fig. 4). The same is the case for the *Vipera* 1 lineage. Here, *Vipera antiqua* from the lower Miocene at Dolnice, Czech Republic, is almost identical to modern *Vipera ammodytes*, *V. maghrebiana* from the middle Miocene of Morocco is another member of this lineage, and *Vipera meotica* from the upper Miocene of Ukraine shows great similarity to *Vipera aspis* (Szyndlar 1987; Szyndlar & Zerova 1992).

The *Vipera* 2 lineage is represented by *V. platyspondyla* from the lowermost Miocene of Dolnice, Czech Republic, and represents an early ancestral member of this lineage. It is believed to have been sympatric with the small *Vipera antiqua*, as the fossil remains of these two taxa have been found together at the same deposits (Szyndlar & Schleich 1993). In the middle Miocene this lineage is represented by *V. ukrainica* from Ukraine (Zerova 1992).

The *Macrovipera* lineage is represented by a comparatively good sample of ancestral taxa. *Vipera sarmatica* has been described from the middle Miocene of Moldavia, followed by *V. burgenlandica* from the upper Miocene of Austria (Bachmayer & Szyndlar 1987; Szyndlar & Zerova 1992). This taxon is today considered to be a member of this lineage, although the fossil remains also showed characteristics both of *Vipera xanthina* (i.e. *Vipera* 2) and of *Vipera palaestinae* (a species whose taxonomic status is not clearly resolved, see above). From the uppermost Miocene, the *Macrovipera* lineage is represented by *V. gedulyi* from Hungary. From the Pliocene, *V. kuchurganica* from Ukraine is referred to this lineage (Zerova, Lungu & Chkhikvadze 1987; Szyndlar 1991).

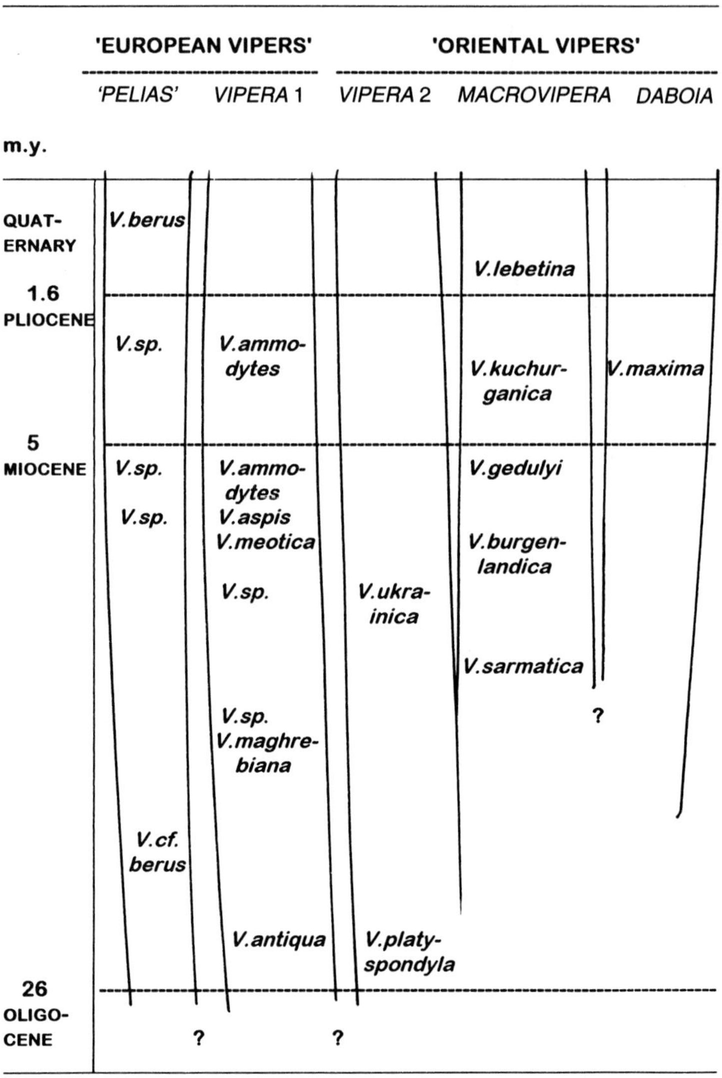

Fig. 4. Neogene and Quaternary lineages within the 'Eurasian vipers'.

The *Daboia* lineage is represented by Pliocene *V. maxima* from Spain (Szyndlar 1988). This is a long distance from the present distribution of *Daboia* (*s. str.*), which is south-east Asia, and may reflect a rather recent (Pliocene/Pleistocene) wide distribution, and indeed a much warmer climate. In addition there are a number of fossil records that very clearly can be referred to one or another of these lineages from most parts of the Neogene. The split of the Eurasian vipers into the lineages characteristic for the living members goes back to the lower Miocene, 26 million years ago (Szyndlar 1991), which is even earlier than indicated by the phylogenetic reconstructions based on molecular methods and/or morphology. Consequently, the seven taxonomic lineages of 'Eurasian vipers' each have a long evolutionary history.

In this perspective, and in combination with the different ecological and reproductive patterns observed, it could be appropriate to treat the lineages as taxonomic entities, carrying their own names.

References

Aldridge, R. D. (1979). Female reproductive cycles of the snakes *Arizona elegans* and *Crotalus viridis*. *Herpetologica* **35**: 256–261.

Ashe, J. S. & Marx, H. (1988). Phylogeny of the viperine snakes (Viperinae). Part II. Cladistic analysis and major lineages. *Fieldiana* (*Zool.*) (*N.S.*) No. 52: 1–23.

Bachmayer, F. & Szyndlar, Z. (1987). A second contribution to the ophidian faunas (Reptilia: Serpentes) of Kohfidisch, Austria. *Annln naturh. Mus. Wien* (*A*) **88**: 25–39.

Brodmann, P. (1987). *Die Giftschlangen Europas und die Gattung* Vipera *in Afrika und Asien.* Kümmerly & Frey, Bern.

Groombridge, B. (1980). *A phyletic analysis of viperine snakes*. PhD thesis: City of London Polytechnic, Council for National Academic Awards.

Groombridge, B. (1986). Phyletic relationships among viperine snakes. In *Studies in herpetology*: 219–222. (Ed. Roček, Z.). Charles University, Prague.

Herrmann, H. -W., Joger, U. & Nilson, G. (1992a). Phylogeny and systematics of viperine snakes III. Resurrection of the genus *Macrovipera* (Reuss, 1927) as suggested by biochemical evidence. *Amphibia-Reptilia* **13**: 375–392.

Herrmann, H. -W., Joger, U. & Nilson, G. (1992b). Molecular phylogeny and systematics of viperine snakes I. General phylogeny of European vipers (*Vipera sensu stricto*). In *Proceedings of the sixth ordinary general meeting of the Societas Europaea Herpetologica*: 219–224. (Eds Korsós, Z. & Kiss, I.). Hungarian Natural History Museum, Budapest.

Herrmann, H. -W., Joger, U., Nilson, G. & Sibley, C. G. (1987). First steps towards a biochemically based reconstruction of the phylogeny of the genus *Vipera*. In *Proceedings of the fourth ordinary general meeting of the Societas Europaea Herpetologica*: 195–200. (Eds van Gelder, J. J., Strijbosch, H. & Bergers, P.J.M.). Societas Europaea Herpetologica, Faculty of Sciences, Nijmegen.

Joger, U. (1984). The venomous snakes of the Near and Middle East. *Beih. Tüb. Atlas vord. Orients Reihe A* No. 12: 1–115.

Joger, U., Herrmann, H. -W. & Nilson, G. (1992). Molecular phylogeny and systematics of viperine snakes II. A revision of the *Vipera ursinii* complex. In *Proceedings of the sixth ordinary general meeting of the Societas Europaea Herpetologica*: 239–244. (Eds Korsós, Z. & Kiss, I.). Hungarian Natural History Museum, Budapest.

Kramer, E. (1961). Variation, Sexualdimorphismus, Wachstum und Taxonomie von *Vipera ursinii* (Bonaparte, 1835) und *Vipera kaznakovi* Nikolskij, 1909. *Rev. suisse Zool.* **68**: 627–725.

Latifi, M. (1991). The snakes of Iran. *Contr. herpet.* No. 7: 1–159.

Marx, H. & Rabb, G. B. (1965). Relationships and zoogeography of the viperine snakes (family Viperidae). *Fieldiana* (*Zool.*) **44**: 161–206.

Nilson, G. (1980). Male reproductive cycle of the European adder, *Vipera berus* and its relation to annual activity periods. *Copeia* **1980**: 729–737.

Nilson, G. (1981). Ovarian cycle and reproductive dynamics in the female adder, *Vipera berus*

(Reptilia: Viperidae). *Amphibia-Reptilia* **2**: 63–82.

Nilson, G. (1989). Book reviews. Hans Schiemenz (1987): Die Kreuzotter/*Vipera berus*. *Amphibia-Reptilia* **10**: 345–346.

Nilson, G. & Andrén, C. (1986). The mountain vipers of the Middle East — the *Vipera xanthina* complex (Reptilia, Viperidae). *Bonn. Zool Monogr.* No. 20: 1–90.

Nilson, G. & Andrén, C. (1988). *Reproductive ecology and demography in Viperinae*. First World Congress of Herpetology, University of Kent, Canterbury. Abstract.

Nilson, G. & Andrén, C. (In preparation). *The meadow and steppe vipers of Europe and Asia, the* Vipera ursinii *complex.*

Nilson, G., Andrén, C. & Joger, U. (1993). A re-evaluation of the taxonomic status of the Moldavian steppe viper based on immunological investigations, with a discussion of the hypothesis of secondary intergradation between *Vipera ursinii rakosiensis* and *Vipera (ursinii) renardi*. *Amphibia-Reptilia* **14**: 45–57.

Nilson, G., Höggren, M., Tuniyev, B. S., Orlov, N. L. & Andrén, C. (1994). Phylogeny of the vipers of the Caucasus (Reptilia, Viperidae). *Zoologica Scr.* **23**: 353–360.

Obst, F. J. (1982) [1983]. Zur Kenntnis der Schlangengattung *Vipera* (Reptilia, Serpentes, Viperidae). *Zool. Abh. st. Mus. Tierk., Dresden* **38**: 229–235.

Orlov, N. L. & Tuniyev, B. S. (1990). Three species in the *Vipera kaznakowi* complex (Eurosiberian group) in the Caucasus: their present distribution, possible genesis, and phylogeny. *Asiat. herpet. Res.* **3**: 1–36.

Saint Girons, H. (1976). Les différents types de cycles sexuels des mâles chez les vipères européennes. *C. r. hebd. Séanc. Acad. Sci., Paris (Sér. D)* **282**: 1017–1019.

Saint Girons, H. (1978). Morphologie externe comparée et systématique des vipères d'Europe (Reptilia, Viperidae). *Rev. suisse Zool.* **85**: 565–595.

Saint Girons, H. (1980). Biogéographie et évolution des vipères européennes. *C. r. Séanc. Soc. Biogéogr.* **56** No. 496: 146–172.

Saint Girons, H. (1982). Reproductive cycles of male snakes and their relationships with climate and female reproductive cycles. *Herpetologica* **38**: 5–16.

Schuett, G. W. (1992). Is long-term sperm storage an important component of the reproductive biology of temperate pitvipers? In *Biology of the pitvipers*: 169–184. (Eds Campbell, J. A. & Brodie, E. D.). Selva, Tyler, Texas.

Szyndlar, Z. (1984). Fossil snakes of Poland. *Acta zool. cracov.* **28**: 1–156.

Szyndlar, Z. (1987). Snakes from the lower Miocene locality of Dolnice (Czechoslovakia). *J. Vertebr. Paleont.* **7**: 55–71.

Szyndlar, Z. (1988). Two new extinct species of the genera *Malpolon* and *Vipera* (Reptilia, Serpentes) from the Pliocene of Layna (Spain). *Acta zool. cracov.* **31**: 687–706.

Szyndlar, Z. (1991). A review of Neogene and Quaternary snakes of central and eastern Europe. Part II: Natricinae, Elapidae, Viperidae. *Estud. geol. (Madrid)* **47**: 237–266.

Szyndlar, Z. & Schleich, H. H. (1993). Description of Miocene snakes from Petersbuch 2 with comments on the lower and middle Miocene ophidian faunas of southern Germany. *Stuttg. Beitr. Naturk. Ser. B* **192**: 1–47.

Szyndlar, Z. & Zerova, G. A. (1992). Miocene snake fauna from Cherevichnoie (Ukraine, USSR), with description of a new species of *Vipera*. *Neues Jb. Geol. Paläont. Abh.* **184**: 87–99.

Wüster, W., Otsuka, S., Malhotra, A. & Thorpe, R. S. (1992). Population systematics of Russell's viper: a multivariate study. *Biol. J. Linn. Soc.* **47**: 97–113.

Zerova, G. A. (1992). *Vipera (Daboia) ukrainica* — a new viper (Serpentes; Viperidae) from the Middle Sarmatian (Upper Miocene) of the Ukraine. *Neues Jb. Geol. Paläont. Abh.* **184**: 235–249.

Zerova, G. A., Lungu, A. N. & Chkhikvadze, V. M. (1986) [1987]. Large fossil vipers from northern Black Sea-side and Transcaucasus. *Trudy zool. Inst. Leningr.* **158**: 89–99.

4 Evolution of viperine snakes

HANS-WERNER HERRMANN and ULRICH JOGER

Synopsis

A quantitative immunological determination of serum albumin differences between 41 viperid species was used to reconstruct the phylogenesis of this group.

The data are currently being supplemented by mitochondrial cytochrome *b* sequences. A comparison of our molecular trees and the morphological trees of Marx, Ashe & Watrous (1988) reveals fundamental differences in branching pattern and groupings, whereas the monophyletic groups of Groombridge (1980), based on morphology and internal anatomy, are essentially the same as ours. After a revision, the data set of Marx *et al.* turned out to be inadequate for a cladistic analysis, whereas the revised characters of Groombridge could be used to compute a cladogram with HENNIG86.

According to our data, the Viperinae most probably arose in Africa with *Causus*-like forms. Within the African viperids, the genera *Causus, Atheris, Bitis* and *Cerastes* were all found to be monophyletic. The monophyly of *Echis* could not be proved. A larger monophyletic group comprises all genera of the 'Eurasian stock' (*Daboia, Eristicophis, Macrovipera, Pseudocerastes, Vipera*). This group is related to *Cerastes* (and probably *Echis*), and reached Eurasia about 20 million years BP, according to palaeontological and molecular evidence.

Introduction

The subfamily Viperinae is generally regarded as the sister group of the Crotalinae. This has recently been corroborated by a comparative study of mitochondrial RNAs (Heise *et al.* 1995). Moreover, these authors confirm earlier results of Cadle (1992) and Knight & Mindell (1993) concerning the phylogenetic position of the East Asian viperid *Azemiops feae*. The genus *Azemiops* had been regarded as the most primitive member of the whole family and given a subfamily of its own (Liem, Marx & Rabb 1971). The new data show that it is in fact a basal branch of the Crotalinae.

The African *Causus*, the second viperid genus with a high proportion of ancestral characters, has also been regarded as a separate subfamily (e.g., Hoge & Romano-Hoge 1978/79; Groombridge 1986). Cadle's (1992) immunological data support this view.

Apart from this, there is little agreement about the relative phylogenetic positions and the taxonomy of the remaining 11 genera of Viperinae: the Ethiopian

Symp. zool. Soc. Lond. (1997) No. 70: 43–61
ISBN 0 19 854986 5

Adenorhinos, Atheris and *Bitis*, the Saharo-Sindian *Cerastes, Echis, Eristicophis* and *Pseudocerastes*, the Palaearctic *Macrovipera* and *Vipera*, and the Oriental *Daboia*. Obst (1982) united *Daboia, Macrovipera, Pseudocerastes* and the *Vipera xanthina* complex under the generic name *Daboia*. Herrmann, Joger & Nilson (1992b) refuted this arrangement. Instead, their comparative immunological analysis favoured the separation of these genera from each other and from *Vipera*. Groombridge (1986) found morphological and anatomical evidence that the African and Eurasian viperines are both monophyletic groups. *Echis* and *Cerastes* were tentatively attached to the African viperines. However, Marx *et al.* (1988) and Ashe & Marx (1988) published contradictory phylogenetic hypotheses. Their 'all-tree' unites *Echis* with *Daboia* (= *Vipera russelli*) and *Bitis, Cerastes* with *Pseudocerastes, Eristicophis, Macrovipera* (= *Vipera lebetina*) and the *Vipera xanthina* group, whereas the different European *Vipera* species, *Adenorhinos* and *Atheris* represent basal branches with no close relationships between each other. In their qualitative trees, Ashe & Marx (1988) again show polyphyly of *Vipera*; *Bitis* is allied either with *Atheris* and *Adenorhinos* or with *Echis*; *Cerastes* becomes the sister genus of *Eristicophis, Vipera xanthina* clusters either with *Echis, Pseudocerastes, Eristicophis* and *Cerastes*, or remains at the base of the tree as the most primitive viperine.

The confusion resulting from these different morphological analyses encouraged us to try to reconstruct a phylogenetic tree of the viperines by a quantitative immunological comparison of blood serum albumins — a technique which we had previously applied successfully to the phylogenies of gekkonid and agamid lizards (Joger 1984a, 1991). Antibody reactions are specific to complex molecular structures of an antigen (here: serum albumin) and in heterologous reactions, their strength is proportional to the degree of amino acid homology between the proteins compared (Maxson 1992). Therefore immunological identity always implies identity of a complex character with nearly indefinite numbers of possible states, thus greatly reducing the danger of being misled by convergent evolution (Joger in press).

Materials and methods

Blood samples for the immunological comparisons of serum albumins were collected from live specimens of the taxa listed in Appendix 1. Immunological methods were carried out as described in Joger (1984a). Serum albumins were purified using vertical discontinuous PAG electrophoresis. For antisera production we used up to six rabbits per antigen. Taxa and populations against which antisera were produced are listed in Appendix 1 and marked by an asterisk.

Quantitative precipitin tests were carried out and immunological distances (IDs) were calculated as percentage of the homologous reaction.

In a few cases, when reciprocal IDs were missing in the data matrices, identical reciprocal values were used to fill the matrices. To obtain a measure for ID reciprocity we calculated the percent non-reciprocity with a correction for non-random elements (Sarich & Cronin 1976).

Two methods of constructing dendrograms were both applied:

1. The Fitch–Margoliash algorithm using the procedure 'FITCH' of the computer program 'PHYLIP' (Felsenstein 1988). This is an additive method which does not require the assumption of a similar evolutionary rate. The outgroup *Boulengerina annulata* (Elapidae) was used for rooting the tree.
2. The UPGMA clustering method which requires ultrametric distances with almost identical evolutionary rates in the different clades. Since albumins of different species may show different rates of evolution, the IDs used for UPGMA were corrected for relative rates (Herrmann 1990). The *F* value (Prager & Wilson 1976) was evaluated to assess the goodness of fit between input and output data sets.

The morphological and anatomical data published by Groombridge (1980) and Marx *et al.* (1988) were used for a cladistic reanalysis with the maximum parsimony method of the computer program HENNIG86 (Farris 1988) after critical checking for phylogenetic information content and independence of the individual characters. Doubtful characters were omitted (Herrmann 1995). The procedures used were *mhennig**, *bb**, *xw s* and *nelsen*.

Results

Morphological/anatomical trees

In the character set of Ashe & Marx (1988) we found many character states that could easily arise by convergence (e.g., scale counts or meristic characters) and thus these characters had to be omitted as doubtful regarding their homology. Other problematical characters were those that have obviously evolved several times independently (e.g., supraocular 'horns' in desert vipers). The remaining characters were not sufficient to reconstruct a reliable cladogram with HENNIG86. Groombridge's (1980) character set, however, comprised enough informative characters for a cladistic analysis.

Two most parsimonious cladogram hypotheses were computed: (1) assuming additivity of character states; (2) assuming non-additivity of states. Those bifurcations that were identical in both trees were included in the consensus tree shown in Figs 1 and 2; those that differed were regarded as unresolved and shown as multiple branching events. For instance, no reliable synapomorphy was found that differentiates *Vipera* species from each other or from *Macrovipera*; on the other hand, six synapomorphies unite *Eristicophis* and *Pseudocerastes* (Fig. 1). The Palaearctic vipers are united by a single character, as are the Afrotropical vipers (except *Causus*). Whether Russell's viper (*Daboia*) shares a common branch with the Palaearctic or with the Afrotropical clade could not be determined. *Bitis* and *Atheris* are confirmed as monophyletic genera. *Cerastes* and *Echis*, which Groombridge (1986) only tentatively assigned to the Afrotropical clade, are now clearly members of the *Atheris* lineage, but not as a monophyletic sister group of *Atheris*. As a genus, *Cerastes* appears monophyletic; *Echis*, however, appears paraphyletic. Intrageneric relations remain largely unresolved.

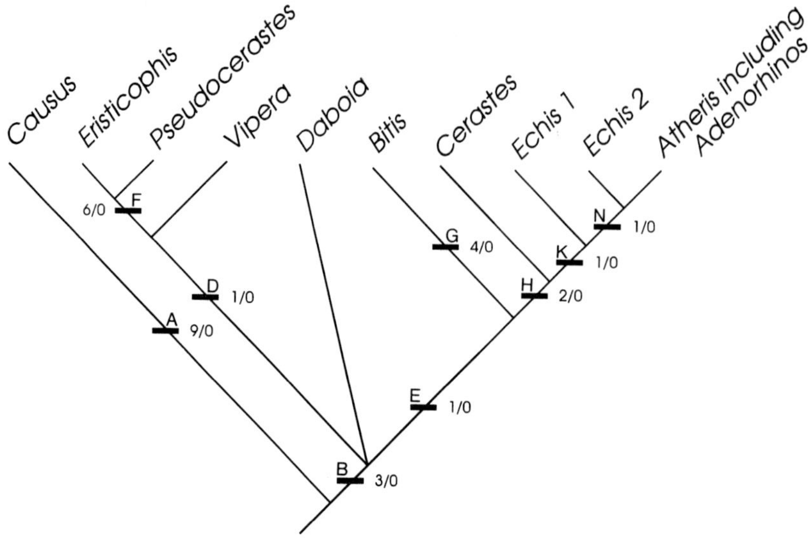

Fig. 1. Consensus cladogram of the Viperinae, computed with HENNIG86 using the data of Groombridge (1980). *Vipera* includes *Macrovipera*. Only branches that were found identical under assumptions of both additivity and non-additivity were accepted. Capital letters denote branches containing derived character states as listed in Appendix 2. First digits indicate the number of synapomorphies, second digits the number of reversals or convergent states found. (Steps = 310, consistency = 0.58, retention = 0.89).

Immunological trees

In order to keep pairwise comparisons in an acceptable order of magnitude, Palaearctic and Afrotropical vipers were investigated separately.

Two alternative immunological trees of Afrotropical vipers (with *Daboia* and *Macrovipera* to represent Eurasian vipers) are shown in Figs 3 and 4. Tables 1 and 2 give the corresponding distance matrices. Both trees show virtually identical orders of branching, except that the level of resolution of intrageneric relations is sometimes different. *Bitis* and *Atheris* are again confirmed as monophyletic, with *Bitis* branching off from the main stem first. Small, southern African *Bitis* and large, tropical African *Bitis* are separate evolutionary units, with the puffadder, *B. arietans*, as a third group without any close affiliation to *B. gabonica* and *B. nasicornis* as previously thought.

Within *Atheris*, the separate position of *A. superciliaris* is obvious, as in the morphological cladogram. *Echis* and *Cerastes*, especially the latter, appear now as relatives of the Palaearctic *Macrovipera* and the tropical Asian *Daboia*.

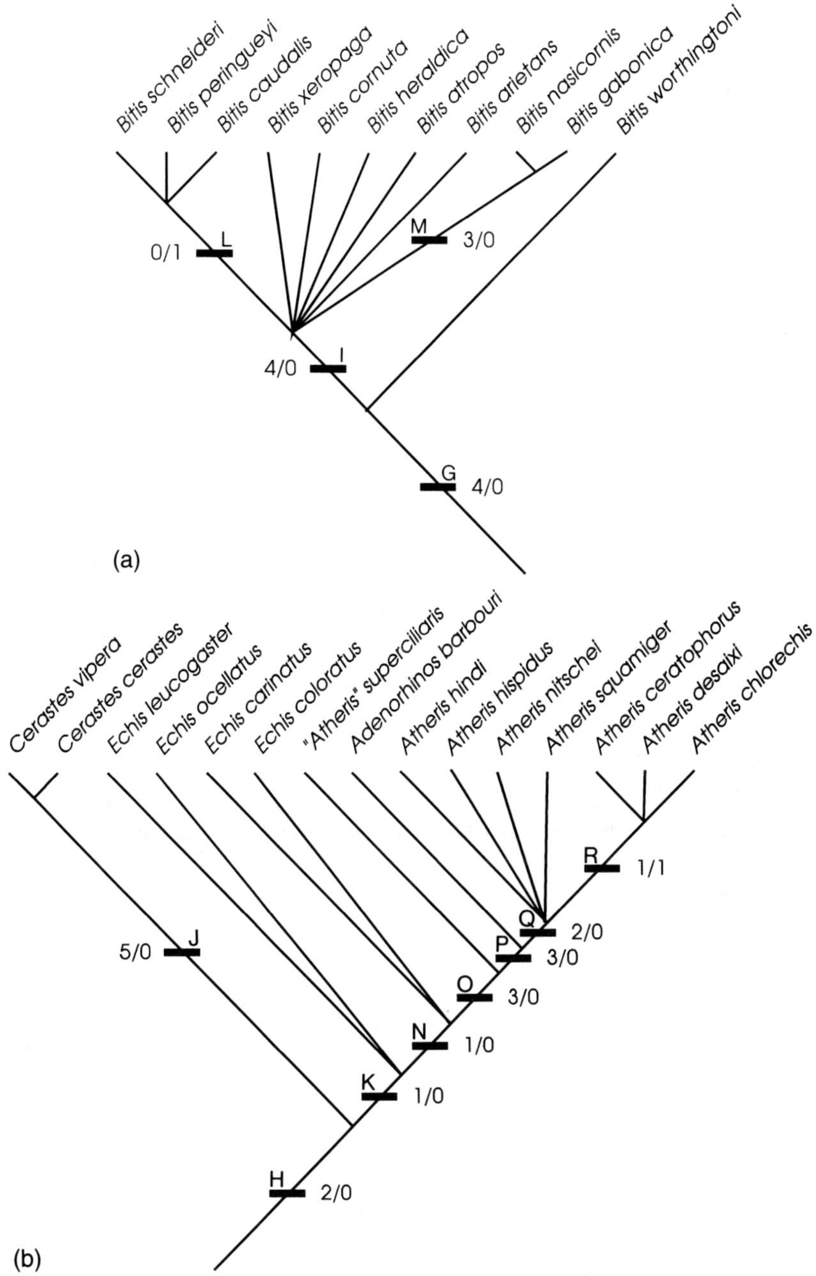

Fig. 2. Species level resolution of the right half of Fig. 1; (a) *Bitis*; (b) *Cerastes*, *Echis* and *Atheris*. No species level resolution was achieved for *Vipera*/*Macrovipera*. Within *Causus*, only *C. defilippii* and *C. lichtensteini* could be set apart from the rest of the genus on the basis of two synapomorphies and one reversal (branch C in Appendix 2, not shown in figs).

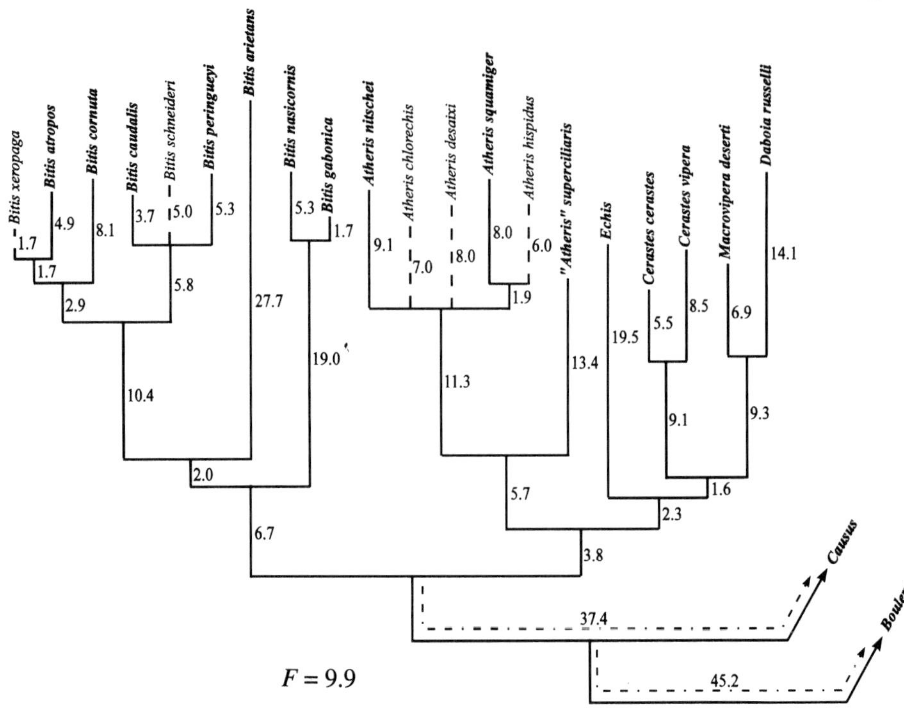

Fig. 3. Fitch–Margoliash dendrogram of mainly African Viperinae, computed from albumin immunological distances. Stippled branches denote species for which no antisera were available (unidirectional IDs) and which were retroactively joined to the tree by applying the method of Beverly & Wilson (1982). The tree was rooted by using *Causus* and *Boulengerina* as outgroups. Numbers indicate branch lengths in %ID.

Table 1. Distance matrix used to compute the Fitch–Margoliash tree depicted in Fig. 3. Lower left: mean reciprocal IDs measured after correction for non-random elements (input); upper right: IDs taken from the Fitch–Margoliash tree (output)

	Bari	*Bnas*	*Bgab*	*Batr*	*Bcor*	*Bcau*	*Bper*	*Asup*	*Anit*	*Asqu*	*Ccer*	*Cvip*	*Echi*	*Mdes*	*Drus*	*Crho*
B. arietans		54	50	46	49	48	49	59	66	67	59	62	62	60	68	74
B. nasicornis	49		7	45	48	46	48	58	65	66	57	60	61	59	66	72
B. gabonica	50	7		41	44	43	44	50	57	58	50	53	53	51	59	65
B. atropos	45	41	34		13	17	19	50	57	58	49	52	53	51	58	64
B. cornuta	57	51	53	13		21	22	53	60	61	52	55	56	54	61	68
B. caudalis	44	47	47	22	20		9	52	59	59	51	54	54	53	60	66
B. peringueyi	49	52	50	19	19	9		53	60	61	53	56	56	54	61	68
A. superciliaris	61	57	48	50	58	50	59		34	35	38	41	41	39	46	53
A. nitschei	76	69	66	80	77	75	71	40		19	45	48	48	46	53	67
A. squamiger	75	64	66	75	74	67	69	56	19		45	48	49	47	54	68
C. cerastes	59	55	51	47	56	49	62	38	54	59		14	36	31	38	60
C. vipera	67	63	56	47	45	43	66	49	56	57	14		39	34	41	63
Echis	67	54	48	49	61	53	56	44	61	59	35	41		37	45	63
M. deserti	67	61	52	51	57	52	53	29	64	66	28	37	48		21	61
D. russelli	65	63	62	60	68	60	58	60	70	65	39	42	37	21		69
C. rhombeatus	71	67	69	64	70	64	68	60	77	75	60	67	65	69	61	

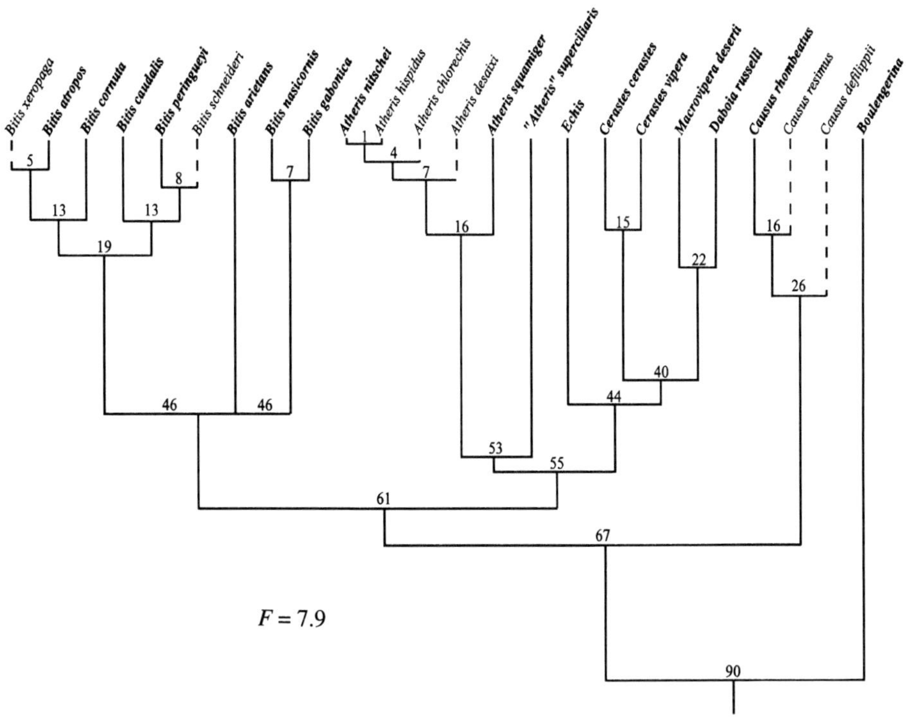

Fig. 4. UPGMA dendrogram of mainly African Viperinae, computed from albumin immunological distances. Distances were corrected for differing relative evolutionary rates. Stippled branches denote species for which no antisera were available (unidirectional IDs). Numbers indicate average ID between the terminal taxa derived from the respective branching point.

Table 2. Distance matrix used to compute the UPGMA tree depicted in Fig. 4. Lower left: mean reciprocal IDs after correction for non-random elements and for differing evolutionary rates of albumins (input); upper right: IDs taken from the UPGMA tree (output). Unilateral distances (included in the tree) not shown

	Bari	*Bnas*	*Bgab*	*Batr*	*Bcor*	*Bcau*	*Bper*	*Asup*	*Anit*	*Asqu*	*Ccer*	*Cvip*	*Echi*	*Mdes*	*Drus*	*Crho*
B. arietans		46	46	46	46	46	46	61	61	61	61	61	61	61	61	67
B. nasicornis	46		7	46	46	46	46	61	61	61	61	61	61	61	61	67
B. gabonica	46	7		46	46	46	46	61	61	61	61	61	61	61	61	67
B. atropos	45	43	34		13	19	19	61	61	61	61	61	61	61	61	67
B. cornuta	51	49	49	13		19	19	61	61	61	61	61	61	61	61	67
B. caudalis	43	50	48	24	20		13	61	61	61	61	61	61	61	61	67
B. peringueyi	46	51	48	20	18	9		61	61	61	61	61	61	61	61	67
A. superciliaris	64	64	52	59	63	58	65		53	53	55	55	55	55	55	67
A. nitschei	63	60	56	73	64	69	60	39		16	55	55	55	55	55	67
A. squamiger	63	57	58	70	63	63	61	56	15		55	55	55	55	55	67
C. cerastes	62	61	56	55	60	58	88	47	53	59		15	44	40	40	67
C. vipera	62	63	54	50	43	46	65	55	49	50	15		44	40	40	67
Echis	64	55	48	53	61	58	57	50	54	54	41	43		44	44	67
M. deserti	61	59	49	52	53	53	51	32	54	57	30	36	48		22	67
D. russelli	67	70	66	70	72	70	63	73	67	64	48	47	42	22		67
C. rhombeatus	67	67	67	67	67	67	67	67	67	67	67	67	67	67	67	

Fitch–Margoliash trees were computed separately for Asian vipers (Fig. 5) and for European vipers (Fig. 6), with *Macrovipera lebetina* and *Vipera xanthina* as links. The respective data sets are shown in Tables 3 and 4. Figure 7 shows a UPGMA tree combined from the data of both sets after correction for differential evolutionary rates and inclusion of unilateral distances (taxa against which no antisera could be produced). More detailed dendrograms and distance matrices have been published previously (Herrmann *et al.* 1992a,b; Joger, Herrmann & Nilson 1992).

Table 3. Distance matrix of albumins of Asian vipers — data base for the Fitch–Margoliash tree (Fig. 5) and for the UPGMA tree (Fig. 7). Lower left: mean reciprocal IDs after correction for non-random elements (input); upper right: IDs taken from the Fitch–Margoliash tree. Unilateral distances (included in the UPGMA tree) not shown

	xanthina	*raddei*	*lebetina*	*mauritanica*	*palaestinae*	*Pseudocerastes*	*Echis*
xanthina	0	8.7	20.4	23.6	11.0	17.5	51.1
raddei	8.8	0	25.7	28.9	16.3	22.8	56.4
lebetina	19.6	23.4	0	17.8	19.2	16.9	62.1
mauritanica	30.3	42.3	17.8	0	22.4	20.1	65.3
palaestinae	11.7	15.1	19.4	18.8	0	16.3	52.7
Pseudocerastes	15.2	23.8	18.7	17.1	20.1	0	59.2
Echis	50.5	57.3	63.0	68.2	54.2	54.8	0

Table 4. Distance matrix of albumins of European vipers — data base for the Fitch-Margoliash tree (Fig. 6) and for the UPGMA tree (Fig. 7). Lower left: mean reciprocal IDs after correction for non-random elements (input); upper right: IDs taken from the Fitch–Margoliash tree. Unilateral distances (included in the UPGMA tree) not shown

	ursinii	*berus*	*kaznakovi*	*aspis*	*ammodytes*	*xanthina*	*lebetina*
ursinii	0	16.0	18.3	22.1	20.6	17.9	30.1
berus	19.5	0	12.3	16.1	14.6	15.1	27.3
kaznakovi	18.4	12.2	0	13.4	11.9	17.4	29.6
aspis	19.6	16.7	14.7	0	13.9	21.2	33.4
ammodytes	19.4	14.6	11.2	13.9	0	19.7	31.9
xanthina	17.3	14.1	18.1	20.8	22.7	0	24.2
lebetina	32.0	24.8	28.7	32.8	38.2	24.2	0

Pseudocerastes/Eristicophis, Daboia and *Macrovipera* form a clade which is clearly separated from the genus *Vipera*. The *V. xanthina* group, contrary to Obst (1982), remains with *Vipera*, albeit as a sister group to the rest of the genus (European vipers *sensu stricto*). The affiliation of *V. palaestinae* to the *V. xanthina* group is only tentative.

Within the European vipers, the *V. ursinii* group branches off first, the *V. berus* group second, and the remaining group comprises not only *V. aspis, V. latastei* and *V. ammodytes*, but also *V. kaznakovi* and *V. dinniki*, previously included in the *V. berus* group. Whether *V. seoanei* is a member of the *V. berus* or the *V. aspis* group could not be decided with reliable probability.

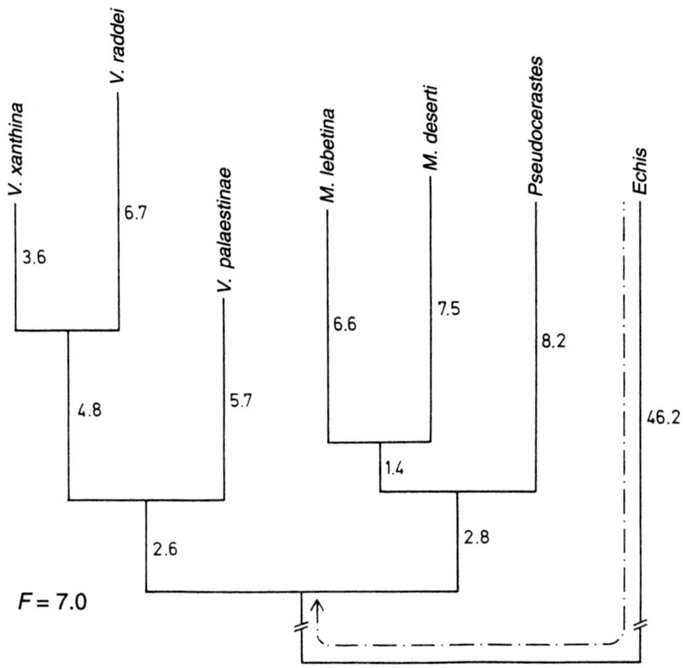

Fig. 5. Fitch–Margoliash dendrogram computed with immunological distances between albumins of Asian vipers (*Vipera*, *Macrovipera*, *Pseudocerastes*). The tree was rooted with *Echis* as outgroup. Numbers denote branch lengths in %ID.

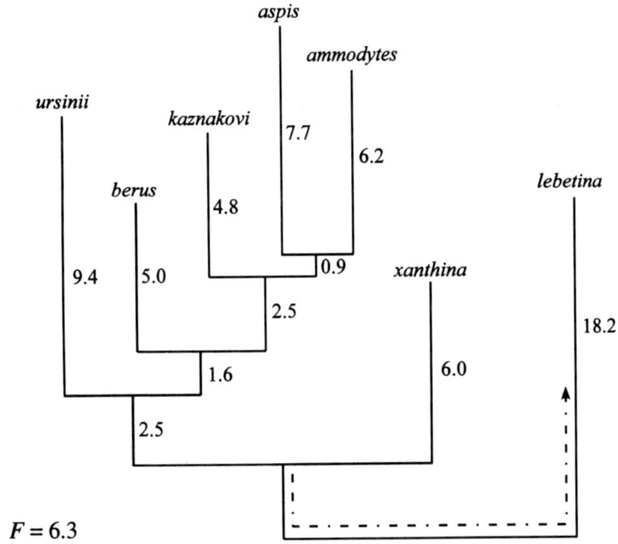

Fig. 6. Fitch–Margoliash dendrogram computed with immunological distances between albumins of European *Vipera*. The tree was rooted with *Macrovipera lebetina* as outgroup. Numbers denote branch lengths in %ID.

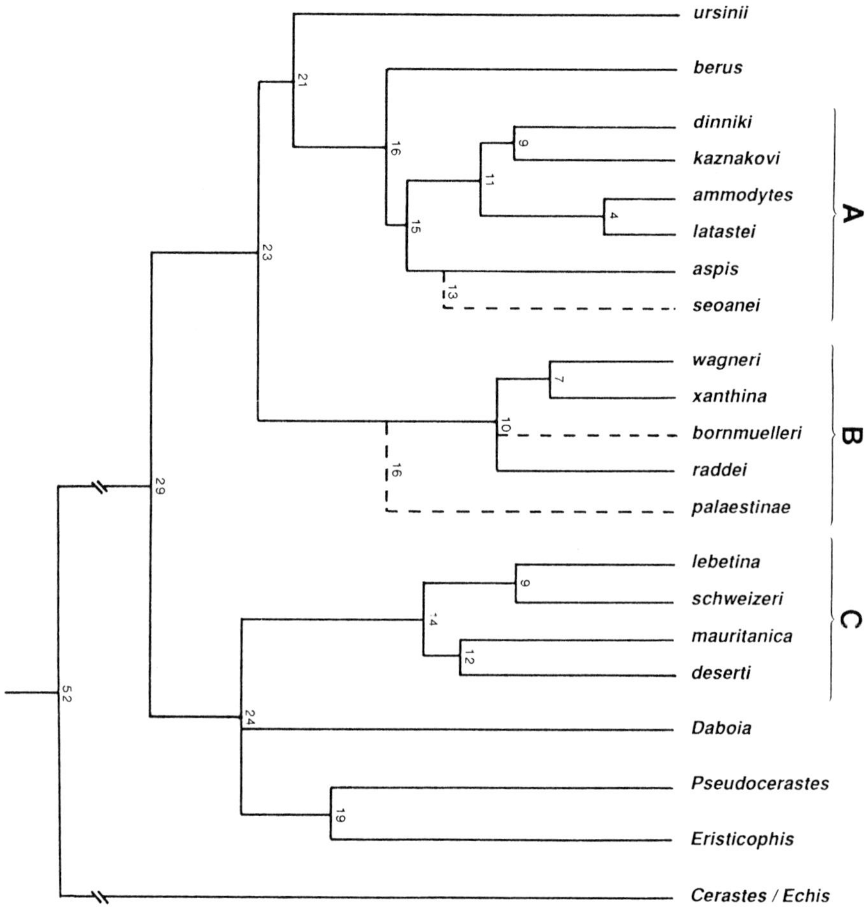

Fig. 7. UPGMA dendrogram of Eurasian vipers, computed from the data of Tables 3 and 4 after addition of several species for which only unidirectional IDs were available, and after correction for differing evolutionary rates. The allocations of stippled branches are not unequivocal. Numbers indicate average ID between the terminal taxa derived from the respective branching point.
A = *Vipera aspis* complex; B = *Vipera xanthina* complex; C = *Macrovipera*.

Discussion

Evolution of Viperinae

In the dispute about the conflicting phylogenetic hypotheses presented by Groombridge (1986) and by Ashe & Marx (1988), our results give strong support to Groombridge regarding monophyly of the Eurasian vipers, and attachment of *Echis* and *Cerastes* to the Afrotropical vipers. The main difference is in the branching point of the Eurasian vipers. The morphological/anatomical data suggest a sister-group relationship between Eurasian vipers (or at least Palaearctic vipers, excluding *Daboia*) and Afrotropical vipers (excluding *Causus*). The immunological trees show *Cerastes* as the genus which is closest to the Eurasian clade, thus making the

Afrotropical group paraphyletic. Moreover, if *Macrovipera* is taxonomically separated from *Vipera*, *Vipera* is monophyletic according to the immunological data, whereas Groombridge regards this genus as paraphyletic, and Ashe & Marx even as polyphyletic. Both of them refer to the low degree of head shield fragmentation in European *Vipera*, a character state considered plesiomorphic in colubroid snakes. Implicitly they suggest that head shield fragmentation occurred gradually along the main stem of viperines. On the other hand, the Crotalinae show that head shield fragmentation must have occurred more than once. Parallel evolutionary trends towards increasing head shield fragmentation in all the main groups of vipers should be considered. However, even a reversal to large head shields in *Vipera* may be possible.

Support for the immunological hypothesis comes from three sides:

1. Historical biogeography. If the Viperinae are of African origin (as suggested by the exclusively Afrotropical *Causus* being undisputedly the most ancestral genus), it makes sense to assume a primarily African evolution before Eurasia was invaded by viperines, and then a fairly recent but diverse radiation in the Palaearctic. Other taxa originating from Africa, like the Proboscidea and the Hominoidea, did not reach Eurasia before the Oligocene/Miocene boundary.
2. Palaeontological evidence. The first viperids in Eurasia stem from the lowermost Miocene (Szyndlar 1991). Both Groombridge (1986) and Ashe & Marx (1988) imply with their published dendrograms that the 'Oriental vipers' (*Macrovipera*, *Pseudocerastes*, *Eristicophis* and *Daboia*) should be much older than the European *Vipera*. Palaeontological evidence (Szyndlar & Schleich 1993) shows that both groups appeared in the same period (about 20 million years BP), which is consistent with the immunological tree depicting them as sister groups (Fig. 7). Temporal evolution of viperine albumins seems to be fairly regular. In the absence of good external calibration events, evidence for this can be taken from the relative branch lengths in the additive Fitch–Margoliash trees (different branches ending in approximately the same level) and from the congruence between Fitch–Margoliash and the non-additive, rate-corrected UPGMA trees. However, there are conservative albumins like those of *Vipera xanthina* and *V. palaestinae*, and progressive ones like those of *Bitis arietans*.

   Unfortunately, in Africa, Tertiary snake fossils are virtually nonexistent. However, correlations between albumin immunological distances and time (Maxson 1992) suggest that the African groups may be much older than the European fossil record. The average precipitin distance of 61 between the two main stems of African viperines (except *Causus*) is equivalent to a microcomplement ID of about 85 (Joger in press). According to the 'molecular clock' calibration established from albumin IDs of mammals, crocodiles and frogs by Maxson (1992: fig. 7), this would mean that the African viperines are at least 50 million years old.
3. Mitochondrial DNA. A preliminary branching order generated with cytochrome *b* sequences of viperids currently in progress in our laboratory is identical with the immunological tree. However, bootstrap values for the basal branches are not yet significant.

Taxonomical conclusions

Snakebite case histories (e.g., Warrell *et al.* 1974; Gillissen *et al.* 1994) make it obvious that a viperid taxonomy reflecting true genetic differences is much needed as a basis for toxinology, antivenom production and medical treatment of snakebite (see Joger 1984b: 7). Descriptions of new species, particularly in the genera *Echis* and *Vipera*, have made traditional species lists obsolete and have questioned the specificity of commercially available antivenins. A stable state of viperine taxonomy has not yet been attained. The number of *Echis* species recognized has risen from two (Klemmer 1963) to 12 (Cherlin 1990), but has subsequently been reduced to eight (Golay *et al.* 1993). The number of species of *Vipera* has also significantly increased (see Nilson & Andrén, this volume pp. 31–42).

The following taxonomical changes and clarifications can be drawn from our data:

1. 'Oriental' vipers. These species, many of which had traditionally been included in *Vipera* and were transferred to *Daboia* by Obst (1982) should be considered different genera: *Daboia* (*D. russelli* which may consist of several species); *Pseudocerastes*; *Eristicophis*; *Macrovipera* (*M. lebetina*, *M. mauritanica*, *M. deserti* and *M. schweizeri*, which had been considered subspecies of *V. lebetina*) (Herrmann *et al.* 1992b).
2. The *V. xanthina* group and *V. palaestinae*. These are provisionally redrawn into the genus *Vipera* from Obst's *Daboia*. However, it should be noted that they show considerable phyletic distances from the European vipers and from each other.
3. European *Vipera*. The *Vipera ursinii* complex seems to be the sister group of all other European vipers. The high diversity within this complex raises the question whether the allopatric subspecies of meadow vipers should be assigned to different species (Joger, Herrmann & Nilson 1992; Nilson, Andrén & Joger 1993). *Vipera dinniki* and *V. kaznakovi* are transferred from the *V. berus* group to the *V. aspis* group (e.g., if subgenera were introduced, from *Pelias* to *Vipera sensu stricto*) (Herrmann *et al.* 1992a).
4. *Bitis*. The puffadder is not a close relative of *B. gabonica* and *B. nasicornis* but forms a group of its own, which might warrant subgenus or even genus rank. The same is true for the small southern African *Bitis*. Within the latter, the following subgroups are discernible: *B. caudalis* (*B. schneideri*/*B. peringueyi*); *B. cornuta* (*B. atropos*/*B. xeropaga*) (Herrmann & Joger 1995).
5. *Atheris*. *A. superciliaris* roots at the very basis of the *Atheris* lineage, morphologically and immunologically very distinct from the rest of this genus. It can certainly be taken out of *Atheris*. On morphological evidence, it could probably be assigned to *Adenorhinos*, a genus with a single known species which we have not yet been able to obtain for molecular analysis.

For the sake of nomenclatural stability, many of the above-named groupings have not yet been given taxonomical ranking. Before drawing such consequences, we are currently repeating the phylogenetic analysis of viperines by applying an independent method, the sequencing of mitochondrial DNA genes. We also hope to clarify thereby a number of still unresolved relationships.

Future research

Among the many open questions of viperine phylogeny (some of which have already been mentioned above) we consider the following the most urgent:

1. Branching point of the Crotalinae. The crotalines may be the sister group of the viperines including *Causus*, or excluding *Causus*, or may even branch off from within the Viperinae *sensu stricto*.
2. Systematics of *Echis*. This genus, of the greatest medical importance, remains totally unresolved phylogenetically and taxonomically. Our data suggest a paraphyly of *Echis*. The clarification of the questions of species numbers and interrelationships requires material from all parts of the enormous distributional range of the genus.
3. *Vipera xanthina* complex. Species recognized in this taxonomically and toxinologically neglected group number from two (Schätti, Baran & Sigg 1991) to seven (Nilson & Andrén 1992) or even eight, if *V. palaestinae* is considered a member of the *V. xanthina* group.
4. *Vipera berus* complex. The taxonomic status of *V. (berus) bosniensis*, *V. seoanei*, *V. nikolskii* and *V. barani* requires further study. A preliminary study using mitochondrial cytochrome *b* sequences (Joger, Lenk *et al.* in press) indicates that *nikolskii* is merely a subspecies of *V. berus*, whereas *seoanei* is a separate species. *V. barani* and *bosniensis* appear closely related to each other, but whether they represent one or two distinct species or just subspecies of *V. berus* remains uncertain.

Acknowledgements

We cannot name all the numerous colleagues who provided live snakes or plasma samples for this study. As representative for all of them, we would like to thank Göran Nilson, Göteborg. Facilities for antiserum production were kindly provided by Professors E. Geyer, Marburg, and U. Stewart, Darmstadt. The theses of H.-W. Herrmann were supervised and supported by Prof. H.-O. von Hagen, Marburg. For their hospitality and help in organizing field trips in Africa, we are deeply indebted to Bill Branch, Port Elizabeth, Wulf Haacke, Pretoria, and Mike Griffin, Windhoek. The co-operation of the nature conservation authorities in the republics of South Africa and Namibia is gratefully acknowledged. The Deutsche Forschungsgemeinschaft provided a research grant (Jo-134/3), the Deutsche Studienstiftung des Deutschen Volkes and the Deutscher Akademischer Austauschdienst provided travel subsidies.

References

Ashe, J. S. & Marx, H. (1988). Phylogeny of the viperine snakes (Viperinae): Part II. Cladistic analysis and major lineages. *Fieldiana (Zool.)* (N.S.) No. 52: 1–23.

Beverly, S. M. & Wilson, A. C. (1982). Molecular evolution in *Drosophila* and higher Diptera. I. Micro-complement fixation studies of a larval hemolymph protein. *J. molec. Evol.* **18**: 251–264.

Cadle, J. E. (1992). Phylogenetic relationships among vipers: immunological evidence. In *Biology of the pitvipers*: 41–48. (Eds Campbell, J. A. & Brodie, E. D., Jr.). Selva, Tyler, Texas.

Cherlin, V. A. (1990). Taxonomic revision of the snake genus *Echis* (Viperidae). II. An analysis of taxonomy and description of new forms. *Tr. zool. Inst. Akad. Nauk SSSR* **207**: 193–223. (In Russian.)

Farris, J. S. (1988). *HENNIG86 software, version 1.5, and reference*. Manual. Privately printed and distributed.

Felsenstein, J. (1988). *PHYLIP (Phylogeny inference package)*. Version 3.1. Manual. Department of Genetics, University of Washington, Seattle.

Gillissen, A., Theakston, R. D. G., Barth, J., May, B., Krieg, M. & Warrell, D. A. (1994). Neurotoxicity, haemostatic disturbances and haemolytic anaemia after a bite by a Tunisian saw-scaled or carpet viper (*Echis 'pyramidum'* complex): failure of antivenom treatment. *Toxicon* **32**: 937–944.

Golay, P., Smith, H. M., Broadley, D. G., Dixon, J. R., McCarthy, C., Rage, J.-C, Schätti, B. & Toriba, M. (1993). *Endoglyphs and other major venomous snakes of the world. A checklist.* Azemiops S. A., Geneva.

Groombridge, B. C. (1980). *A phyletic analysis of viperine snakes*. PhD thesis: City of London Polytechnic, Council for National Academic Awards.

Groombridge, B. (1986). Phyletic relationships among viperine snakes. In *Studies in herpetology*: 219–222. (Ed. Roček, Z.). Charles University, Prague.

Heise, P. J., Maxson, L. R., Dowling, H. G. & Hedges, S. B. (1995). Higher-level snake phylogeny inferred from mitochondrial DNA sequences of 12S rRNA and 16S rRNA genes. *Mol. Biol. Evol.* **12**: 259–265.

Herrmann, H.-W. (1990). *Albumin-immunologische Untersuchungen zur Phylogenie der paläarktischen Großvipern (Reptilia: Serpentes: Viperidae)*. Diploma thesis: University of Marburg, Germany.

Herrmann, H.-W. (1995). *Zur Phylogenie afrikanischer Vipern (Reptilia: Serpentes: Viperidae)*. PhD thesis: University of Marburg, Germany.

Herrmann, H.-W. & Joger, U. (1995). Molecular data on the phylogeny of African vipers — preliminary results. In *Scientia herpetologica*: 92–97. (Eds Llorente, A., Montori, A., Santos, X. & Carretero, M. A.). Associacíon Herpetologica Española, Salamanca.

Herrmann, H.-W., Joger, U. & Nilson, G. (1992a). Molecular phylogeny and systematics of viperine snakes I. General phylogeny of European vipers. In *Proceedings of the sixth ordinary general meeting of the Societas Europaea Herpetologica*: 219–224. (Eds Korsos, Z. & Kiss, I.). Hungarian Natural History Museum, Budapest.

Herrmann, H.-W., Joger, U. & Nilson, G. (1992b). Phylogeny and systematics of viperine snakes. III: resurrection of the genus *Macrovipera* (Reuss, 1927) as suggested by biochemical evidence. *Amphibia-Reptilia* **13**: 375–392.

Hoge, A. R. & Romano-Hoge, S. A. R. W. L. (1978/79) [1981]. Poisonous snakes of the world. Part I. Check list of the pit vipers Viperoidea, Viperidae, Crotalinae. *Mems Inst. Butantan* **42/43**: 179–310.

Joger, U. (1984a). Morphologische und biochemisch-immunologische Untersuchungen zur Systematik und Evolution der Gattung *Tarentola* (Reptilia: Gekkonidae). *Zool. Jb. (Anat.)* **112**: 137–256.

Joger, U. (1984b). *The venomous snakes of the Near and Middle East.* Beihefte Tübinger Atlas des Vorderen Orients, **A 12**, Reichert, Wiesbaden.

Joger, U. (1991). A molecular phylogeny of agamid lizards. *Copeia* **1991**: 616–622.

Joger, U. (In press). Phylogenetische Rekonstruktion mit biochemischen Methoden. *Zool. Beitr.*

Joger, U., Herrmann, H.-W. & Nilson, G. (1992). Molecular phylogeny and systematics of viperine snakes II. A revision of the *Vipera ursinii* complex. In *Proceedings of the sixth ordinary general meeting of the Societas Europaea Herpetologica*: 239–244. (Eds Korsos, Z. & Kiss, I.). Hungarian Natural History Museum, Budapest.

Joger, U., Lenk, P., Baran, I., Böhme, W., Heidrich, P. & Wink, M. (In press). The phylogenetic position of *Vipera barani* and *V. nikolskii* within the *Vipera berus* complex. In *Proceedings of the eighth ordinary general meeting of the Societas Europaea Herpetologica*, Bonn 1995.

Klemmer, K. (1963). Liste der rezenten Giftschlangen. In *Die Giftschlangen der Erde*: 255–464. (Ed. Behringwerke AG). Behringwerk-Mitteilungen, Sonderband, Marburg.

Knight, A. & Mindell, D. P. (1993). Substitution bias, weighting of DNA sequence evolution, and the phylogenetic position of Fea's viper. *Syst. Biol.* **42**: 18–31.

Liem, K. F., Marx, H. & Rabb, G. B. (1971). The viperid snake *Azemiops*: its comparative cephalic anatomy and phylogenetic position in relation to Viperinae and Crotalinae. *Fieldiana (Zool.)* No. 59: 64–126.

Marx, H., Ashe, J. S. & Watrous, L. E. (1988). Phylogeny of the viperine snakes (Viperinae): Part I. Character analysis. *Fieldiana (Zool.)* (N. S.) No. 51: 1–16.

Maxson, L. R. (1992). Tempo and pattern in anuran speciation and phylogeny: an albumin perspective. In *Herpetology. Current research on the biology of amphibians and reptiles*: 41–58. (Ed. Adler, K.). Society for the Study of Amphibians and Reptiles, Oxford (Ohio).

Nilson, G. & Andrén, C. (1992). The species concept in the *Vipera xanthina* complex: reflecting evolutionary history or hiding biological diversity? *Amphibia-Reptilia* **13**: 421–424.

Nilson, G., Andrén, C. & Joger, U. (1993). A re-evaluation of the taxonomic status of the Moldavian steppe viper based on immunological investigations, with a discussion of the hypothesis of secondary intergradation between *Vipera ursinii rakosiensis* and *Vipera (ursinii) renardi*. *Amphibia-Reptilia* **14**: 45–57.

Obst, F. J. (1982) [1983]. Zur Kenntnis der Schlangengattung *Vipera* (Reptilia, Serpentes, Viperidae). *Zool. Abh. st. Mus. Tierk., Dresden* **38**: 229–235.

Prager, E. M. & Wilson, A. C. (1976). Congruency of phylogenies derived from different proteins. *J. molec. Evol.* **9**: 45–57.

Sarich, V. M. & Cronin, J. E. (1976). Molecular systematics of the Primates. In *Molecular anthropology: genes and proteins in the evolutionary ascent of the Primates:* 141–170. (Eds Goodman, M., Tashian, R. E. & Tashian, J. H.). Plenum Press, New York & London.

Schätti, B., Baran, I. & Sigg, H. (1991). Rediscovery of the Bolkar viper: morphological variation and systematic implications on the '*Vipera xanthina* complex'. *Amphibia-Reptilia* **12**: 305–327.

Szyndlar, Z. (1991). A review of Neogene and Quaternary snakes of central and eastern Europe. Part II. Natricinae, Elapidae, Viperidae. *Estud. geol. (Madrid)* **47**: 237–266.

Szyndlar, Z. & Schleich, H. H. (1993). Description of Miocene snakes from Petersbuch 2 with comments on the lower and middle Miocene ophidian faunas of southern Germany. *Stuttg. Beitr. Naturk. Ser. B* **192**: 1–47.

Warrell, D. A., Davidson, N. McD., Ormerod, L. D., Pope, H. M., Watkins, B. J., Greenwood, B. M. & Reid, H. A. (1974). Bites by the saw-scaled or carpet viper (*Echis carinatus*): trial of two specific antivenoms. *Br. med. J.* **4**: 437–439.

Appendix 1

Taxa and populations examined (albumins used for production of antisera are marked with an asterisk):

Viperidae:

Atheris chlorechis	Togo
Atheris desaixi	Kenya: Mt. Kenya
Atheris hispidus	Kenya: Kagamega forest
*Atheris nitschei nitschei**	Rwanda: Cyangugu
*Atheris squamiger squamiger**	Togo
*Atheris superciliaris**	Malawi: near Lake Malawi
*Bitis arietans**	Namibia: Kaokoveld
	Togo*
	South Africa: Cape Province, Port Elizabeth
*Bitis atropos**	South Africa: Transvaal, Drakensberg area, Barberton
	South Africa: Transvaal, Drakensberg area, Haenertsburg*
*Bitis caudalis**	Namibia: Kaokoveld
	Namibia: Swakopmund
	South Africa: Transvaal, Magaliesberge area
	South Africa: Transvaal, Messina*
*Bitis cornuta cornuta**	Namibia: Lüderitz
	South Africa: Cape Province, Saldanah Bay*
	South Africa: captive-bred specimen
*Bitis gabonica gabonica**	South Africa: Natal, Mtubatuba*
	Zaire: Kivu, Irangi
*Bitis gabonica rhinoceros**	Togo
*Bitis nasicornis**	Togo*
	Zaire: Kivu, Irangi*
*Bitis peringueyi**	Namibia: Swakopmund
Bitis schneideri	South Africa: Cape Province, Kleinsee
Bitis xeropaga	South Africa: Cape Province, Aggenys
Causus defilippii	South Africa: Natal
Causus resimus	Burundi
*Causus rhombeatus**	South Africa: Transvaal
*Cerastes cerastes**	Jordan: Wadi Araba*
	Morocco: Zagora*
*Cerastes vipera**	Egypt: El Fayum*
	Morocco: Erfoud*

*Daboia russelli russelli**	Pakistan
Daboia russelli siamensis	Thailand
*Echis leucogaster**	Algeria: Hoggar
Echis multisquamatus	Turkmenia
Echis ocellatus	Mali
Echis pyramidum	Egypt: El Fayum
Eristicophis macmahoni	Pakistan
*Macrovipera deserti**	Libya: Djebel Nefusa
*Macrovipera lebetina lebetina**	Cyprus: Limassol
Macrovipera lebetina obtusa	Turkey: Ararat
	Turkey: Gaziantep
Macrovipera lebetina turanica	Turkmenia: Kopet-Dagh
Macrovipera mauritanica	Morocco: Sousse, Aoulouz
Macrovipera schweizeri	Greece: Cyclades, Milos
*Pseudocerastes persicus persicus**	Pakistan
Pseudocerastes persicus fieldi	Israel
Vipera aspis aspis	France: La Licamarie, south Saint Etienne
*Vipera aspis francisciredi**	Italy: Monte Geove, Terracina
Vipera aspis atra	Switzerland: Maggiatel, Corgello
Vipera ammodytes ammodytes	Croatia: Gracac
*Vipera ammodytes meridionalis**	Greece: Euboea
Vipera ammodytes transcaucasiana	Georgia: Boschumi
	Turkey: Borcka, Artvin
*Vipera berus berus**	Sweden: Göteborg
Vipera berus sachalinensis	Russia: Sachalin
Vipera dinniki	Georgia
*Vipera kaznakovi**	Turkey: Hopa
Vipera latastei	Spain: Valencia
	Spain: Zamora
	Spain: Pueblo, Rioconejos
*Vipera palaestinae**	Israel: Tel Aviv
*Vipera raddei raddei**	Turkey: Ararat*
	Turkey: Digor
*Vipera ursinii graeca**	Greece: Pindos mountains
*Vipera xanthina**	Turkey: Ciclikara*
	Turkey: Xanthos

Elapidae:

*Boulengerina annulata**	Burundi

Appendix 2

Synapomorphic characters (taken from Groombridge 1980) defining clades in the HENNIG86 cladogram (Figs 1 and 2).

A. (*Causus*) — *Causus*-type supranasal sac; *Causus*-type scale surface micro-ornamentation; *Causus*-type hemipenis; 'maxillo'-postorbital ligament runs only to spine on prefrontal; venom gland primitive (elongated); musculus add. ext. profundus divided; musculus pterygoideus with attachment extending anteriorly along compound bone; moderate or low lateral and medial laminae of compound bone present; ectopterygoid slender, rounded postero-lateral projection.

B. (*Vipera*, *Macrovipera*, *Pseudocerastes*, *Eristicophis*, *Daboia*, *Bitis*, *Cerastes*, *Echis*, *Adenorhinos*, *Atheris*) — facial carotid artery ventral to V_3; musculus hyotrachealis shifts attachment off hyoid to musculus neurocostomandibularis complex; latero-dorsal process of prefrontal–frontal articulation reduced or absent.

C. (*Causus defilippii*, *lichtensteini*) — tracheal lung present, anterior pulmonary artery (tracheal artery) present and in right-dorsal position, heart–liver gap variably present, left lung absent, antero-lateral pocket of right lung absent, thin and short tracheal artery present, posterior pulmonary artery reduced, long intra-pulmonary bronchus, extending to the posterior half of the liver, left-ventral anterior pulmonary artery present, right posterior pulmonary artery absent; venom gland normal type ('reversal'); muscle 'X' absent.

D. (*Vipera*, *Macrovipera*, *Pseudocerastes*, *Eristicophis*) — frontal and parietal scales well-formed.

E. (*Bitis*, *Cerastes*, *Echis*, *Adenorhinos*, *Atheris*) — medial head of musculus retractor pterygoideus large.

F. (*Pseudocerastes*, *Eristicophis*) — *Pseudocerastes*-type supranasal sac; *Pseudocerastes*- and *Eristicophis*-type micro-ornamentation of scale surface; narrow horizontal anterior lamina of septomaxilla present; latero-dorsal process of septomaxilla prominently curved; prefrontal robust and deeply concave; postorbital 'twisted' distally.

G. (*Bitis*) — parallel elongation of rostral; *Bitis*-type supranasal sac; *Bitis*-type (not *B. peringueyi*) micro-ornamentation of scale surface; ectopterygoid with broad head and pointed posterolateral projection.

H. (*Cerastes*, *Echis*, *Adenorhinos*, *Atheris*) — *Echis*- and *Cerastes*-type micro-ornamentation of scale surface; wide horizontal anterior lamina of septomaxilla moderately developed.

I. (*Bitis schneideri*, *peringueyi*, *caudalis*, *xeropaga*, *cornuta*, *heraldica*, *atropos*, *arietans*, *nasicornis*, *gabonica*) — several small scales separate nasal and rostral (parallelism); anterior ridge of septomaxilla present; maxilla–ectopterygoid articulation with very prominent dorsal transverse ridge; postorbital contacts prootic.

J. (*Cerastes*) — elongation of rostral (parallelism); several small scales separate nasal and rostral (parallelism); *Cerastes*-type hemipenis; tracheal lung absent; anterior pulmonary artery (tracheal artery) absent; heart–liver gap present; left lung absent, antero-lateral pocket of right lung present, thin and short tracheal artery present; parietal extended postero-laterally.

K. (*Echis*, *Adenorhinos*, *Atheris*) — maxilla–ectopterygoid articular fossa rather triangular, ligaments attached to knob on diagonal ridge.

L. (*Bitis schneideri*, *peringueyi*, *caudalis*) — prootic foramen: separate V_4 foramen.

M. (*Bitis nasicornis*, *gabonica*) — double or multiple snout horns; transverse scale rows regularly duplicated; origin of musculus retractor pterygoideus: prootic modified.

N. (*Echis carinatus*, *coloratus*, *Adenorhinos*, *Atheris*) — prootic foramina: anterior twig of V_4 widely exposed.

O. (*Adenorhinos*, *Atheris*) — *A. superciliaris*-, *Adenorhinos*- and *Atheris s.str.*-type micro-ornamentation of scale surface; medial head of musculus retractor pterygoideus absent; wide horizontal anterior lamina of septomaxilla strongly developed.

P. (*Adenorhinos*, *Atheris hindi*, *Atheris s.str.*) — naso-rostral present (parallelism); *Adenorhinos*- and *Atheris s.str.*-type hemipenis; vomerine process of premaxilla much reduced or virtually absent.

Q. (*Atheris hindi*, *hispidus*, *nitschei*, *squamiger*, *ceratophorus*, *desaixi*, *chlorechis*) — gular scales strongly keeled; lower scales of transverse scale rows frequently duplicated.

R. (*Atheris ceratophorus*, *desaixi*, *chlorechis*) — upper 'rostral' with vertical division; prootic foramen: separate V_4 foramen (convergence).

5 Phylogenetic relationships of the '*Agkistrodon* complex' based on mitochondrial DNA sequence data

CHRISTOPHER L. PARKINSON, SCOTT M. MOODY and JON E. AHLQUIST

Synopsis

We investigated phylogenetic relationships among 15 species of pitvipers commonly known as the '*Agkistrodon* complex' or the tribe Agkistrodontini. We sequenced a 505 base-pair region of the mitochondrial 16S ribosomal DNA for all genera of this group, including all three New World species of the genus *Agkistrodon* and four members of the Old World *Agkistrodon*, to test the monophyly of this genus. Results indicate that *Agkistrodon* is paraphyletic, therefore the genus *Gloydius* should be used for the Old World members. The genera *Calloselasma, Deinagkistrodon* and *Hypnale* are separate lineages, thus their status as genera is upheld. The tribe Agkistrodontini is not valid and the name should be suppressed.

Introduction

Snakes of the family Viperidae are united by possession of a solenoglyphous dentition used for venom delivery. These vipers are usually grouped in the Caenophidia or advanced snakes. The family Viperidae has three recognized subfamilies: Azemiopinae, for the monotypic species *Azemiops feae*; Viperinae, for the true vipers and the enigmatic genus *Causus*; and the Crotalinae, for the pitvipers (Liem, Marx & Rabb 1971; Cadle 1988, 1992). The Viperinae are found in Africa and Eurasia, and are argued to be a monophyletic assemblage (Ashe & Marx 1988). The Crotalinae are recognized by most to be monophyletic, although this has recently been questioned by Cadle (1992).

The main synapomorphy of the subfamily Crotalinae is the heat-sensing loreal pit. However, this organ or a functionally similar system has evolved more than once (e.g. in boas and pythons) (Cock Buning 1985). Moreover, Cadle (1992: 44), using immunological data, suggested that *Azemiops* might fall within the subfamily Crotalinae, thus suggesting that the Crotalinae may be paraphyletic. If this is the case, then either a secondary loss of the loreal pit occurred in *Azemiops* or multiple

Symp. zool. Soc. Lond. (1997) No. 70: 63–78
ISBN 0 19 8549865 5

evolution of this character is required to explain these relationships. Recently two papers based on sequence data from the mitochondrial (mt) 12S and 16S ribosomal DNA genes (rDNA) suggested that *Azemiops* is the sister group to the Crotalinae, rather than being a member of the Crotalinae (Knight & Mindell 1993; Heise *et al.* 1995). The same DNA sequences for *Azemiops, Vipera* and *Coluber* were used in both studies; the only differences were in alignment, analyses and number of taxa (larger in the Heise *et al.* 1995 study). Thus the phylogenetic position of *Azemiops* is still in question. Here we subscribe to the conservative view that *Azemiops* is not part of the Crotalinae, but could be their sister taxon. In this light we believe the Crotalinae to be a monophyletic group composed of 17–20 genera, depending on taxonomic judgments.

Generic descriptions and systematic problems

Snakes of the genus *Agkistrodon* and the putatively allied genera (*Calloselasma, Deinagkistrodon* and *Hypnale*) were grouped in the subfamily Agkistrodontinae of the family Crotalidae by Burger in his unpublished dissertation (1971). This group was later changed to the tribe Agkistrodontini of the Crotalinae by Hoge & Romano-Hoge (1978/79). This group has been loosely known as the *Agkistrodon* complex (Gloyd & Conant 1990). The Agkistrodontini have recently received much study, both by traditional morphological methods (Gloyd & Conant 1990; Kardong 1990; Malnate 1990) and by molecular techniques (Campbell & Whitmore 1989; Minton 1990; Knight, Densmore & Rael 1992; Porter, Haiduk & de Queiroz 1994). Although these studies have been important, a comprehensive molecular phylogenetic analysis is warranted. Using nucleotide sequence data generated from a region of the mitochondrial 16S rDNA we address the following questions. (1) Is the genus *Agkistrodon* (*sensu stricto*) monophyletic, i.e. what is the status of the genus *Gloydius*? (2) Are the genera *Calloselasma, Deinagkistrodon* and *Hypnale* valid, and if so, what are their relationships to each other and to other pitvipers? (3) What is the relationship of the genus *Trimeresurus* (representing other Old World taxa) to the Agkistrodontini? (4) What is the status of the tribe Agkistrodontini?

The genus *Agkistrodon* (*sensu lato*) presently contains three New World species (*A. bilineatus, A. contortrix* and *A. piscivorus*) while the Old World group contains approximately 10 species (depending on personal preferences): *A. blomhoffii, A. caliginosus, A. halys, A. himalayanus, A. intermedius, A. monticola, A. saxatilis, A. shedaoensis, A. strauchi, A. tsushimaensis, A. ussuriensis* (Gloyd & Conant 1990; Zhao & Adler 1993 and references therein; Isogawa, Moriya & Mitsui 1994). Three taxa which were traditionally placed within this genus have recently been removed and given generic rank: *Calloselasma* (Cope 1860; revived by Chernov 1957), *Deinagkistrodon* (Gloyd 1979), *Hypnale* (Fitzinger 1843; resurrected by Gloyd 1977). *Calloselasma* and *Deinagkistrodon* are monotypic while *Hypnale* contains three species.

The genus *Agkistrodon* Palisot de Beauvois is diagnosed by the following characters (summarized from Gloyd & Conant 1990); (1) deep loreal pit, (2) large venom glands in temporal region of both sides of the head, (3) maxillary bone bearing large canaliculate fangs, (4) hemipenis deeply bifurcate, (5) choanal process wide, bearing a keel-like process on the dorsal side of the palatine bone, (6) postorbitals in contact or near the frontals, (7) nostril centred between two nasal scales, (8) postoculars and suboculars combined range from 2–5, (9) ventrals range from 127–178, (10) subcaudals range from 26–71, generally paired in the Old World forms whereas they are single in New World forms.

Chernov (1957) resurrected the generic name *Calloselasma* Cope for *Agkistrodon rhodostoma* on the basis of osteological characters. This genus is characterized by the following: (1), (2) and (3) same as above, (4) hemipenis deeply bifurcate, almost entirely covered with enlarged spines on the base of the lobes, and distal two-thirds of the lobes covered with densely packed flounces, (5) choanal process of the palatine very elongate, usually with one tooth, (6) dorsal scales smooth and not keeled, (7) m. retractor pterygoidei inserting at just two sites: anterior pterygoid and medial process of the ectopterygoid, (8) a distinct m. pterygoideus glandulae present.

Gloyd (1979) described the genus *Deinagkistrodon* for the species known as *Agkistrodon acutus* on the basis of the following characters: (1), (2) and (3) are the same as in *Agkistrodon*, (4) hemipenis deeply bifurcate, (5) choanal process of the palatine strongly curved inward and elongate, (6) dorsal scales strongly keeled, (7) tip of snout upturned, formed by greatly enlarged rostral and internasal scales, (8) origin of the m. levator anguli oris lying posterior to that of the m. pseudotemporalis with no overlap of their origins, (9) m. pterygoideus glandulae partially distinct.

The fourth genus, *Hypnale*, was resurrected by Gloyd in 1977 and has the following characteristics: (1), (2) and (3) are the same as in *Agkistrodon*, (4) hemipenis deeply bifurcate, may or may not have proximal spines, closely resembles that of *Calloselasma*, (5) choanal process of the palatine low or absent, (6) dorsal scales weakly keeled, (7) tip of snout upturned in two species and slightly in the third, (8) m. pterygoideus glandulae distinct.

Based on the above morphological characters and some molecular data, numerous phylogenetic hypotheses have been proposed. Gloyd & Conant (1990) in their monograph on the *Agkistrodon* complex did not favour separating Old and New World species; however, they did suggest some relationships between taxa. They listed many characters (Gloyd & Conant 1990: 408) for an affinity between *Calloselasma* and *Hypnale*, although they also stated that there are many differences, especially in the shape of the palatine bone. Brattstrom (1964) did not differentiate the group (*Agkistrodon, sensu lato*) into separate genera, instead classifying all as *Agkistrodon*. He hypothesized that the New World *Agkistrodon* are polyphyletic, having a lineage of *A. bilineatus, A. piscivorus* and the Old World taxon *A. acutus* (*Deinagkistrodon acutus*), while *A. contortrix* is grouped with the rest of the Old World taxa (i.e. all other *Agkistrodon, Calloselasma* and *Hypnale*). Burger (1971) in his doctoral dissertation revised the Crotalinae and placed them as the family Crotalidae. He suggested that all of the Old World species be included in the genus *Hypnale*.

He referred to the New World *Agkistrodon* as a group, while lumping *acutus* and *rhodostoma* into *Calloselasma*. Burger suggested, in a phylogenetic sense, that his *Hypnale* (members of the Old World *Agkistrodon* and *Hypnale*) gave rise to *Calloselasma* and other pitviper genera. *Hypnale* was hypothesized to have crossed the Bering land-bridge, giving rise to the New World *Agkistrodon* and *Porthidium*. Burger, however, never published his dissertation.

Recently, researchers have turned to molecular systematics to help them to understand the questions which have arisen from the morphological hypotheses. Campbell & Whitmore (1989), using epidermal keratins, hypothesized that the agkistrodontines form a monophyletic group. Relationships within the group, however, were uncertain. Their data suggest that New World *Agkistrodon* is polyphyletic; that *Calloselasma* and *Deinagkistrodon* are not closely related; and that Old World *Agkistrodon* may be the sister group of *A. bilineatus*. They did not study *Hypnale*. Knight *et al.* (1992) examined restriction fragment polymorphisms and a partial sequence of the 16S rDNA. Their results indicate monophyly of the New World *Agkistrodon* (*sensu stricto*) and monophyly of the Old World *Agkistrodon* (*sensu stricto*) but not monophyly of the *Agkistrodon* (*sensu lato*) nor a sister-group relationship. They did not, however, suggest that the genus should be split, owing to problems in their data.

The generic status of *Calloselasma* and *Deinagkistrodon* is supported, although *Hypnale* was not included in the Knight *et al.* (1992) study. The validity of *Calloselasma, Deinagkistrodon* and *Hypnale* is supported by immunoelectrophoretic results of Minton (1990), although he does not provide a scenario of relationships among the three genera. Minton states that the New World *Agkistrodon* are closer to each other than to members of the Old World *Agkistrodon*.

Materials and methods

DNA purification, amplification and sequencing

DNA was extracted from red blood cells, soft tissues (i.e., liver, heart), or from shed skins (see specimens examined for voucher data) following procedures found in Knight & Mindell (1993). The 16S ribosomal sequences were amplified with the primers L2510 (5′-CGCCTGTTTATCAAAAACAT-3′) and H3059 (5′-CCGGTCTGAACTCAGATCACGT-3′). The primers are designated L for light strand and H for heavy strand and their positions identified with the 3′ ends corresponding to published human sequences (Anderson *et al.* 1981). A third primer, L2932 (5′-ATCGAAACAAGTTACTCCAG-3′), is an internal primer used only in the cycle sequencing reactions. Polymerase chain reaction (PCR) was accomplished by using *Taq* polymerase and its reaction buffers from Perkin Elmer/Cetus. Final $MgCl_2$ concentration was 3.4–3.9 mM. Thermal cycling was carried out in a Perkin Elmer/Cetus 480 thermal cycler. Thermal cycle parameters were 94°C, 35 s/45°C, 35 s/72°C, 1 min for 35 cycles. Samples were then held at 4°C until removed and frozen.

Twenty-five microlitres of the PCR products from amplification were purified by electrophoresis in 2.0% low-melting agarose (Sea Plaque) and visualized by ethidium bromide staining. The bands of interest were cut out with a sterile razor blade and purified by using Wizard PCR Prep kits (Promega) following the manufacturer's instructions.

Direct sequencing of the PCR product was carried out with the CircumVent thermal cycle sequencing kit (New England BioLabs) using internal incorporation of $\alpha\ ^{35}S$ nucleotides. After cycle sequencing, 2.5 μl of the products were separated at *c.* 1800 V in 6% poly-acrylamide, 7 M urea, 50 cm × 21 cm 0.4 mm gels (Jordan Scientific). They were dried and autoradiographs were exposed for 2–4 days.

Data alignment and sequence statistics

DNA alignment is tedious and difficult to accomplish with a non-coding gene owing to the fact that one can not use the protein sequence as the template. The 16S region was aligned by using the multiple sequence alignment program CLUSTAL V with the default settings (Higgins & Sharp 1989; Higgins, Bleasby & Fuchs 1992). After alignment, changes were made by eye to minimize gaps and transversions using ESEE (Cabot & Breckenback 1989). Nucleotide composition, number of supposed transitions (ts), supposed transversions (tv), both pairwise (p distance) and corrected pairwise distances (both Jukes and Cantor and Kimura's 2 parameter) of the L-strand were calculated by using MEGA (Kumar, Tumara & Nei 1993). We plotted the transition/transversion ratio against both uncorrected and corrected pairwise distances to ascertain the role of substitutions in our data set.

Phylogenetic analyses

The aligned data were subjected to maximum parsimony (MP) analyses using PAUP 3.1 (Swofford 1993). Numerous weighting schemes have been developed for MP analyses of rDNA sequences. These range from weighting transversions greater than transitions (Brown *et al.* 1982; Knight & Mindell 1993, 1994), weighting stems differently than loops (Wheeler & Honeycutt 1988; Hillis & Dixon 1991; Dixon & Hillis 1993), using the transversions only (Cracraft & Helm-Bychowski 1991), and not weighting at all (Nedbal, Allard & Honeycutt 1994; Heise *et al.* 1995). For this study we chose to analyse all the data unweighted and by weighting the transversions twice as much as the transitions (tv 2 : ts 1). Gaps were either excluded, or added as a fifth character, or coded at the end of the sequence matrix as a non-additive binary matrix (Swofford 1993). A heuristic search with additive tree bisection (10 replications) was performed on the different data sets (three gap methods and weighted data tv 2 : ts 1). Bootstrap (Felsenstein 1985) resampling (500 replications) was also performed to test node support on the weighted data, again using the heuristic option with additive tree bisection (10 replications).

Results

The aligned data are shown in Fig. 1. A total of 138 sites were variable out of a total of 505 bases. Of these 68 sites were informative under the criteria of parsimony. The largest gap was two sites. The region between sites 300–338 is highly variable: 17 of 138 variable sites occur here. Base composition is: adenine (A) ranged from 33.9–36.3,

$\bar{x}$ = 35.2; thymine (T) 22.9–27.1, $\bar{x}$ = 25.2; cytosine (C) 20.2–22.2, $\bar{x}$ = 21.5; guanine (G) 17.3–18.8, $\bar{x}$ = 18.1. The number of supposed transitions ranged from 10 to 28; transversions from four to 36. The highest values were between the outgroups and certain members of the ingroup. The lowest values were between the two species of *Crotalus*. The transition: transversion ratio generally decreases as divergence between taxa increases. This follows the Brown *et al.* (1982) theory that transition saturation increases as divergence increases.

```
                                                                            6
                                                                            0
Bitis            ATATTAAAGG  CAACGCTGCC  CAGTGAACAA  TTAAACGGCC  GCGGTATCCT  AACCGTGCAA
Echis            G......G..  ..........  .......T..  ..........  ..........  ..........
C. adamanteus    G.........  ..........  .......A..  ..T.......  ......C...  ..........
C. tigris        G.........  ..........  .......A..  ..T.......  ......C...  ..........
S. catenatus     G.........  ..........  .......A..  ..T.......  ......C...  ..........
B. asper         G.........  .G........  .......A..  ..........  ..........  ..........
A. contortrix    G.........  .TT.......  .......A..  ..........  ..........  ..........
A. bilineatus    G.........  .G........  .......A..  ..........  ..........  ..........
A. piscivorus    G.........  .G........  .......A..  ..........  ..........  ..........
A. intermedius   G.........  ..........  .......T..  ..........  ..........  ..........
A. shedaoensis   G.........  ..........  .......T..  ..........  ..........  ..........
A. strauchi      G.........  ..........  .......T..  ..........  ..........  ..........
A. ussuriensis   G.........  ..........  .......T..  ..........  ..........  ..........
Deinagkistrodon  G.........  .G........  .......T..  ..........  ......C...  ..........
H. hypnale       G.........  .G........  .......T..  ..........  ..........  ..........
Calloselasma     G.........  .G........  .......T..  ..........  ......C...  ..........
T. flavoviridis  G.........  .G........  ........T.  ..........  ..........  ..........

                                                                            1
                                                                            2
                                                                            0
Bitis            AGGTAGCATA  ATCATTTGTC  TATTAATTGT  AGACCCGTAT  GAAAGGCAAT  ATGAGAGCCT
Echis            ..........  ..........  ..........  .....T....  ........CC  ..........
C. adamanteus    ..........  ..........  ..........  .....A....  .........A  .........C
C. tigris        ..........  ..........  ..........  .....A....  A........A  .........C
S. catenatus     ..........  ..........  ..........  .....A....  .........A  .........C
B. asper         ........C.  ..........  ..........  .....T....  .........A  .........C
A. contortrix    ..........  ..........  ..........  .....A....  .........A  .........C
A. bilineatus    ..........  ..........  ....CC....  .....T....  .........A  .......T.C
A. piscivorus    ..........  ........A.  ..........  .....T....  .........A  .......T.C
A. intermedius   ..........  ..........  ..........  .....T....  .........A  .........C
A. shedaoensis   ..........  ..........  ..........  ..T..T....  .........A  .........C
A. strauchi      ..........  ..........  ..........  .....T....  .........A  .........C
A. ussuriensis   ..........  ..........  ..........  .....T....  .........A  .........C
Deinagkistrodon  ..........  ..........  ..........  .....T....  .........A  .....G..T.
H. hypnale       ..........  ..........  ..........  .....T....  .........A  .....G.T.C
Calloselasma     ........C.  ....C.....  ..........  .....T....  .........A  ....A....C
T. flavoviridis  ..........  ..........  ..........  .....T....  .........A  .........C

                                                                            1
                                                                            8
                                                                            0
Bitis            AACTGTCTCT  TATAACAAAT  CAATTAAACT  GATCTCCCAG  TACAAAAGCT  GGGATACCCA
Echis            ..........  ..........  ..........  .......T..  ..........  A.A..TA.AC
C. adamanteus    ..........  .....T....  ..........  .......T..  ..........  A.A....TA.
C. tigris        ..........  .....T....  ..........  .......T..  ..........  A.A....TA.
S. catenatus     G.........  .....T....  ..........  .......T..  ..........  AA...TTTAT
B. asper         ..........  ..........  ..........  .......T..  ..........  A.A....TA.
A. contortrix    ..........  ..........  ..........  .......T..  ..........  A.T...TTAT
A. bilineatus    ..........  ..........  ..........  .......T..  ..........  A.A...TTA.
A. piscivorus    ..........  ..........  ..........  .......T..  ..........  A.A...TTA.
A. intermedius   ..........  ..........  ..........  .......T..  .C........  A.A....TA.
A. shedaoensis   ..........  ..........  ..........  .......T..  .C........  A.A....TA.
A. strauchi      ..........  ..........  G.........  .....T.T..  .C........  A.A.....A.
A. ussuriensis   ..........  ..........  ..........  .....T.T..  .C........  A.A...TTT.
Deinagkistrodon  ..........  ...GG.....  ..........  .......T..  ..........  A.A.....A.
H. hypnale       ..........  ..........  ..........  .......T..  ..........  A.A.....A.
Calloselasma     G.........  ..........  ....A.....  .......T..  ..........  A.A..GTTA.
T. flavoviridis  ..........  ..........  ..........  .......T..  ..........  A.A....TA.
```

```
                                                                                 2
                                                                                 4
                                                                                 0
Bitis             CATAAGACCA GAAGACCCTG TGAAGCTTTA ACTAACCTAT TAAACCTCAT AATAGCTAGT
Echis             .......... .......... .......... .T...A.... ......C... ....A...A.
C. adamanteus     T......... .......... .......... .......... .......AG. ....A...C.
C. tigris         .......... .......... .......... .T...A.... .......A.. ....AT..C.
S. catenatus      .......... .TT....... .......... .......... ......CT.. ....A...C.
B. asper          .......... .......... .......... .........C ......CA.. ........C.
A. contortrix     .......... .......... .......... .......... ...C...AT. ....A...C.
A. bilineatus     .......... .......... .......... .......... ......AA.. ........C.
A. piscivorus     .........T .......... .......... .....A.... ......AAG. ........CC
A. intermedius    .......... .......... .......... .......... ......CA.. ....A...C.
A. shedaoensis    .......... .......... .......... .......... ......CA.. ....A...C.
A. strauchi       .......... .......... .......... .....A.... ......CA.. ........C.
A. ussuriensis    .......... .......... .......... .......... .......A.. ....A...C.
Deinagkistrodon   .......... .......... .......... .AA....... ....T.AA.. ....A...C.
H. hypnale        .......... .......... .......... ..A....... ......AA.. ........C.
Calloselasma      .......... .......... .......... .TA....... ....T.AA.. ....A..TT.
T. flavoviridis   .......... .......... .......... ..A...T... ......CA.. ....AT..C.

                                                                                 3
                                                                                 0
                                                                                 0
Bitis             TTCGGTTGGG GCGACCTTGG AATAAAAAAG AACTTCCAAA AAACATGACC TTCCTCATAT
Echis             ..T....... .......... ..-...G..A .........- -.CA.GC... .--...C.GC
C. adamanteus     .......... .......... .......... .........C --.T....A- ..-......A
C. tigris         ..A....... .......... .......... .........T --.T.....- ..A......A
S. catenatus      .......... .......... ..C....... .........T --.TG....- ........TA
B. asper          ..A....... .......... ..C....... .........C --CT.....- .C.A.....A
A. contortrix     .......... .......... .......... .........T --.T.....- .........A
A. bilineatus     .......... .......... .......... .........T --.T.....- .........A
A. piscivorus     ..T....... .......... .......... .........T --.T.....- .........A
A. intermedius    .......... .......... .......... .........T --.T.....- -........G
A. shedaoensis    .......... .......... .......... .........T --.T.....- -........A
A. strauchi       .......... .......... ........GA .........- C-.T.....- -........A
A. ussuriensis    .......... .......... .......... .........T C-.T.....- -.T......A
Deinagkistrodon   .......... .......... ..C....... .........C --.......- ........TA
H. hypnale        ..A....... .......... .......... .........C ..TT.....- -........A
Calloselasma      ..T....... ..A....... .......... .........C ..-T.....- -........A
T. flavoviridis   .......... .......... .......... .........C --.T.....- ....A....A

                                                                                 3
                                                                                 6
                                                                                 0
Bitis             AA--AAGGCG GACAGGCCAC AATTA-GACC CAGAACTTCT GATAAATGAA CCAAGTTACT
Echis             .CCT.....T A...A...T. C.AATA.... ...C.AAG.. ...TG..... ..........
C. adamanteus     ..TT-....A A...A....A C.CC...... ...C..AG.. ..C..T.... A.........
C. tigris         ..T.-....A ....A....A C.CC...... ...C.TAG.. .....T.... A.........
S. catenatus      ..C.-..... .........A T.CA.....A ...C..AG.. .....T.... AT........
B. asper          .........A A...A....A C.C....... ...C..AG.. ..C..TC... A.........
A. contortrix     .-.C.....A A...A....A T......... ...C.TAG.. .....T.... AT........
A. bilineatus     ..TC.....A A...A....A T......... ...T.AAG.. .....TC... AT........
A. piscivorus     .-CC...... A...A....A T......... ...T.TAA.. .....TC... AT........
A. intermedius    G.C..-...A A...A..... .C-....... ...T.T.A.. .....T.... AT........
A. shedaoensis    G.C..-...A A...A..... .C-....... ...T.T.A.. .....T.... AT........
A. strauchi       ..C......A ....A..... .C-A...... ...T.T.A.. .....T.... AT........
A. ussuriensis    ..T......A A...A..... .C-....... ...T.T.A.. ...G.T.... AT........
Deinagkistrodon   ..T.-....A A..GA....A C.CCT...T. ...T..AA.. .......... A.........
H. hypnale        ..CT.....A A...C..... ..AC...... ...T.TAG.. ..A..T.... A.........
Calloselasma      ..T......A A...C...G. ..A..T.... ...C..GG.. ..GT.C.... AT........
T. flavoviridis   ....C....A T...A...TA C.C....... ...C..AG.. .....T.... AT.......A
```

```
                                                                                    4
                                                                                    2
                                                                                    0
Bitis             CCAGGGATAA  CAGCGCTATC  TTCTTTAAGA  GCCCATATCA  AA-AAGAAGG  TTTACGACCT
Echis             ..........  ..........  .....C....  ..........  ..........  ..........
C. adamanteus     ..........  ..........  .....C....  ..........  ..C.......  ..........
C. tigris         ..........  ..........  .....C....  ..........  ..C.......  ..........
S. catenatus      ..........  ..........  .....C....  .T........  ..........  ...C......
B. asper          ..........  ..........  ..........  ..........  ..........  ..........
A. contortrix     ..........  ..........  ..........  ..........  ..........  ..........
A. bilineatus     ..........  ..........  ..........  ..........  ..........  ..........
A. piscivorus     ..........  ..........  ..........  ..........  ..........  ..........
A. intermedius    ..........  ..........  ......T...  ..........  ..........  ..........
A. shedaoensis    ..........  ..........  ......T...  ..........  ..........  ..........
A. strauchi       ..........  ..........  ......T...  ..........  ..........  ..........
A. ussuriensis    ..........  ..........  ......T...  ..........  ..........  ..........
Deinagkistrodon   ..........  ..........  .....C....  ..........  ..........  ..........
H. hypnale        ..........  ..........  .....C....  .T........  ..........  ..........
Calloselasma      ..........  ..........  .....C....  ..........  ..........  ..........
T. flavoviridis   ..........  ......C...  ..........  ..........  ..........  ..........

                                                                                    4
                                                                                    8
                                                                                    0
Bitis             CGATGTTGGA  TCAGGACATC  CTAGCAATGC  AACAGTTACT  AAGGGTTCGT  TTGTTCAACG
Echis             ..........  ..........  ....T.....  ..........  ..A.......  ..........
C. adamanteus     ..........  ..........  .A..T.....  .G.C......  ..A.......  ..........
C. tigris         ..........  .......G..  .A..T.....  .G.C......  ..A.......  ..........
S. catenatus      ..........  ..........  .C..T.....  ...C......  ..A.......  ..........
B. asper          ..........  ........C.  .C..T.G...  ...C.C....  ..A.......  ..........
A. contortrix     ..........  ..........  .C..T.....  ...C......  ..A.......  ..........
A. bilineatus     ..........  ..........  .C..T.....  .G.C......  ..A.......  ..........
A. piscivorus     ..........  ..........  ....T.....  ...C......  ..A..AA...  ..........
A. intermedius    ..........  ..........  .C..T.....  .C.C......  ..A.......  ..........
A. shedaoensis    ..........  ..........  .C..T.....  .C.C......  ..A.......  ..........
A. strauchi       ..........  ..........  .C..T.....  ...C......  ..A.......  ..........
A. ussuriensis    ..........  ..........  .C..T.....  ...C......  ..A.......  ..........
Deinagkistrodon   ..........  ........C.  .C..T.....  .C.C......  ..A.......  ..........
H. hypnale        ..........  ..........  .C..T.....  .G.C......  ..A.......  ..........
Calloselasma      ..........  ........C.  .C..T.....  .G.C......  ..A.......  ..C.......
T. flavoviridis   ..........  ..........  .C..T.....  .G.C......  ..A.......  ..........

                                              5
                                              0
                                              5
Bitis             ATTAATAGTC  CTACGTGATC  TGAGT
Echis             ..........  ......A...  .....
C. adamanteus     .....C....  ..........  .....
C. tigris         .....C....  ......T...  .....
S. catenatus      .....C....  .C........  .C...
B. asper          ..........  ..........  .....
A. contortrix     .....C....  ..........  .....
A. bilineatus     .....C....  ..........  .....
A. piscivorus     .....C....  ..........  .....
A. intermedius    ..........  ..........  .....
A. shedaoensis    ..........  ....T.....  .CT.A
A. strauchi       .....C....  ..........  .C...
A. ussuriensis    ..........  ..........  .T...
Deinagkistrodon   ..........  ..........  .....
H. hypnale        .....C....  ..........  .....
Calloselasma      ..........  ..........  .....
T. flavoviridis   .....C....  ..........  .T...
```

Fig. 1. Aligned mitochondrial 16S rDNA sequences (L-strand) of *Bitis arietans, Echis ocellatus, Crotalus adamanteus, C. tigris, Sistrurus catenatus, Bothrops asper, Agkistrodon contortrix, A. bilineatus, A. piscivorus, A. intermedius, A. shedaoensis, A. strauchi, A. ussuriensis, Deinagkistrodon acutus, Hypnale hypnale, Calloselasma rhodostoma* and *Trimeresurus flavoviridis*. A dot(.) indicates identity with *Bitis*. Dashes (–) indicate gaps inserted to maintain alignment. Sequences will be deposited in GENBANK.

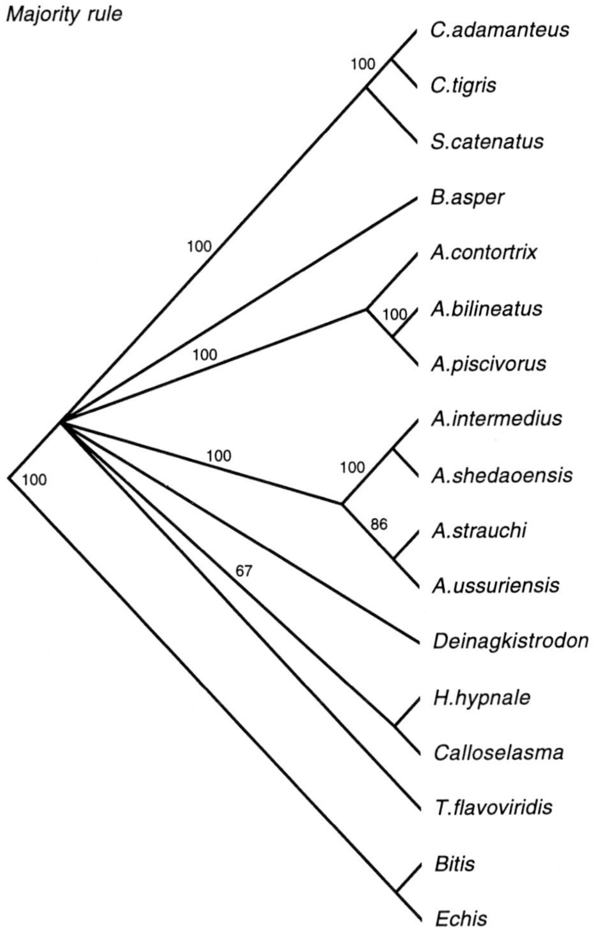

Fig. 2. Majority rules consensus tree of the 21 most parsimonious trees found treating gaps as: (1) missing = seven trees; (2) new-state = four trees; (3) nonadditive binary matrix = 10 trees. Nodal values indicate the percentage of times that node was represented out of 21 trees. *Bitis* and *Echis* were used as outgroups.

Comparing heuristic trees with no weighting for the three different gap methods reveals no clear-cut differences. Figure 2 is a 50% majority rules consensus tree of the 21 most parsimonious (MP) trees generated from the unweighted heuristic searches. Gaps were treated as: (1) missing; seven MP trees, length 280; (2) new-state; four MP trees, length 309; (3) nonadditive binary; 10 MP trees, length 296. For the weighted parsimony analyses we did not include gaps in the analyses. By weighting ts:tv 1:2 and using a heuristic search with additive tree bisection (10 replications) a single most parsimonious tree was found (Fig. 3) having a length of 400. A bootstrap analysis of the weighted data resulted in Fig. 4. The bootstrap values are low on certain nodes but groupings are consistent with the other trees presented here.

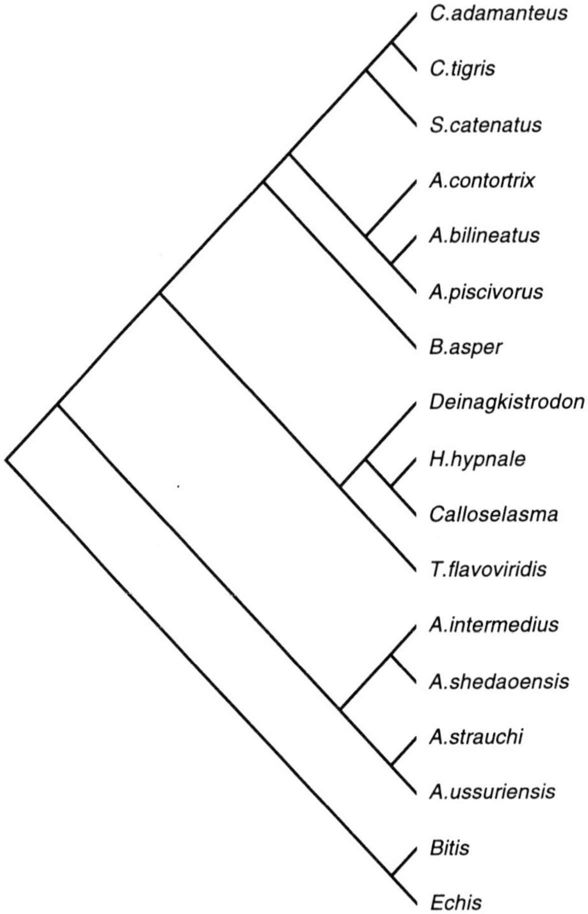

Fig. 3. Heuristic search, using random addition sequence and 10 replications, of the weighted data (tv 2 : ts 1) with *Bitis* and *Echis* as outgroups. Total length = 400.

Discussion

Comparing the three different trees we see that the main difference is in the placement of *Trimeresurus flavoviridis* and *Deinagkistrodon acutus*. The placement of *T. flavoviridis* (representing other Old World taxa) is not certain. In the weighted analysis it falls as the sister group of *Calloselasma*, *Deinagkistrodon* and *Hypnale*, but in the consensus trees its placement is in the basal polytomy of the pitvipers. *Deinagkistrodon* follows the same pattern: in the weighted analysis it is the sister group to the *Calloselasma/Hypnale* clade and in the consensus trees it falls within the basal polytomy of the pitvipers.

Unweighted heuristic searches of the data yielded 21 most parsimonious trees, depending on how gaps were considered. Seven MP trees resulted when gaps were treated as missing; four MP trees when gaps were added as a fifth character state; and 10 MP trees resulted when gaps were coded as non-additive binary characters. When these trees were compared, most of the discordance came from certain taxa

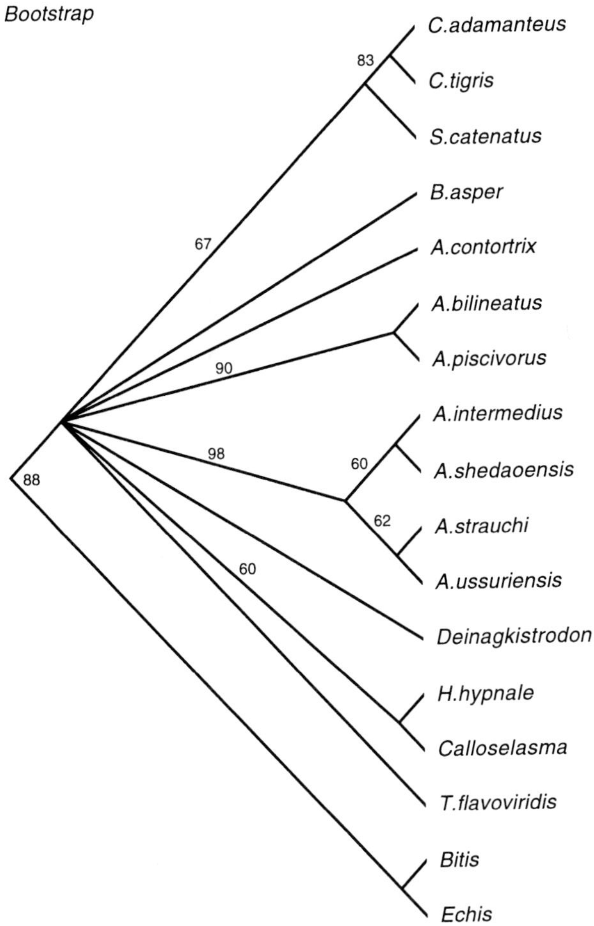

Fig. 4. Bootstrap analysis (using heuristic search, random addition sequence, and 10 replications per resampling) of the weighted data (tv 2 : ts 1) with *Bitis* and *Echis* as outgroups. Values at a node indicate percentage of times the node is represented out of 500 bootstrap replications.

(i.e. *Trimeresurus flavoviridis*, *Deinagkistrodon acutus*), but differences also arose from the placement of monophyletic lineages. By testing the three different gap modes for our data and building a consensus tree (Fig. 4) a better understanding of species groups is attained. Owing to the nature of consensus trees, resolution is lost when polytomies are formed. Thus, Figs 2 and 4 do not yield much information on higher phylogenetic relationships, with the exceptions of the *Crotalus/Sistrurus* and *Calloselasma/Hypnale* clades.

Many authors have used different weighting schemes for transversions and transitions (Brown *et al.* 1982; Knight & Mindell 1993; Alves-Gomes *et al.* 1995). The weighted tree gave greater resolution for deeper branches than the unweighted consensus trees. The possibility of near-transition saturation is evident by looking at the transition:transversion ratios. Thus by weighting the transversion higher, a more robust phylogeny is attained.

In traditional classifications the crotalids form a monophyletic group, and all *Agkistrodon* (*sensu lato*) are monophyletic. However, our data indicate (Fig. 3) that the New and Old World *Agkistrodon* do not form a monophyletic group nor are they sister groups. The New World pitvipers form a monophyletic group in relation to the Old World taxa. The data thus suggest that there was only a single immigration into the New World, but since all crotaline genera were not examined, they do not demonstrate it unequivocally.

Status of the genus *Gloydius*

Hoge & Romano-Hoge (1978/79) proposed a new generic name *Gloydius* for the Old World *Agkistrodon* based on morphological characters which they felt separated them from their New World congeners. Kardong (1990) argued that the characters which were used to diagnose the new genus might reflect allometric differences rather than taxonomic differences. A recent study by Knight *et al.* (1992), however, found taxonomic differences on a molecular level. Knight nonetheless stated that more conclusive data are needed to make the taxonomic change, thus did not side with Hoge & Romano-Hoge (1978/79). The data presented here and by Knight *et al.* (1992) are conclusive evidence of the validity of the genus *Gloydius*; we therefore support the change proposed by Hoge & Romano-Hoge (1978/79) of the generic name for the Old World *Agkistrodon* (*sensu stricto*) to *Gloydius*.

Validity of *Calloselasma*, *Deinagkistrodon* and *Hypnale*

In all analyses *Calloselasma*, *Deinagkistrodon* and *Hypnale* form separate evolutionary lineages, thus their validity as genera, according to these data, is supported. An interesting point to note is that in all analyses *Calloselasma* and *Hypnale* formed a clade. These data and the morphological characters presented by Groombridge (1986), Gloyd & Conant (1990), and Malnate (1990) suggest that *Calloselasma* is the sister taxon to *Hypnale*. This is very peculiar biogeographically as their ranges do not overlap at all, and indeed the two are separated by many miles and various mountain ranges. *Calloselasma* is found on the Malayan peninsula while *Hypnale* is found in the Western Ghats of India and on the island of Sri Lanka. Many other organisms have been found to share this distributional pattern (Smith 1931; Moody 1980; Nussbaum & Gans 1980). Hora (1949, 1953) proposed the 'Satpura hypothesis' as an explanation of these enigmatic distributions: migrating fauna followed the Satpura trend of mountains through peninsular India down into the Western Ghats region of India and onto the island of Sri Lanka during the Pleistocene. However, there are problems with his hypothesis, the major one being the Garo–Rajmahal Gap, in which two great rivers flow. Hora (1953) countered this assertion by stating that during the many glaciation events uplifting occurred, generating a bridge. The data presented here lend credence to the Satpura hypothesis, but more testing is needed.

Relationships of *Trimeresurus* to the Agkistrodontini and the status of the tribe Agkistrodontini

We used *Trimeresurus flavoviridis* to represent Old World crotalids not considered part of the '*Agkistrodon* complex'. This is very risky as we are using only one species to represent many taxa, thus these results lack stringency. In the weighted analysis (Fig. 3), *T. flavoviridis* forms the sister taxon to ((*Calloselasma, Hypnale*) *Deinagkistrodon*); in the consensus trees it falls within the basal polytomy. We suggest that the New World taxa are monophyletic with respect to the Old World taxa, and the tribe Agkistrodontini is invalid. If *Calloselasma*, *Deinagkistrodon* and *Hypnale* are the sister group to the other Old World genera (*Ovophis, Protobothrops, Trimeresurus, Tropidolaemus*) rather than to the genus *Gloydius*, the Agkistrodontini of Hoge & Romano-Hoge (1978/79) is paraphyletic. Additional data (C. L. Parkinson unpubl.) support these findings. F. Kraus (pers. comm.) is also in agreement with these results, supporting both the validity of the genus *Gloydius* and the paraphyly of the tribe Agkistrodontini.

The genus *Agkistrodon* formerly (Brattstrom 1964), included *Calloselasma, Deinagkistrodon, Hypnale* and *Gloydius* (Burger 1971; Chernov 1957; Gloyd 1977; and Hoge & Romano-Hoge 1978/79). It was diagnosed on the basis of a pitviper which has nine large symmetrical head shields and lacks a rattle. We suggest that these characters are pleisiomorphic at the generic level, and therefore new characters and character systems are needed. Molecular sequence data are just one system, which may be used as long as it is in conjunction with morphology. Many more data are needed in order to ascertain the phylogenetic relationships within this group of crotalids.

Specimens examined

Institutional abbreviations are listed in Leviton *et al.* (1985). Personal abbreviations are: CLP (Chris Parkinson, field tag), MOODY (Scott Moody, field tag).

Agkistrodon bilineatus — CLP 140. Mexico: Tamaulipas.
Agkistrodon contortrix — MOODY 338. US: OH, Athens Co., Waterloo Tsp.
Agkistrodon piscivorus — CLP 30. US: SC.
Bitis arietans — West Africa; Togo.
Bothrops asper — CLP 50. Costa Rica.
Calloselasma rhodostoma — UTA-R 22247. No locality data.
Crotalus adamanteus — CLP 4. US: FL, St. Johns Co., Flagler Estates.
Crotalus tigris — CLP169. US: AZ, Pima Co.
Deinagkistrodon acutus — CLP 28. China.
Echis ocellatus — MVZ 174468. West Africa: Ghana, Upper West Region.
Gloydius intermedius — UTA-R 16966. No locality data.
Gloydius shedaoensis — ROM 20468. PRC: Liaoning, Snake Is.
Gloydius strauchi — ROM 20473. PRC: Sichuan, Waqie.
Gloydius ussuriensis — ROM 20452. PRC: Jilin, Kouqian.

Hypnale hypnale — CLP 164. Sri Lanka: near Colombo.
Sistrurus catenatus — MOODY 502. US: TX, Haskell Co. near Stanford Lake.
Trimeresurus flavoviridis — UMMZ 199973. Japan: Ryukyu Is., Tokunoshima.

Acknowledgements

For tissues or DNAs used in this study, we thank John Cadle, Instituto Clodomiro Picado, F. Kraus, A. Knight, W. Lamar, R. Murphy, L. Porras and R. W. VanDevender. For the care and housing of live snakes we thank W. McMahan and the Louisville Zoological Gardens. For comments on the manuscript we thank D. Rountree, S. Akkaraju and H. Dessauer. C. Parkinson wishes to thank Professor Roger Thorpe, Dr Anita Malhotra, Dr Wolfgang Wüster and the Zoological Society of London for the invitation to be part of the symposium. This work was supported by the University of Louisville's Department of Biology, Dr P. W. Flanagan, Vice President of Research and Dean of Graduate programs, Sigma Xi, and SSAR Grants-in-Herpetology to CLP.

References

Anderson, S., Bankier, A. T., Barrell, B. G., de Bruijn, M. H. L., Coulson, A. R., Drouin, J., Eperon, I. C., Nierlich, D. P., Roe, B. A., Sanger, F., Schreier, P. H., Smith, A. J. H., Staden, R. & Young, I. G. (1981). Sequence and organization of the human mitochondrial genome. *Nature, Lond.* **290**: 457–465.

Alves-Gomes, J. A., Orti, G., Haygood, M., Heiligenberg, W. & Meyer, A. (1995). Phylogenetic analysis of the South American electric fishes (Order Gymnotiformes) and the evolution of their electrogenic system: a synthesis based on morphology, electrophysiology, and mitochondrial sequence data. *Mol. Biol. Evol.* **12**: 298–318.

Ashe, J. S. & Marx, H. (1988). Phylogeny of the viperine snakes (Viperinae): Part II. Cladistic analysis and major lineages. *Fieldiana, Zool. (N.S.)* No.52: 1–23.

Brattstrom, B. H. (1964). Evolution of the pit vipers. *Trans. S. Diego Soc. nat. Hist.* **13**: 185–268.

Brown, W. M., Prager, E. M., Wang, A. & Wilson, A. C. (1982). Mitochondrial DNA sequences of primates: tempo and mode of evolution. *J. molec. Evol.* **18**: 225–239.

Burger, L. W. (1971). Genera of pitvipers (Serpentes: Crotalidae). PhD Diss., University of Kansas. [*Diss. Abst. Int.* B32: 6119.]

Cabot, E. L. & Breckenback, A. T. (1989). Simultaneous editing of multiple nucleic acid and protein sequences with ESEE. *Comput. Appl. Biosci* **5**: 233–234.

Cadle, J. E. (1988). Phylogenetic relationships among advanced snakes: a molecular perspective. *Univ. Calif. Publs Zool.* **119**: 1–177.

Cadle, J. E. (1992). Phylogenetic relationships among vipers: immunological evidence. In *Biology of the pitvipers*: 41–48. (Eds Campbell, J. A. & Brodie, E. D.). Selva, Tyler, Texas.

Campbell, J. A. & Whitmore, D. H. (1989). A comparison of the skin keratin biochemistry in vipers with comments on its systematic value. *Herpetologica* **45**: 242–249.

Chernov, S. A. (1957). [Systematic position of the poisonous snake *Ancistrodon rhodostoma* (Boie) (Serpentes, Crotalidae) in connection with its craniology.] *Zool. Zh.* **36**: 790–792. [In Russian, English summary.]

Cock Buning, T. de (1985). Qualitative and quantitative explanation of the forms of heat sensitive organs in snakes. *Acta biotheor.* **34**: 193–206.

Cope, E. D. (1860). Catalogue of the venomous serpents in the museum of the Academy of Natural Sciences of Philadelphia, with notes on the families, genera, and species. *Proc. Acad. nat. Sci. Philad.* **11**: 332–347.

Cracraft, J. & Helm-Bychowski, K. (1991). Parsimony and phylogenetic inference using DNA sequences: some methodological strategies. In *Phylogenetic analysis of DNA sequences*: 184–220. (Eds Miyamoto, M. M. & Cracraft, J.). Oxford University Press, New York & Oxford.

Dixon, M. T. & Hillis, D. M. (1993). Ribosomal RNA secondary structure: compensatory mutations and implications for phylogenetic analysis. *Mol. Biol. Evol.* **10**: 256–267.

Felsenstein, J. (1985). Confidence limits on phylogenies: an approach using the bootstrap. *Evolution* **39**: 783–791.

Fitzinger, L. J. F. J. (1843). *Systema reptilium . . . Fasciculus primus: Amblyglossae.* Braümuller et Seidel Bibliopolas, Vindobonae. [Reprinted in 1973 by Society for the Study of Amphibians and Reptiles.]

Gloyd, H. K. (1977). Descriptions of new taxa of crotalid snakes from China and Ceylon (Sri Lanka). *Proc. biol. Soc. Wash* **90**: 1002–1015.

Gloyd, H. K. (1979). A new generic name for the hundred-pace viper. *Proc. biol. Soc. Wash.* **91**: 963–964.

Gloyd, H. K. & Conant, R. (Eds) (1990). Snakes of the *Agkistrodon* complex: a monographic review. *Contr. Herpet.* No. 6: 1–614.

Groombridge, B. C. (1986). Phyletic relationships among viperine snakes. In *Studies in herpetology:* 219–222. (Ed. Roček, Z.). Societas Europaea Herpetologica; Charles University, Prague.

Higgins, D. G., Bleasby, A. J. & Fuchs, R. (1992). CLUSTAL V: improved software for multiple sequence alignment. *Comput. Appl. Biosci.* **8**: 189–191.

Higgins, D. G. & Sharp, P. M. (1989). Fast and sensitive multiple sequence alignments on a microcomputer. *Comput. Appl. Biosci.* **5**: 151–153.

Heise, P. J., Maxson, L. R., Dowling, H. G. & Hedges, S. B. (1995). Higher-level snake phylogeny inferred from mitochondrial DNA sequences of 12S rRNA and 16S rRNA genes. *Mol. Biol. Evol.* **12**: 259–265.

Hillis, D. M. & Dixon, M. T. (1991). Ribosomal DNA: molecular evolution and phylogenetic inference. *Q. Rev. Biol.* **66**: 411–453.

Hoge, A. R. & Romano-Hoge, S. A. R. W. L. (1978/79)[1981]. Poisonous snakes of the world. Part I. Check list of the pit vipers, Viperoidea, Viperidae, Crotalinae. *Mems Inst. Butantan* **42/43**: 179–309.

Hora, S. L. (1949). Satpura hypothesis of the distribution of the Malayan fauna and flora to peninsular India. *Proc. natn. Inst. Sci. India* **15**: 310–314.

Hora, S. L. (1953). The Satpura hypothesis. *Sci. Prog., Lond.* **41**: 245–255.

Isogawa, K., Moriya, A. & Mitsui, S. (1994). A new snake of the genus *Agkistrodon* (Serpentes: Viperidae) from Tsushima Island, Nagasaki Prefecture, Japan. *Jap. J. Herpet.* **15**: 101–111.

Kardong, K. V. (1990). General skull, bone and muscle variation in *Agkistrodon* and related genera. In *Snakes of the* Agkistrodon *complex: a monographic review.* (Eds Gloyd, H. K. & Conant, R.). *Contr. Herpet.* No. 6: 573–581.

Knight, A., Densmore III, L. D. & Rael, E. D. (1992). Molecular systematics of the *Agkistrodon* complex. In *Biology of the pitvipers*: 49–70. (Eds Campbell, J. A. & Brodie, E. D.). Selva, Tyler, Texas.

Knight, A. & Mindell, D. P. (1993). Substitution bias, weighting of DNA sequence evolution, and the phylogenetic position of Fea's viper. *Syst. Biol.* **42**: 18–31.

Knight, A. & Mindell, D. P. (1994). On the phylogenetic relationship of Colubrinae, Elapidae, and Viperidae and the evolution of front-fanged venom systems in snakes. *Copeia* **1994**: 1–9.

Kumar, S., Tumara, K. & Nei, M. (1993). *MEGA: molecular evolutionary genetics analysis, version 1.01*. Pennsylvania State University., University Park, PA.

Leviton, A. E., Gibbs, R. H., Heal, E. & Dawson, C. E. (1985). Standards in herpetology and ichthyology: Part 1. Standard symbolic codes for institutional resource collections in herpetology and ichthyology. *Copeia* **1985**: 802–832.

Liem, K. F., Marx, H. & Rabb, G. B. (1971). The viperid snake *Azemiops*: its comparative cephalic anatomy and phylogenetic position in relation to Viperinae and Crotalinae. *Fieldiana* (*Zool.*) No. 59: 67–126.

Malnate, E. (1990). A review and comparison of hemipenial structure in the genus *Agkistrodon* (*sensu lato*). In *Snakes of the* Agkistrodon *complex: a monographic review*. (Eds Gloyd, H. K. & Conant, R.). *Contr. Herpet.* No. 6: 583–588.

Minton, S. (1990). Immunologic relationships in *Agkistrodon* and related genera. In *Snakes of the* Agkistrodon *complex: a monographic review*. (Eds Gloyd, H. K. & Conant, R.). *Contr. Herpet.* No. 6: 589–600.

Moody, S. M. (1980). *Phylogenetic and historical biogeographical relationships of the genera in the family Agamidae (Reptilia: Lacertilia)*. PhD thesis: University of Michigan.

Nedball, M. A., Allard, M. W. & Honeycutt, R. L. (1994). Molecular systematics of hystricognath rodents: evidence from the mitochondrial 12S r RNA gene. *Mol. Phylog. Evol.* 3: 206–220.

Nussbaum, R. A. & Gans, C. (1980). On the *Ichthyophis* (Amphibia: Gymnophiona) of Sri Lanka. *Spolia zeylan.* **35**: 137–154.

Porter, C. A., Haiduk, M. W. & de Queiroz, K. (1994). Evolution and phylogenetic significance of ribosomal gene location in chromosomes of squamate reptiles. *Copeia* **1994**: 302–313.

Smith, M. A. (1931). *The fauna of British India including Ceylon and Burma. Reptilia and Amphibia* 1. *Loricata, Testudines*. Taylor & Francis, London.

Swofford, D. L. (1993). *PAUP: phylogenetic analysis using parsimony, Version 3.1*. Illinois Natural History Survey, Champaign, Ill.

Wheeler, W. C. & Honeycutt, R. L. (1988). Paired sequence difference in ribosomal RNAs: evolutionary and phylogenetic implications. *Mol. Biol. Evol.* **5**: 90–96.

Zhao, E. & Adler, K. (1993). Herpetology of China. *Contr. Herpet.* No. 10: 1–522.

6 Systematic implications of lactate dehydrogenase isozyme phenotypes in Neotropical pitvipers (Viperidae: Crotalinae)

STEVEN D. WERMAN

Synopsis

Lactate dehydrogenase isozyme phenotypes were surveyed from kidney tissues of pitvipers by means of horizontal starch-gel electrophoresis. Individuals of 23 species, representing the general *Agkistrodon*, *Atropoides*, *Bothriechis*, *Bothriopsis*, *Bothrops*, *Cerrophidion*, *Crotalus*, *Lachesis*, *Porthidium*, *Sistrurus* and *Trimeresurus*, were examined. Phenotypic variation based on the mobility class of LDH-A, the loss of specific isozyme expression and relative isozyme intensity differences were used to evaluate existing systematic and phylogenetic issues.

The observed variation of LDH phenotypes was found to have systematic utility. Mobility class defines the relationships: (((*Bothrops*, *Bothriopsis*) *Porthidium*) *Bothriechis*), with the remaining genera as sister lineages. Species groups based on the presumed apomorphic loss of isozymes and expression differences include (*Bothrops* spp.), (*Bothriechis schlegelii–B. lateralis–B. nigroviridis*), (*Porthidium nasutum–P. ophryomegas*) and (*Atropoides nummifer–A. picadoi*). The genera previously assigned to *Bothrops* (*sensu lato*) appear to be valid and the arboreal species assigned to *Bothriechis* and *Bothriopsis* are probably independent lineages. The general conclusions of this study are in concordance with previous cladistic analyses and biogeographic information of Neotropical pitvipers.

Introduction

The techniques of protein electrophoresis and the identification of isozyme phenotypes have been widely applied to questions and issues in evolution and systematics (Avise 1974; Ferguson 1980; Wake 1981; Buth 1984; Hillis 1987). While allozyme information is most frequently utilized in genetic studies at the population and species levels, isozymes have emerged as characters useful in resolving systematic problems of species and higher taxa (Ferris & Whitt 1978; Whitt 1981). Murphy *et al.* (1990) provide an in-depth description of enzyme data and their range of potential application.

Symp. zool. Soc. Lond. (1997) No. 70: 79–88
ISBN 0 19 854986 5

Lactate dehydrogenase (LDH, E.C. 1.1.1.27) is a tetrameric enzyme composed of subunits usually coded by two and sometimes three genetic loci (Whitt 1970a,b; Markert 1983). If in tissues where two loci are expressed equally (i.e., the 'A' and 'B' loci), and there is no restriction on subunit assembly, five isozymes are usually formed by the random association of the two different subunits into two homotetramers and three heterotetramers [AAAA, (A_4); AAAB, (A_3B_1); AABB, (A_2B_2); BBBB, (B_4)] in a respective ratio of 1: 4: 6: 4: 1. This pattern is also observed in phenotypes of heterozygous individuals where a single locus is expressed (Fig. 1).

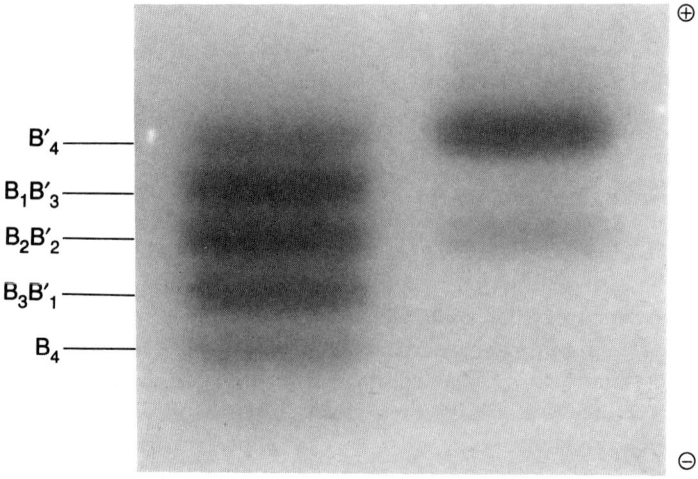

Fig. 1. A five-banded pattern (left) for an individual heterozygous for two alleles (B/B') at the LDH-B locus. Note the approximate 1 : 4 : 6 : 4 : 1 ratio of isozyme intensity. The individual on the right is a homozygote (B'/B') with a small sub-band trailing behind the B'_4 homotetramer. Migration is in the direction of the anode.

Several studies (Whitt 1970b; Gorman 1971; Frankel 1982; Sites, Bezy & Thompson 1986) show that in many tissues, interlocus heterotetramers may be absent. The loss of heterotetramer formation is presumably under genetic control either by the restriction of formation of functional tetramers due to sequence divergence between loci (Whitt 1970b; Frankel 1982; Murphy *et al.* 1990), or by the rapid degradation of some of these isozymes in specific tissues (Fritz *et al.* 1969).

Observed variation, in the presence or absence of specific LDH isozymes, intensity of their expression and mobility differences, exhibits taxon-specific patterns (Gorman 1971; Toledo & Ribeiro 1978) and is thus of systematic and evolutionary interest. Therefore, it follows that character information from isozyme studies is of value not only for taxonomic studies, but also for phylogenetic analysis.

This report focuses on kidney LDH variation in Neotropical pitvipers. Pitvipers are venomous snakes characterized by the possession of paired infra-red detecting pit organs embedded in the maxillae, with an external opening visible between the external naris and the orbit. New World pitvipers range from southern Canada to southern Argentina (Hoge

& Romano-Hoge 1978/79), inhabit areas ranging from deserts to wet tropical forests and include both terrestrial and arboreal forms. In this study, species comparisons were made with representatives of *Agkistrodon*, *Atropoides*, *Bothriechis*, *Bothriopsis*, *Bothrops*, *Cerrophidion*, *Crotalus*, *Lachesis*, *Porthidium* and *Sistrurus*. In addition, one Old World representative from tropical South-East Asia, *Trimeresurus gramineus*, was also assayed for an outgroup reference comparison (Maddison, Donoghue & Maddison 1984).

The features of LDH variation explored here include mobility class differences (*sensu* Morizot & Cranmer 1985) of the A_4 homotetramer, the relative loss of specific isozyme expression or activity, and the relative inter-isozyme intensity differences within an individual as described below.

Materials and methods

Kidney tissues were dissected from 108 individual pitvipers (Werman 1992) and frozen in liquid nitrogen prior to laboratory freezer storage at -70°C. Tissue extracts were prepared by mechanical homogenization in an equal volume of 1.0 M Tris–HCl (pH 7.0) and centrifuged at 19 250 g at 4.0°C for at least 30 min. Supernatants were removed and subjected to electrophoresis in 9.0 mm thick horizontal starch gels. All gels were 14.0% w/v, starch/buffer. Assays for LDH were run with a Tris–citrate/borate, pH 8.7, buffer system (Poulik 1957; Murphy *et al.* 1990) at 14.8 V/cm for 6 h. Following electrophoresis, gels were cut horizontally into 1.5 mm slices and histochemically stained for LDH (Shaw & Prasad 1970; Murphy *et al.* 1990). LDH phenotypes were scored qualitatively following the scheme listed in Table 1.

Table 1. Qualitative scoring of kidney LDH phenotypes as characters and character states for pitvipers. Alphabetic designations for character states are arbitrary and do not imply evolutionary polarity

1. Mobility class of LDH-A
   - A — fast (anodal)
   - B — medium
   - C — slow (cathodal)
2. Isozyme loss
   - A — five isozymes present
   - B — A_3B_1 absent
   - C — A_3B_1 and A_1B_3 absent
3. Expression intensity of homotetramers in kidney
   - A — $A_4 = B_4$
   - B — $A_4 > B_4$
   - C — $A_4 < B_4$
4. Intensity pattern of symmetrical isozyme tetramers
   - A — $A_4 < A_2B_2 > B_4$
   - B — $A_4 < A_2B_2 = B_4$
   - C — $A_4 > A_2B_2 = B_4$
   - D — $A_4 = A_2B_2 > B_4$
   - E — $A_4 > A_2B_2 < B_4$
   - F — $A_4 < A_2B_2 < B_4$
   - G — $A_4 > A_2 B_2 > B_4$

Results and discussion

Specimens representing 23 species in 11 genera were assayed for LDH isozyme pattern and expression in kidney tissue (Fig. 5). In all cases the LDH-B homotetramer (B_4) migrated faster (anodally) than the LDH-A homotetramer (A_4). Intraspecific allelic variation was rarely observed in LDH phenotypes (Werman 1992). This is exemplified in the isozyme patterns of *Bothrops jararaca* (Fig. 2). Analysis of allelic variation was not within the scope of this study.

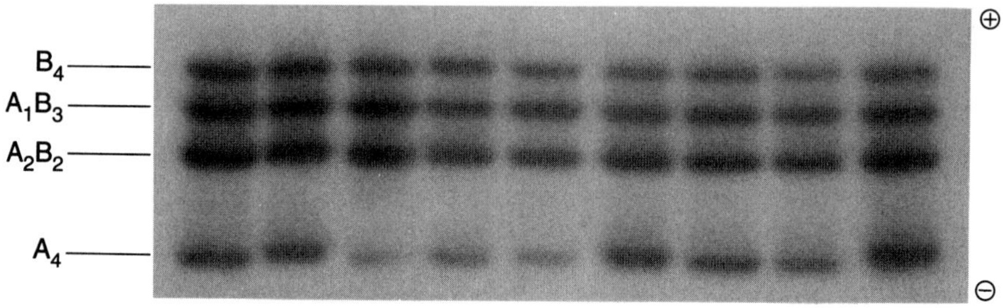

Fig. 2. Intraspecific variation in the kidney LDH phenotypes of nine individuals of *Bothrops jararaca*. Differences in intensities of the isozymes among individuals are due to minor differences in tissue dilution prior to homogenization.

Mobility class

The LDH A_4 isozymes fall into three general mobility zones or classes (Figs 3, 5 and Table 1, character 1). A cathodal or 'slow' class (state C) was unique to *Bothrops brazili*. A 'medium' class (state B) was observed for the other members of *Bothrops*, as well as *Bothriechis*, *Bothriopsis taeniata* and *Porthidium nasutum*. The 'fast' (state A), anodal mobility class of A_4 isozymes appears to be a primitive condition. This class was found in *Agkistrodon piscivorus*, *Atropoides*, *Cerrophidion godmani*, *Crotalus durissus*, *Lachesis muta*, *Sistrurus miliarius* and *Trimeresurus gramineus*. Within each of the latter two classes minor allelic differences in A_4 mobility were

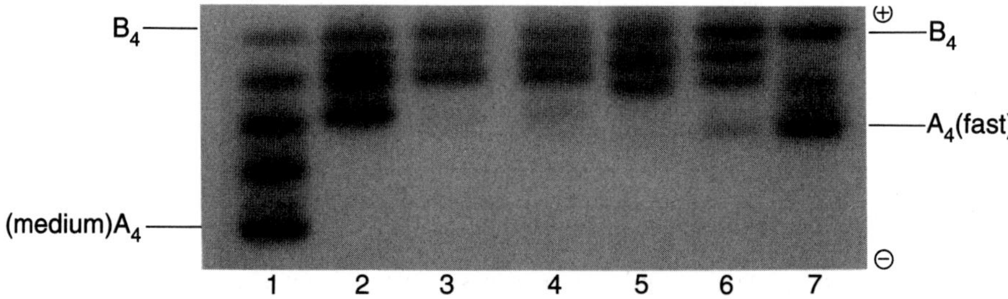

Fig. 3. Differences in 'medium' and 'fast' mobility class characterizations of the A_4 isozymes in pitvipers. The 'slow' class can be seen in *Bothrops brazili* of Fig. 5. The taxa are: 1, *Bothrops moojeni*; 2, *Lachesis muta*; 3, *Cerrophidion godmani*; 4, *Atropoides picadoi*; 5, *Agkistrodon piscivorus*; 6, *Sistrurus miliarius* and 7, *Crotalus durissus*.

evident and in my view warrant the characterization of mobility classes in LDH A_4 isozymes in pitvipers. Others, however, support the conclusion that mobility classes simply represent allelic variation with extreme differences in electrophoretic migration (D. G. Buth & D. M. Hillis pers. comm.). In spite of these opinions, the 'medium' mobility class phenotype does provide a potentially important synapomorphy defining a major clade among Neotropical pitvipers (Werman 1992).

Patterns of isozyme loss

Typically, in tissues where two LDH loci are expressed and the genetic divergence between the loci is small, five isozymes are usually formed based on random assembly (described above). However, many pitvipers in this study exhibited a loss of specific isozymes (Figs 4, 5 and Table 1, character 2). All five isozymes (state A) were observed in *Bothrops moojeni* and *B. pradoi* (also the unidentified *Bothrops* sp., see Fig. 5). All other pitvipers surveyed, including *T. gramineus*, were missing at least one isozyme. The possibility exists that the phenotype of five isozymes reflects an apomorphic reversal from a primitive four-banded (absent A_3B_1 isozyme) pattern. Thus, it is possible that *B. moojeni* and *B. pradoi* are closely related species.

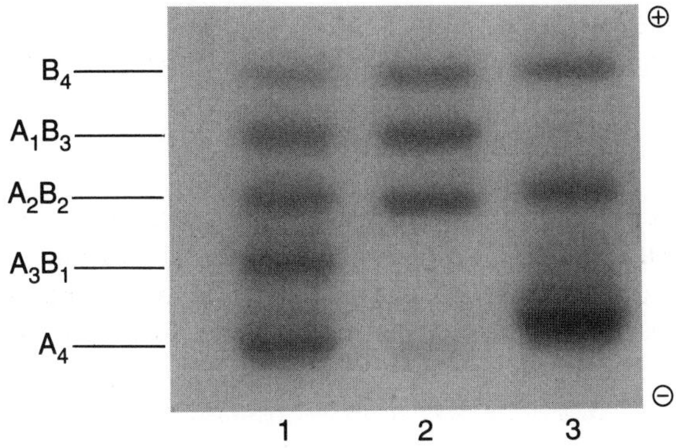

Fig. 4. Loss of asymmetrical isozymes in pitvipers. 1, *Bothrops moojeni*, with all five isozymes; 2, *Bothrops alternatus*, missing the A_3B_1 isozyme; 3, *Bothriechis lateralis*, missing both the A_3B_1 and the A_1B_3 isozymes.

Three-banded patterns (state C) consisting of A_4, A_2B_2, and B_4 isozymes were observed in *Agkistrodon piscivorus*, *Bothriechis lateralis*, *B. nigroviridis* (A_1B_3 very faint) and *Crotalus durissus*. *C. durissus* varied intraspecifically with either a three-banded pattern as above (data not shown), or a nearly two-banded pattern with a very faint A_2B_2 isozyme. Because *B. lateralis* and *B. nigroviridis* show other LDH isozyme similarities to *B. schlegelii*, and are clearly separated from the other genera by a different mobility class A_4 isozyme, it is likely that the absence of the asymmetrical heterotetramers is a convergent loss for these taxa (i.e. *Bothriechis* and *Crotalus*).

All other species examined exhibited a four-banded pattern, missing the A_3B_1 heterotetramer (state B). This feature is also found in other groups of snakes and lizards (Gorman 1971; Sites *et al.* 1986; S. D. Werman unpubl.) and is possibly symplesiomorphic. No individual examined in the present study had lost more than two of the five isozymes (Fig. 5).

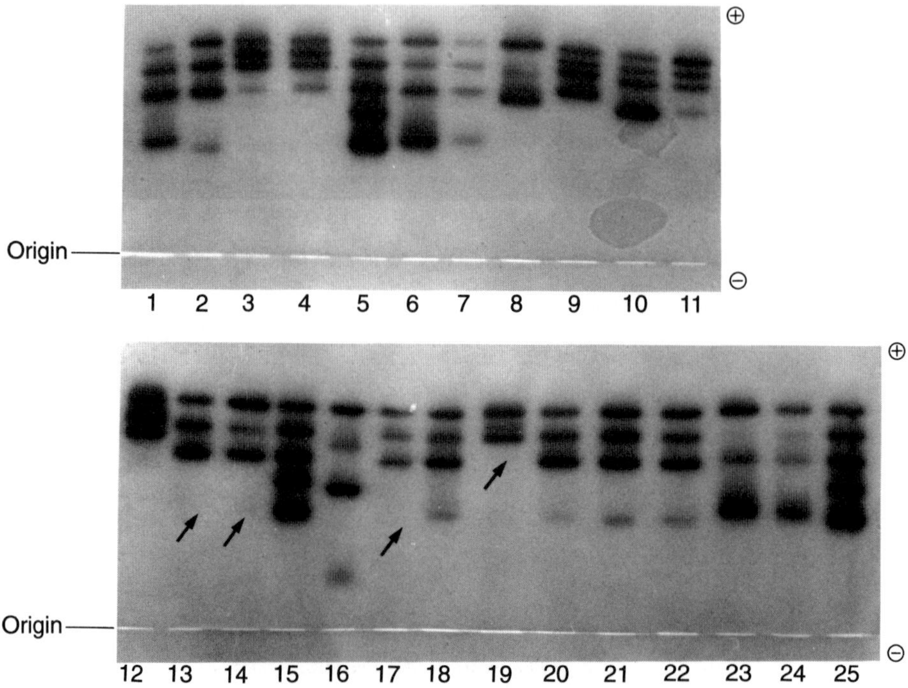

Fig. 5. Representative patterns of kidney LDH expression in pitvipers. The arrows indicate faint A_4 isozyme bands that may not be evident in the photograph. The taxa are: 1, *Porthidium nasutum*; 2, *Bothrops neuwiedi*; 3, *Atropoides nummifer*; 4, *Atropoides picadoi*; 5, *Bothrops pradoi*; 6 & 7, *Bothriechis schlegelii*; 8, *Crotalus durissus*; 9, *Lachesis muta*; 10, *Trimeresurus gramineus*; 11, *Sistrurus miliarius*; 12, *Agkistrodon piscivorus*; 13, *Bothrops alternatus*; 14, *Bothrops atrox*; 15, *Bothrops* sp.; 16, *Bothrops brazili*; 17, *Bothriopsis taeniata*; 18, *Bothrops erythromelas*; 19, *Cerrophidion godmani*; 20, *Bothrops itapetininga*e; 21, *Bothrops jararaca*; 22, *Bothrops jararacussu*; 23, *Bothriechis lateralis*; 24, *Bothriechis nigroviridis*; 25, *Bothrops moojeni*.

Isozyme intensity

The patterns of relative staining intensity of LDH isozymes within kidney varied considerably. Differences were observed most readily between and among distantly related pitvipers. Several patterns were either species-specific or diagnostic for closely related taxa or genera (Fig. 5 and Table 2).

In general, there were two criteria used to distinguish LDH isozyme patterns of differential isozyme intensity in the pitviper kidney (Table 1). The first was the relative intensities between the A_4 and B_4 isozymes (character 3). Either the A_4 was expressed equally to the B_4 (state A), or the A_4 was expressed strongly over the B_4 (state B), or the B_4 isozyme was dominately expressed over the A_4 (state C). *Agkistrodon piscivorus* and *Crotalus durissus* both had homotetramers of equal intensity (state A). *Atropoides*, *Bothriopsis taeniata*, *Cerrophidion godmani*, *S. miliarius* and most of the species of *Bothrops* had a weaker A_4 as compared to the B_4 (state C), whereas *Bothriechis*, *Bothrops moojeni* and *B. pradoi*, *L. muta*, *P. nasutum* and *T. gramineus* all had strongly expressed B_4 isozymes relative to the A_4 (state B).

Table 2. Character state arrays for representative pitviper taxa based on Table 1. Superscript denotes additional taxa with an identical array

Taxon	Character 1	2	3	4
Agkistrodon piscivorus	A	C	A	A
Atropoides nummifer[a]	A	B	C	B
Bothriechis schlegelii	B	B	B	C
B. lateralis[b]	B	C	B	E
Bothrops alternatus[c]	B	B	C	B
B. brazili	C	B	C	B
B. moojeni[d]	B	A	B	C
Cerrophidion godmani	A	B	C	B
Crotalus durissus	A	C	A	E
Lachesis muta	A	B	B	C
Porthidium nasutum[e]	B	B	B	D
Sistrurus miliarius	A	B	C	F
Trimeresurus gramineus	A	B	B	G

[a] *Atropoides picadoi*
[b] *Bothriechis nigroviridis*
[c] *Bothrops atrox*, *B. erythromelas*, *B. itapetiningae*, *B. jararaca*, *B. jararacussu*, *B. neuwiedi* and *Bothriopsis taeniata*.
[d] *Bothrops pradoi* and *Bothrops* sp.
[e] *Porthidium ophryomegas* (data not shown).

The second was the intensity of the A_2B_2 isozyme relative to the two homotetramers, the A_4 and B_4. This character was further differentiated into seven patterns or character states. The patterns, described as follows, do not include the complete loss of a particular isozyme because this is regarded as a separate character. Thus, the relative intensities of the A_4 to A_2B_2 to B_4 isozymes (bands) were used as independent qualitative character states. *Agkistrodon piscivorus* was unique (state A) in that it showed a symmetrical phenotype of a strong A_2B_2 band flanked by two less intense, but equal A_4 and B_4 isozymes. *Bothriechis schlegelii*, *Bothrops moojeni* and *L. muta* were typified by a strong A_4 band with weaker but equal A_2B_2 and B_4 bands (state C). *Porthidium nasutum* (and *P. ophryomegas*, data not shown) had a strong A_4 band, a weak B_4 band and an intermediate A_2B_2 band (state D). *Bothriechis lateralis*, *B. nigroviridis* and *Crotalus durissus* had equivalent A_4 and B_4 bands with a weaker A_2B_2 in between (state E). The pigmy rattlesnake, *S. miliarius*, had a strong B_4 band with a stepwise decrease in intensity of the A_2B_2 and A_4 bands respectively (state F). *Trimeresurus gramineus* had the opposite condition to state F, in that the A_4 was the strongest with a decrease in intensity to the A_2B_2 and B_4 bands, respectively (state G).

The remaining taxa were characterized by a weak A_4 band and stronger but equal A_2B_2 and B_4 bands (state B). These included most *Bothrops*, *Bothriopsis taeniata* and *Atropoides*.

Systematic considerations

The diagnostic value of the data here, for general systematic purposes, is useful and can be applied at interspecies and intergeneric levels. Lactate dehydrogenase patterns provide specific 'key' characters (*sensu* Mayr 1969) for the identification of taxa if methods of protein electrophoresis are available. Many of the pitvipers examined here have LDH patterns in kidney that are either unique or shared only with closely related forms (Table 2). For example, the five-banded kidney isozyme pattern was found in only three species of pitviper assayed here, all of South American distribution (*Bothrops moojeni*, *B. pradoi* and *Bothrops* sp.). In addition, characteristic isozyme phenotypes for the arboreal species *Bothriechis schlegelii*, *B. nigroviridis* and *B. lateralis* can be used not only to diagnose these Central American forms from all other species in this study, but also to differentiate the latter two from *B. schlegelii*. In addition, the phenotypes of this arboreal genus are distinct from that of *Bothriopsis taeniata*, an arboreal South American form.

The mobility class feature of the A_4 isozyme separates the pitvipers into two major groups. The 'medium' class, for example, is characteristic of many *Bothrops*, especially those restricted to South America. The 'fast' class is dominated by many Central American genera as well as divergent lineages (i.e. *Trimeresurus*, *Crotalus* and *Lachesis*). In general, many species herein can be easily identified or placed into a particular phenotypic group based on their particular LDH pattern as observed under the electrophoretic conditions employed here.

For phylogenetic reconstruction, LDH isozyme data may have limits to its value as an independent source. This stems primarily from the potential homoplasious nature of the pattern differences, specifically the loss or reduction events. Losses due to either the restriction of subunit assembly (Whitt 1970b) or the differential degradation of various isozymes (Fritz *et al.* 1969) which may have direct ancestor–descendant transmission are difficult to distinguish from parallel independent loss in the absence of additional information. Thus, singular isozyme losses can corroborate a phylogenetic hypothesis but they may have little power to test it. A possible example may include the Central American arboreal pitvipers (*Bothriechis*), because other studies (Crother, Campbell & Hillis 1992; Werman 1992) suggest that this genus comprises a distinct clade. The overall phenotypic similarities of LDH support these authors' conclusions in that all three taxa assayed here share a strongly expressed, 'medium' class A_4 isozyme and *Bothriechis nigroviridis* and *B. lateralis* show a loss, or great reduction, of the A_1B_3 heterotetramer. It is also noteworthy that the LDH phenotypes of the jumping vipers, *Atropoides nummifer* and *A. picadoi*, are identical, as are those of *Porthidium nasutum* and *P. ophryomegas* (data not shown) of the hognose pitviper group.

Thus, isozyme evolution may be appropriately evaluated if enzyme character information can be superimposed on a previous and stable phylogeny. In cases involving the loss of tissue specific expression (Ferris & Whitt 1978), the silencing of duplicate genes (Whitt 1981) and the restriction of heterotetramer assembly (Sites *et al.* 1986; Murphy *et al.* 1990) this approach may be productive and interesting.

In general, the LDH phenotypes are consistent with the cladistic hypothesis of Neotropical pitviper relationships as outlined in Werman (1992). This phylogeny, in concert with biogeographic information, allows for the recognition of the many genera previously assigned to *Bothrops (sensu lato)*. These relationships are: (((((*Bothriopsis–Bothrops*) *Porthidium*) (*Bothriechis–Ophryacus*)) *Atropoides*) *Cerrophidion*?). While the LDH phenotypes herein failed to distinguish *Bothriopsis* from *Bothrops*, the remaining genera were identified and are likely to be natural groupings. Ongoing efforts of DNA sequence analysis for genes of systematic utility will surely illuminate these issues.

Acknowledgements

Special thanks are extended to Shari Reeder, who helped in the preparation of early drafts of the manuscript, and to Dr Edward Hurlbut for his assistance in preparing the figures. I also thank Mesa State College, the School of Natural Sciences and faculty of the Department of Biology for financial support and encouragement.

References

Avise, J. C. (1974). Systematic value of electrophoretic data. *Syst. Zool.* **23**: 465–481.

Buth, D. G. (1984). The application of electrophoretic data in systematic studies. *A. Rev. Ecol. Syst.* **15**: 501–522.

Crother, B. I., Campbell, J. A. & Hillis, D. M. (1992). Phylogeny and historical biogeography of the palm-pitvipers, genus *Bothriechis*: biochemical and morphological evidence. In *The biology of pitvipers*: 1–19. (Eds Campbell, J. A. & Brodie Jr., E. D.). Selva, Tyler, Texas.

Ferguson, A. (1980). *Biochemical systematics and evolution*. Wiley, New York.

Ferris, S. D. & Whitt, G. S. (1978). Phylogeny of tetraploid catostomid fishes based on the loss of duplicate gene expression. *Syst. Zool.* **27**: 189–203.

Frankel, J. S. (1982). Lactate dehydrogenase specificity and subunit assembly in neural tissues of the teleost *Phallichthys amates*. *Experientia* **38**: 673–674.

Fritz, P. J., Vesell, E. S., White, E. L. & Pruitt, K. M. (1969). The roles of synthesis and degradation in determining tissue concentration of lactate dehydrogenate-5. *Proc. natn. Acad. Sci. USA* **62**: 558–565.

Gorman, G. C. (1971). Evolutionary genetics of island lizard populations. *Yb. Am. phil. Soc.* **1971**: 318–319.

Hillis, D. M. (1987). Molecular versus morphological approaches to systematics. *A. Rev. Ecol. Syst.* **18**: 23–42.

Hoge, A. R. & Romano-Hoge, S. A. R. W. L. (1978/79) [1981]. Poisonous snakes of the world. Part 1. Checklist of the pit vipers Viperoidea, Viperidae, Crotalinae. *Mems Inst. Butantan* **42/43**: 179–309.

Maddison, W. P., Donoghue, M. J. & Maddison, D. R. (1984). Outgroup analysis and parsimony. *Syst. Zool.* **33**: 83–104.

Markert, C. L. (1983). Isozymes: conceptual history and biological significance. In *Isozymes curr. Top. biol. med. Res.* 7: 1–17.

Mayr, E. (1969). *Principles of systematic zoology*. McGraw-Hill, New York.

Morizot, D. C. & Cranmer, K. (1985). Genetic distance estimation from electrophoretic mobility class data. *Isozyme Bull.* **18**: 83.

Murphy, R. W., Sites Jr., J. W., Buth, D. G. & Haufler, C. H. (1990). Proteins I: isozyme electrophoresis. In *Molecular systematics*: 45–126. (Eds Hillis, D. M. & Moritz, C. M.). Sinauer, Sunderland, Massachusetts, USA.

Poulik, M. D. (1957). Starch gel electrophoresis in a discontinuous system of buffers. *Nature, Lond.* **180**: 1477–1479.

Shaw, C. R. & Prasad, R. (1970). Starch gel electrophoresis of enzymes — a compilation of recipes. *Biochem. Genet.* **4**: 297–320.

Sites Jr., J. W., Bezy, R. L. & Thompson, P. (1986). Nonrandom expression of lactate dehydrogenase isozymes in the lizard family Xantusiidae. *Biochem. Syst. Ecol.* **14**: 539–545.

Toledo, S. A. & Ribeiro, A. F. (1978) Directional reduction of lactate dehydrogenase isozymes in teleosts. *Evolution* **32**: 212–216.

Wake, D. B. (1981). The application of allozyme evidence to problems in the evolution of morphology. In *Evolution today. Proceedings of the second international congress of systematic and evolutionary biology*: 257–270. (Eds Scudder, G. G. E. & Reveal, J. L.). Hunt Institute for Botanical Documentation, Carnegie-Mellon University, Pittsburgh.

Werman, S. D. (1992). Phylogenetic relationships of Central and South American pitvipers of the genus *Bothrops* (*sensu lato*): cladistic analyses of biochemical and anatomical characters. In *The biology of pitvipers*: 21–40. (Eds Campbell, J. A. & Brodie, Jr., E. D.). Selva, Tyler, Texas.

Whitt, G. S. (1970a). Developmental genetics of the lactate dehydrogenase isozymes of fish. *J. exp. zool.* **175**: 1–36.

Whitt, G. S. (1970b). Directed assembly of polypeptides of the isozymes of lactate dehydrogenase. *Archs Biochem. Biophys.* **138**: 352–354.

Whitt, G. S. (1981). Evolution of isozyme loci and their differential regulation. In *Evolution today. Proceedings of the second international congress of systematic and evolutionary biology*: 271–289. (Eds Scudder, G. G. E & Reveal, J. L.). Hunt Institute for Botanical Documentation, Carnegie-Mellon University, Pittsburgh.

7 DNA evolution of South American pitvipers of the genus *Bothrops* (Reptilia: Serpentes: Viperidae)

MARIA DA GRAÇA SALOMÃO, WOLFGANG WÜSTER, ROGER S. THORPE, JEAN-MARC TOUZET and BBBSP*

Synopsis

The Neotropical pitvipers currently and formerly included in the genus *Bothrops* comprise a large number (> 60) of species. The phylogenetic relationships among the taxa included in this group remain incompletely understood. Here, we analyse the relationships among various species of *Bothrops* (*B. alternatus*, *atrox*, *caribbaeus*, *cotiara*, *insularis*, *isabelae*, *jararaca*, *jararacussu*, *leucurus*, *marajoensis* and *moojeni*) and *Bothriopsis* (*B. taeniata*) by means of comparative mitochondrial DNA sequencing. Portions of the cytochrome *b* gene were amplified by means of the polymerase chain reaction (PCR), and sequenced. A 565 bp region of the cytochrome *b* gene was subjected to phylogenetic analysis, using maximum parsimony, Fitch–Margoliash and maximum likelihood algorithms. All trees confirmed the monophyly of the *Bothrops atrox* group exclusive of *Bothrops jararacussu*, the rooting of *Bothriopsis taeniata* within the genus *Bothrops*, the basal dichotomy between species with and without a lacunolabial scale, and the sister-species relationships between *B. jararaca* and *B. insularis*. On the basis of these results, we synonymize the genus *Bothriopsis* with *Bothrops*. Our results are discussed with reference to previous analyses of the phylogeny of Neotropical pitvipers, and points of consensus and remaining problem areas are identified.

Introduction

The snakes of the genus *Bothrops* are the most important cause of snakebite accidents in Latin America (Bolaños 1984; Campbell & Lamar 1989; Ministério da Saúde 1990; Cardoso 1992; Otero *et al.* 1992). In spite of this, the systematics and the relationships among members of this group remain controversial and unclear. Problematic areas include the status of some of the genera resurrected by Burger (1971) and the population systematics of the different species of the *B. atrox* group.

Symp. zool. Soc. Lond. (1997) No. 70: 89–98
ISBN 0 19 854986 5

* Butantan–British *Bothrops* Systematics Project — M.F.D. Furtado, S.A. Hoge, G. Puorto, M.G. Salomão, R.D.G. Theakston, R.S. Thorpe, D.A. Warrell & W. Wüster

Morphology has been widely used to infer the affinities among the species of *Bothrops* (Amaral 1921; Brattstrom 1964; Hoge & Romano-Hoge 1978/79), and to divide the genus into the genera *Bothriechis*, *Bothriopsis*, *Bothrops* and *Porthidium* (Burger 1971; Schätti, Kramer & Touzet 1990; Schätti & Kramer 1993). However, these studies lacked rigorous phylogenetic analysis of the data obtained.

Recently other approaches have been used in an attempt to resolve different levels of the branching hierarchy in these groups. These include analysis of hemipenis morphology and electrophoresis of plasma proteins (Pesantes 1989), and combined allozyme, isozyme and morphological analysis (Werman 1992; this volume pp. 79–88). However, despite these studies, a number of problems remain unresolved.

The use of molecular systematic techniques has contributed considerably to the understanding of the systematics of many groups where morphological data have proved inconclusive (Patterson, Williams & Humphries 1993). The use of mitochondrial DNA sequence analysis has recently increased enormously through the development of the polymerase chain reaction (PCR) (Mullis *et al.* 1986; Saiki *et al.* 1988) and of direct sequencing of PCR-amplified DNA fragments using conserved primers (Kocher *et al.* 1989). Among venomous snakes, mtDNA sequence information has been used to formulate phylogenetic hypotheses at various taxonomic levels (Knight & Mindell 1993, 1994; Knight *et al.* 1993; Malhotra & Thorpe, this volume pp. 114–128). MtDNA sequence information has also been used for the identification of cryptic species in morphologically complex groups, such as the Asiatic cobras (Wüster & Thorpe 1994; Wüster *et al.* 1995).

Among the mitochondrial genes, cytochrome *b* has been used extensively in vertebrates not only because of the availability of conserved primer sites, but also because the rate of base-pair substitution makes it a useful source of information about relationships among closely related taxa (Irwin, Kocher & Wilson 1991; Simon *et al.* 1994).

This study aims to use mitochondrial DNA sequence information to provide additional evidence towards a robust phylogenetic hypothesis for the genus *Bothrops* and related South American pitvipers. Particular attention has been paid to the following problems: (1) Is there a basal dichotomy in the genus *Bothrops sensu* Burger (1971) between species with and without a lacunolabial scale? (2) Is the genus *Bothrops* paraphyletic with respect to *Bothriopsis*, as suggested by the results of Werman (1992)? (3) Is the *B. atrox* species complex monophyletic, especially with respect to *B. jararacussu* (see Werman 1992)?

Materials and methods

Tissue and/or blood samples were obtained from road-killed specimens or live, anaesthetized wild-caught or captive animals. The specimens used in this study and their localities are shown in Table 1. Tissue samples were collected and stored in 70–100% ethanol.

Table 1. List of specimens examined in this study and their localities

Taxon	Locality	Sample size
B. taeniata	Shell Macuma, Centro Pañyai, Morona Santiago, Ecuador	1
B. alternatus	Santa Rita do Passaquatro, São Paulo, Brazil	2
B. atrox[a]	UHE Balbina, Presidente Figueiredo, Amazonas, Brazil	2
B. caribbaeus	Saint Lucia	1
B. cotiara	Herval D'Oeste, Santa Catarina, Brazil	1
B. insularis	Ilha da Queimada Grande, São Paulo, Brazil	2
B. isabelae[a]	Guanare, Portuguesa, Venezuela	1
B. jararaca	São Bento do Sul, Santa Catarina, Brazil	2
B. jararacussu	Miracatu, São Paulo, Brazil	3
B. leucurus[a]	Porto Seguro, Bahia, Brazil	1
B. marajoensis[a]	Marajó Island, Pará, Brazil	1
B. moojeni[a]	Lençois Paulista, São Paulo, Brazil	2
C. durissus	Pindamonhangaba, São Paulo, Brazil	1
T. albolabris	Pantar Island, Indonesia	1

[a] The samples of the 'species' *atrox, isabelae, leucurus, marajoensis* and *moojeni* are simply of representative populations normally assigned to these taxa, and should not be taken to imply recognition of these taxa as valid species.

Genomic DNA was obtained through digestion with proteinase K (20 mg/ml). RNase solution (10 mg/ml) was added to the lysate and protein precipitation was carried out by adding 5 M ammonium acetate solution to the RNase-treated cell lysate. A DNA pellet was obtained by ethanol precipitation which was eluted in TE buffer (Puregene DNA isolation kit — Gentra Systems Inc.).

A 767 bp region of the cytochrome *b* gene was amplified using *Thermus aquaticus* DNA polymerase (GIBCO). The primers were MVZ 16 (Moritz, Schneider & Wake 1992), and modified versions of the universal cytochrome *b* primers of Kocher *et al.* (1989). PCR product was purified from the reaction solution through the use of SpinBind DNA Extraction Unit (FMC BioProducts).

Sequencing was carried out by the double-stranded dideoxy chain termination method (Sanger, Nicklen & Coulson 1977). Sequencing reaction products for both short and long fragments of cytochrome *b* gene were separated at 2000 V in 6% polyacrylamide, 7 M urea, 50 cm × 30 cm × 0.4 mm gels and were visualized by autoradiography. The sequences were aligned against the published human cytochrome *b* sequence (Anderson *et al.* 1981). Both strands were sequenced for most parts of the sequence analysed here.

A homologous region of 565 base-pair positions was subjected to phylogenetic analysis. In order to assess the robustness of the phylogenetic hypotheses generated here, we used four different algorithms: maximum parsimony analysis (heuristic search method — PAUP 3.0 — Swofford 1989), Fitch–Margoliash with and without the assumption of a molecular clock (FITCH and KITSCH programs on PHYLIP 3.5 — Felsenstein 1993), and the maximum likelihood method (DNAML — PHYLIP 3.5 — Felsenstein 1993). The outgroup sequences for tree rooting were from *Trimeresurus albolabris* (A. Malhotra pers. comm.) and *Crotalus durissus terrificus*.

Results

Sequence data

A 565 bp section of the cytochrome *b* gene was available for the taxa of interest in this study, and was subjected to phylogenetic analysis. Of these 565 sites, 142 (25.13%) were variable among the ingroup taxa. The level of base-pair divergence between *Crotalus durissus* and the ingroup taxa varied from 15.1% to 17.1%, between *Trimeresurus albolabris* and the ingroup taxa from 13.8% to 16.1%, and among the ingroup taxa from 2.0% to 13.3%. The transition: transversion (TS:TV) ratio between *C. durissus* and the ingroup taxa varied from 1.22 to 1.71, between *T. albolabris* and the ingroup taxa from 1.66 to 2.71, and among the ingroup taxa from 1.73 to 14.50. Among the latter, there was no indication of a negative correlation between total base-pair divergence and the TS:TV ratio, which suggested that saturation of transitions is not a problem with this database. However, the uniformly low TS:TV between the outgroups and the ingroup taxa suggests considerable saturation of transition sites. As a compromise measure, taking into account both the saturation of transition sites in the outgroup comparisons and the fact that the transitions are likely to carry a useful phylogenetic signal within the ingroup, transversions were weighted twice as heavily as transitions in the phylogenetic analyses.

Phylogenetic analysis

The cladograms obtained by the four algorithms used in this study are shown in Fig. 1, and a semistrict consensus tree of the trees obtained by the four algorithms is shown in Fig. 2. All trees support the placement of the clade containing the two

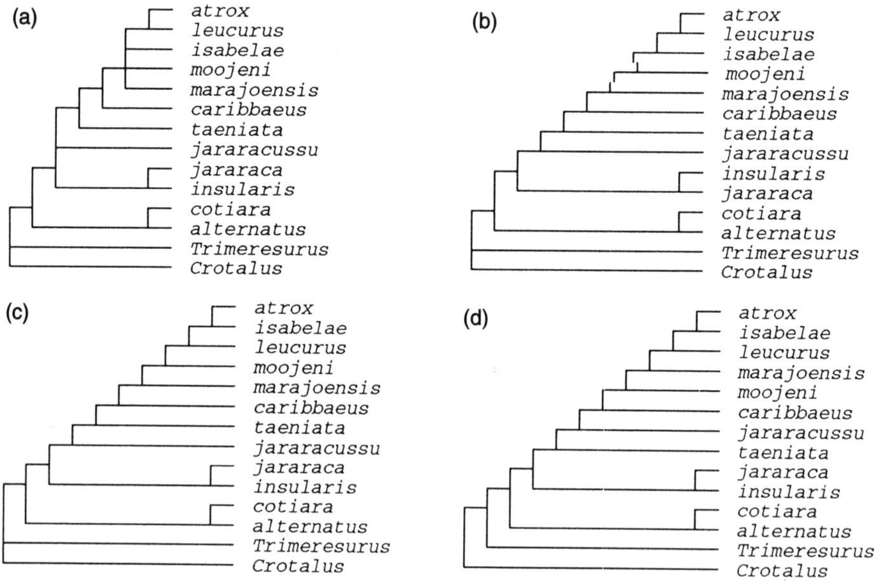

Fig. 1. Phylogenetic trees for the species of *Bothrops* and *Bothriopsis* obtained through the use of four different algorithms. (a) Maximum parsimony algorithm. This is a 75% majority rule consensus tree resulting from 11 equally parsimonious trees revealed by the analysis. (b) Maximum likelihood method. (c) Fitch–Margoliash method. (d) Fitch–Margoliash method with molecular clock assumption.

species with a divided lacunolabial scale (*Bothrops alternatus* and *B. cotiara*) as a sister clade to the remaining taxa, which have an undivided lacunolabial scale. All trees support the monophyly of the South American representatives of the *B. atrox* group, to the exclusion of *B. jararacussu*, and all support the status of *B. caribbaeus* as a sister taxon to this clade. The nesting of *B. jararacussu* and *Bothriopsis taeniata* varies depending on the algorithm used. *Bothriopsis taeniata* is consistently nested within the genus *Bothrops*, usually as a sister taxon to the *B. atrox* group (including *B. caribbaeus*), except in the KITSCH tree. *Bothrops jararaca* and *B. insularis* consistently form a monophyletic group, and represent the sister clade to all other species with an undivided lacunolabial scale in all but the parsimony tree, where the nesting of this clade, *B. jararacussu* and *B. taeniata*, are unresolved. *Crotalus* was rooted as a sister taxon to a clade consisting of *Trimeresurus* and the *Bothrops*/*Bothriopsis* clade in the unrooted KITSCH tree.

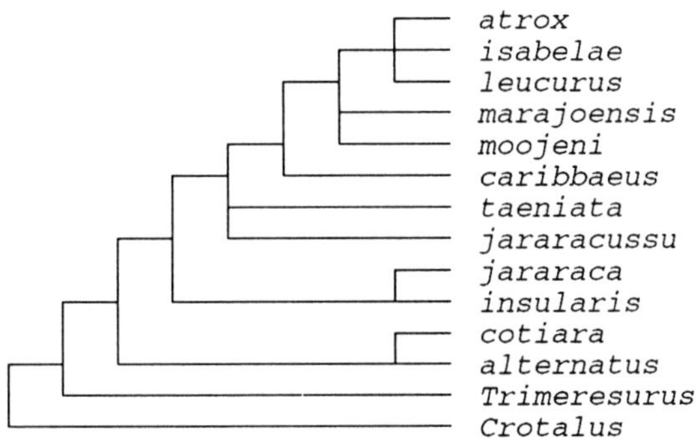

Fig. 2. Semistrict consensus tree of the trees obtained by the four algorithms used in this study. Note that no conclusions regarding the status of the taxa *atrox, isabelae, leucurus, marajoensis* and *moojeni* should be inferred from this analysis, as there is no evidence that these taxa do in fact represent real species. See Wüster *et al.* (this volume pp. 99–113) for a discussion of the systematics of the *Bothrops atrox* species group.

Discussion

Although the evolutionary relationships among pitvipers have formed the subject of intensive research for many decades, studies using advanced methods of phylogenetic inference and molecular systematics have been performed only recently.

The phylogenetic trees resulting from the analyses of our cytochrome *b* sequence data provide a new perspective on the systematics of this group. Comparison with the analyses of Pesantes (1989) and Werman (1992, this volume pp. 79–88) allows the identification of points of agreement and disagreement. The principal points to emerge from our analyses are discussed below with reference to these previous studies.

The status of the *Bothrops atrox* group and *B. jararacussu*

In all our analyses, the sequences representing the nominal species of the *B. atrox* group (*B. atrox*, *isabelae*, *leucurus*, *marajoensis* and *moojeni*) formed a monophyletic group to the exclusion of all other taxa included in the study. This agrees with Pesantes (1989), but conflicts with the results of Werman (1992), in which a clade consisting of *B. jararacussu* and *B. brazili* was rooted within the *B. atrox* group. Our results consistently show *B. jararacussu* as an outgroup to the clade consisting of the *B. atrox* group and *B. caribbaeus*. The mtDNA results are unsurprising, as *B. jararacussu* is morphologically highly distinct from the relatively homogeneous *B. atrox* group. Unfortunately, we do not have sufficient sequence information for *B. brazili* to include this species in the present study. However, preliminary phylogenetic analysis of a 400 bp region of the cytochrome *b* gene suggests that this species is also rooted outside the *Bothrops atrox* complex (unpublished data). In Werman's (1992) study, which spanned all Neotropical crotalines as opposed to just *Bothrops sensu stricto*, the branch lengths were extremely short, indicating that the characters used perhaps lack the resolution required for a phylogenetic analysis of these relatively closely related species. Our results also differ from Pesantes' (1989) results of UPGMA analysis of hemipenis morphology and total plasma protein electrophoresis in that *B. jararacussu* does not form a sister taxon to *B. jararaca*.

The status of *Bothrops insularis*

In all analyses, *B. insularis* was identified as the sister taxon to our sample of *B. jararaca*, from the southern Brazilian state of Santa Catarina. The same sister-group relationship was obtained by Pesantes (1989) through the use of total plasma protein electrophoresis and hemipenial morphology. This indicates a relatively recent common ancestor for the two species, despite profound morphological (Amaral 1921), ecological (Duarte, Puorto & Franco 1995) and venom differences between these taxa. It should be noted that *B. jararaca* is a morphologically very heterogeneous species which displays considerable geographic variation; it is entirely conceivable that it may represent a complex of several species, and that *B. insularis* may be rooted within this complex, leaving *B. jararaca*, as currently understood, paraphyletic.

The status of the genus *Bothriopsis*

All our analyses show *B. taeniata*, the type species of the genus *Bothriopsis*, to be rooted within the genus *Bothrops*, although the precise rooting point differed between analyses. This is in agreement with the results of Werman (1992; this volume pp. 79–88), who found an arrangement which can be represented as (*alternatus* group (*jararaca* (*taeniata*(*atrox* group, *brazili*, *jararacussu*)))). Recognition of *Bothriopsis* would therefore leave the genus *Bothrops* paraphyletic. Albumin immunological distance analysis (Cadle 1992) also showed *Bothriopsis taeniata* to be closer to *B. atrox* than *B. alternatus* was.

The avoidance of paraphyly of the genus *Bothrops* would require either the splitting of that genus, or synonymizing *Bothriopsis* with *Bothrops*. We argue that the latter solution would be preferable for a classification which is consistent with the inferred phylogeny of the group as well as with the requirement for stability in the zoological nomenclature. We base this conclusion on the following considerations:

1. The precise rooting of *B. taeniata* within *Bothrops* is unclear. It would therefore be impossible to propose a genus-level classification for the species currently assigned to *Bothrops*.
2. No phylogenetic analysis has ever been carried out to test whether *Bothriopsis* is monophyletic. The various species of this genus form a heterogeneous group, and no phylogenetic analysis has ever included more than one species of the genus. Assigning all species of this genus to *Bothrops* avoids the creation of further instability which would result from the possible future discovery that the genus may be para- or polyphyletic.
3. A number of hitherto unstudied species of *Bothrops* (e.g., *B. lojanus*, *B. andianus*, *B. pictus*, *B. barnetti*) have uncertain affinities, and any further splitting of *Bothrops* would leave them *incertae sedis*.
4. The various species of *Bothriopsis* are relatively obscure and rarely discussed outside herpetological systematics, and were in any case generally classified as *Bothrops* until the 1980s. On the other hand, many of the species of *Bothrops* which would be involved in name changes in the event of a split of this genus to retain *Bothriopsis* (e.g., *B. alternatus*, *B. jararaca*, *B. neuwiedi*) are of considerable medical importance, and are frequently discussed in the biomedical literature. Changing the generic status of these species would therefore create considerable confusion. Researchers working with animals of medical importance have a particular responsibility to maintain the greatest possible stability compatible with systematic principles.

Finally, it is appropriate at this point to discuss the work of Schätti *et al.* (1990) and Schätti & Kramer (1993), who regarded *Bothriopsis* as a synonym of *Bothriechis*. No evidence of synapomorphies uniting the species of *Bothriopsis* and *Bothriechis* has ever been presented. A superposition of the phylogenetic trees for Central and South American pitvipers published in Werman (1992) clearly shows the branching order (*Bothriechis*(*Porthidium*(*Bothrops*, *Bothriopsis*))). This clearly contradicts the synonymization of *Bothriopsis* with *Bothriechis*. Unfortunately, we do not yet have any sequence information for any species of *Bothriechis*. However, in view of Werman's results, we do not question the status of the genus *Bothriechis* as a genus distinct from *Bothrops*, whereas we hereby synonymize *Bothriopsis* with *Bothrops*.

The *Bothrops alternatus* group

Werman (1992) found a basal dichotomy in the genus *Bothrops*, which separates the species with a lacunolabial scale (e.g., *B. atrox*, *B. jararaca*, *Bothriopsis taeniata*) from species without a lacunolabial (e.g., *B. alternatus*, *B. neuwiedi*). The same results, with different species, were obtained by Pesantes (1989). Our analysis only included two species without a lacunolabial, *B. alternatus* and *B. cotiara*, which consistently formed a monophyletic group, and were consistently rooted as the sister group of all species with an entire lacunolabial scale.

Conclusions: consensus and problems in Neotropical pitviper phylogeny

The present study is the third recent analysis of the systematics and phylogeny of Neotropical pitvipers of the genus *Bothrops* and related genera, after those of Pesantes (1989) and Werman (1992; this volume pp. 79–88). As these three studies have been carried out using different character systems (hemipenis morphology and plasma serum electrophoresis — Pesantes 1989; isozymes and morphology — Werman 1992 and this volume pp. 79–88), a comparison of the results provides an excellent opportunity to take stock of the situation and identify points of agreement as well as remaining problem areas within the systematics of the group.

Principal points of agreement:

—Within the genus *Bothrops*, *sensu* Burger (1971) and Campbell & Lamar (1989), a basal dichotomy between those species with a lacunolabial scale and those without has emerged in all three studies under consideration.
—The *Bothrops atrox* group, defined here as *B. atrox*, *B. isabelae*, *B. leucurus*, *B. moojeni* and *B. marajoensis*, is almost certainly monophyletic. The inclusion of *B. jararacussu* and *B. brazili* in this group in Werman's analysis appears unlikely in view of our results and those of Pesantes (1989), and also on the basis of the morphological distinctness of both these species from the *B. atrox* group.
—*Bothrops insularis* and *B. jararaca* group as sister species in Pesantes' and our analyses.
—Werman's results (1992; this volume pp. 79–88) and ours both show *Bothriopsis taeniata* to be rooted within *Bothrops sensu stricto*, and there is therefore strong evidence in support of our decision to synonymize *Bothriopsis* with *Bothrops*.

Remaining problematic areas:

—Monophyly of the New World pitvipers. Our KITSCH tree suggests that *Trimeresurus* + *Bothrops* form a clade, with *Crotalus* as a sister taxon, suggesting multiple colonization of the New World by pitvipers.
—The phylogenetic affinities of the remaining species formerly classified as *Bothriopsis*.
—The phylogenetic position of *B. jararacussu*, on which there was disagreement between the three studies discussed here.
—The phylogenetic position of a number of other species of *Bothrops* not included here, including *B. lojanus*, *B. venezuelensis*, *B. pictus*, *B. barnetti* and *B. andianus*.
—The phylogenetic position of some species recently excluded from *Bothrops* by Schätti & Kramer (1993), namely *B. pulcher*/*B. campbelli*/*Porthidium almawebi* and *B.*/*Porthidium microphthalmus*/*um*, as well as the superficially similar *Porthidium hyoprora*, which, in terms of pattern and scalation, appears more similar to *Bothrops* than to *Porthidium*.

The resolution of these problems will come from an integration of further multi-disciplinary studies using a number of different character systems and techniques.

With the ever-increasing interest in pitviper systematics and phylogeny, there is now considerable hope that this will eventually be realized.

Acknowledgements

We thank the staff of the Museu Paraense Emílio Goeldi for their help in acquiring specimens for this study. The *Trimeresurus* sequence for outgroup comparison was provided by A. Malhotra. Funded by the Science and Technology for Development Programme of the European Community (Contract number TS3-CT91-0024), the Wellcome Trust (Research Career Development Fellowship to WW), Fundação Banco do Brasil and the British Council (Fellowship to MGS).

References

Amaral, A. (1921). Contribuição para o conhecimento dos ofídios do Brasil — A. Parte II. Biologia da nova espécie, *Lachesis insularis*. *Anex. Mems Inst. Butantan Secc. Ofiologia* **1**: 39–44.

Anderson, S., Bankier, A. T., Barrell, B. G., de Bruijn, M. H. L., Coulson, A. R., Drouin, J., Eperon, I. C., Nierlich, D. P., Roe, B. A., Sanger, F., Schreier, P. H., Smith, A. J. H., Staden, R. & Young, I. G. (1981). Sequence and organization of the human mitochondrial genome. *Nature, Lond.* **290**: 457–465.

Bolaños, R. (1984). *Serpientes, venenos y ofidismo en Centroamerica.* Editorial Universidad de Costa Rica, San José.

Brattstrom, B. H. (1964). Evolution of the pit vipers. *Trans. S. Diego Soc. nat. Hist.* **13**: 185–268.

Burger, W. L. (1971). *Genera of pitvipers. (Serpentes: Crotalidae).* PhD thesis: University of Kansas.

Cadle, J. E. (1992). Phylogenetic relationships among vipers: immunological evidence. In *Biology of the pitvipers*: 41–48. (Eds Campbell, J. A. & Brodie Jr., E. D.). Selva, Tyler, Texas.

Campbell, J. A. & Lamar, W. W. (1989). *The venomous reptiles of Latin America.* Cornell University Press, Ithaca, New York.

Cardoso, J. L. C. (1992). Introdução ao estudo dos acidentes por animais peçonhentos. In *Plantas venenosas e animais peçonhentos*: 139–142. (Ed. Schvartsman, S.). Sarvier, São Paulo.

Duarte, M. R., Puorto, G. & Franco, F. L. (1995). A biological survey of the pitviper *Bothrops insularis* Amaral (Serpentes, Viperidae) — an endemic and threatened offshore island snake of southeastern Brazil. *Stud. neotrop. Fauna Envir.* **30**: 1–13.

Felsenstein, J. (1993). *Phylogeny inference package (PHYLIP), version 3.5c.* Department of Genetics, University of Washington, Seattle.

Hoge, A. R. & Romano-Hoge, S.A.R.W.L. (1978/79) [1981]. Poisonous snakes of the world. Part I. Checklist of the pit vipers, Viperoidea, Viperidae, Crotalinae. *Mems Inst. Butantan* **42/43**: 179–310.

Irwin, D. M., Kocher, T. D. & Wilson, A. C. (1991). Evolution of cytochrome *b* gene of mammals. *J. molec. Evol.* **32**: 128–144.

Knight, A. & Mindell, D. P. (1993). Substitution bias, weighting of DNA sequence evolution, and the phylogenetic position of Fea's viper. *Syst. Biol.* **42**: 18–31.

Knight, A. & Mindell, D. P. (1994). On the phylogenetic relationship of Colubrinae, Elapidae, and Viperidae and the evolution of front-fanged venom systems in snakes. *Copeia* **1994**: 1–9.

Knight, A., Styer, D., Pelikan, S., Campbell, J. A., Densmore III, L. D. & Mindell, D. P. (1993). Choosing among hypotheses of rattlesnake phylogeny: a best-fit rate test for DNA sequence data. *Syst. Biol.* **42**: 356–367.

Kocher, T. D., Thomas, W. K., Meyer, A., Edwards, S. V., Pääbo, S., Villablanca, F. X. & Wilson, A. C. (1989). Dynamics of mitochondrial DNA evolution in animals: amplification and sequencing with conserved primers. *Proc. natn. Acad. Sci. USA* **86**: 6196–6200.

Ministério da Saúde (1990). *Informações sobre o Programa Nacional de Ofidismo.* Fundação Nacional de Saúde, Brasília.

Moritz, C., Schneider, C. J. & Wake, D. B. (1992). Evolutionary relationships within the *Ensatina eschscholtzii* complex confirm the ring species interpretation. *Syst. Biol.* **41**: 273–291.

Mullis, K., Faloona, F., Scharf, S., Saiki, R., Horn, G. & Erlich, H. (1986). Specific enzymatic amplification of DNA *in vitro*: the polymerase chain reaction. *Cold Spring Harbor Symp. quant. Biol.* **51**: 263–273.

Otero, R., Tobon, G. S., Fernando Gómez, L., Osorio, R., Valderrama, R., Hoyos, D., Urreta, J. E., Molina, S. & Arboleda, J. J. (1992). Accidente ofídico en Antioquia y Chocó. Aspectos clínicos y epidemiológicos (marzo de 1989–febrero de 1990). *Acta med. Colombiana* **17**: 229–249.

Patterson, C., Williams, D. M. & Humphries, C. J. (1993). Congruence between molecular and morphological phylogenies. *A. Rev. Ecol. Syst.* **24**: 153–188.

Pesantes, O. C. (1989). *Relações entre algumas espécies do género* Bothrops, *pela eletroforese do plasma e morfologia do hemipênis (Serpentes: Viperidae).* MSc dissertation: Universidade Estadual Paulista Julio de Mesquita Filho, Rio Claro, São Paulo.

Saiki, R. K., Gelfand, D. H., Stoffel, S., Scharf, S. J., Higuchi, R., Horn, G. T., Mullis, K. B. & Erlich, H. A. (1988). Primer directed enzymatic amplification of DNA with a thermostable DNA polymerase. *Science* **239**: 487–491.

Sanger, F., Nicklen, S. & Coulson, A. R. (1977). DNA sequencing with chain-terminating inhibitors. *Proc. natn. Acad. Sci. USA* **74**: 5463–5467.

Schätti, B. & Kramer, E. (1993). Ecuadorianische Grubenottern der Gattungen *Bothriechis*, *Bothrops* und *Porthidium* (Serpentes: Viperidae). *Rev. suisse Zool.* **100**: 235–278.

Schätti, B., Kramer, E. & Touzet, J.-M. (1990). Systematic remarks on a rare crotalid snake from Ecuador, *Bothriechis albocarinata* (Shreve), with some comments on the generic arrangement of arboreal Neotropical pitvipers. *Rev. suisse Zool.* **97**: 877–885.

Simon, C., Frati, F., Beckenbach, A., Crespi, B., Liu, H. & Flook, P. (1994). Evolution, weighting and phylogenetic utility of mitochondrial gene sequences and a compilation of conserved polymerase chain reaction primers. *Ann. ent. Soc. Am.* **87**: 651–701.

Swofford, D. L. (1989). *PAUP — phylogenetic analysis using parsimony, version 3.0.* Illinois Natural History Survey, Champaign, Illinois.

Werman, S. D. (1992). Phylogenetic relationships of Central and South American pitvipers of the genus *Bothrops* (*sensu lato*): cladistic analyses of biochemical and anatomical characters. In *Biology of the pitvipers*: 21–40. (Eds Campbell, J. A. & Brodie Jr., E. D.). Selva, Tyler, Texas.

Wüster, W. & Thorpe, R. S. (1994). *Naja siamensis*, a cryptic species of venomous snake revealed by mtDNA sequencing. *Experientia* **50**: 75–79.

Wüster, W., Thorpe, R. S., Cox, M. J., Jintakune, P. & Nabhitabhata, J. (1995). Population systematics of the snake genus *Naja* (Reptilia: Serpentes: Elapidae) in Indochina: multivariate morphometrics and comparative mitochondrial DNA sequencing (cytochrome oxidase I). *J. evol. Biol.* **8**: 493–510.

8 Systematics of the *Bothrops atrox* complex: new insights from multivariate analysis and mitochondrial DNA sequence information

W. WÜSTER, M. G. SALOMÃO, R. S. THORPE, G. PUORTO, M. F. D. FURTADO, S. A. HOGE, R. D. G. THEAKSTON and D. A. WARRELL

Synopsis

We use multivariate analysis of morphological characters and comparative sequencing of a 520 bp fragment of the cytochrome *b* gene (mitochondrial DNA) to investigate patterns of geographic variation in the *Bothrops atrox* species complex, and the population phylogeny of the group in parts of South America. Populations conventionally assigned to *B. atrox* and *B. moojeni* constitute morphologically distinct groupings, with a zone of phenetically intermediate specimens where their ranges meet. *Bothrops marajoensis*, *B. isabelae*, *B. leucurus* and *B. pradoi* are poorly differentiated or undifferentiated from *B. atrox* and each other. Phylogenetic analysis of the cytochrome *b* sequence information reveals that *B. caribbaeus* constitutes an outgroup to the South American populations of the *B. atrox* complex. Within the South American radiation, the conventional species, as generally understood, largely lack historical reality. Two well-corroborated basal clades are revealed: one includes the populations from the lower Amazon and Tocantins–Araguaia drainages, as well as Marajó Island. The other comprises two further basal clades within it, the first consisting of the populations conventionally assigned to *B. moojeni*, the other a number of populations from the eastern coast of Brazil, various places in western and central Amazonia and western Venezuela. Mantel matrix correlation tests show morphological variation among the populations of the complex to be most strongly associated with vegetation type. This suggests that morphological variation represents primarily an adaptation to current selection pressures, probably for crypsis, and not a reflection of population phylogeny.

Symp. zool. Soc. Lond. (1997) No. 70: 99–113
ISBN 0 19 854986 5

Introduction

The *Bothrops atrox* species group constitutes a complex group of venomous snakes with a long history of taxonomic confusion. The group displays a complex pattern of among- and within-population variation. Over the last 50 years, this has led to the description as separate species of several groups of populations subjectively perceived as distinct (*B. pradoi* — Hoge 1947; *B. marajoensis* and *B. moojeni* — Hoge 1965; *B. isabelae* — Sandner Montilla 1979; *B. atrox aidae* — Sandner Montilla 1981). However, the status of these forms, as well as that of the populations from the eastern coast of Brazil assigned to *B. leucurus*, remains uncertain (Campbell & Lamar 1989; Cunha & Nascimento 1993). Additional problems concern the relationship between these Amazonian and Brazilian taxa with populations from northern South America, Central America and Caribbean islands, which are variously assigned to *B. asper*, *B. xanthogrammus*, *B. caribbaeus* and *B. lanceolatus*. The species of this group are the principal cause of snakebite mortality and morbidity in much of their range (Haad 1980/81; Bolaños 1984; Kouyoumdjian & Polizelli 1988; Cardoso 1992; Nishioka & Silveira 1992; Otero *et al.* 1992). Understanding the population systematics of this group is therefore of potential importance for the treatment of their bites.

Previous attempts to resolve the systematics of this complex used standard taxonomic methods (i.e. the use of single characters to attempt to diagnose taxa defined *a priori* on the basis of subjective perceptions of differences). However, the *B. atrox* complex displays an intricate pattern of geographic, sexual and ontogenetic variation. In such complex groups, the conventional approach to systematics almost inevitably results in seriously flawed systematic conclusions. Recent analyses of the systematics and phylogeny of the genus *Bothrops* and related taxa, based on hemipenis morphology and plasma protein electrophoresis (Pesantes 1989), and a combination of morphology, anatomy, isozymes and allozymes (Werman 1992), have clarified many problems posed by the genus. In both these studies, the nominal species in the *B. atrox* group were considered as units. However, it has never been tested whether the populations making up these units do in fact constitute natural groups. These previous studies therefore cannot resolve the problem of species definitions within this complex.

Three different approaches have been found useful in investigating the systematics of such complex groups. Multivariate morphometrics allows the simultaneous analysis of patterns of variation in large numbers of characters; comparative mitochondrial DNA sequencing can reveal cryptic species as well as population phylogeny without the confounding effects of natural selection on morphological traits; finally, partial Mantel tests can reveal significant associations between observed variation in the phenotype of the animals and hypothesized causes of this variation, such as phylogeny (inferred from DNA sequence information, or hypotheses derived from other sources of information, such as geological, climatic or vegetational history) and current natural selection for present-day ecological conditions (e.g., vegetation, climate), and can thus further the understanding of the underlying phenomena under investigation. These three approaches have been useful in elucidating the population systematics of other complex groups of venomous snakes, such as Asiatic cobras (Wüster & Thorpe 1989, 1990, 1991, 1992, 1994; Wüster, Thorpe, Cox *et al.* 1995) and Russell's viper (Wüster, Otsuka, Malhotra & Thorpe 1992; Wüster, Otsuka, Thorpe & Malhotra 1992), and for investigating the phenomenon of geographic variation in general (Thorpe, Brown *et al.* 1994; Thorpe, Malhotra *et al.* 1995).

Here, we use a combination of multivariate morphometrics, comparative mitochondrial DNA sequencing and Mantel tests of association between morphological variation and hypothesized causes to analyse the population systematics of the populations normally assigned to the species *B. atrox*, *B. moojeni*, *B. marajoensis*, *B. isabelae*, *B. leucurus* and *B. pradoi*, and to attempt to gain an insight into the causes of the pattern of morphological variation in this complex. Comparative sequencing of the same region of cytochrome *b* in a variety of *Bothrops* species and related genera (Salomão *et al.*, this volume pp. 89–98) has shown that the populations assigned to these nominal species constitute a monophyletic group, with *B. caribbaeus* as the sister group.

Materials and methods

Multivariate morphometrics

Approximately 260 preserved specimens were examined by the first author, and 87 morphological characters relating to the scalation, colour pattern, head and body proportions and visceral topography were recorded for each specimen. Details of character selection and recording can be found in Wüster, Thorpe, Puorto & BBBSP (in press).

Specimens were grouped into operational taxonomic units (OTUs) on the basis of collecting gaps and patterns of variation in the generalized phenotype as revealed by principal components analysis (PCA) of individual specimens. The OTUs used in this study are tabulated in Table 1. All characters were tested for significant among-OTU variation by means of analysis of variance (ANOVA) and those characters showing significant variation were selected for further analysis. Canonical variate analysis (CVA) was used to investigate patterns of geographic variation in the populations of the complex. In order to negate the effect of sexual dimorphism, male and female specimens were analysed separately.

Molecular methods

Tissue and blood samples for DNA analysis were collected from field-caught, roadkilled or captive specimens of *Bothrops* and stored in 75% ethanol. The localities of origin of the specimens are shown in Fig. 1. DNA extraction was carried out using the Puregene™ DNA isolation kit (Gentra Systems Inc.). A 767 bp fragment of the cytochrome *b* gene was amplified by means of the polymerase chain reaction (PCR). The amplification was carried out using modified versions of the cytochrome *b* primers of Kocher *et al.* (1989) and primer MVZ 16 of Moritz, Schneider & Wake (1992). The PCR product was purified with the SpinBind® DNA Extraction Unit (FMC BioProducts). Sequencing was carried out by the chain termination method (Sanger *et al.* 1977). Both strands were sequenced wherever possible. Sequences were aligned against the published human mitochondrial DNA sequence (Anderson *et al.* 1981).

Table 1. OTUs used in multivariate analysis and sample sizes for each sex

OTU Locality	Sample size Males	Females
110. Guyana	2	2
120. Northern Peru and neighbouring Colombia	7	6
121. Departamento de Cuzco, Perú	1	0
122. Manú National Park, Departamento Madre de Dios, Perú	3	2
123. Panguana, Rio Llullapichis, Rio Pachitea, Huánuco, Perú	2	1
130. Pastaza Province, Ecuador	5	12
150. Balbina Dam, Pres. Figueiredo, Amazonas, Brazil	8	8
151. Itacoatiara, Amazonas, Brazil	5	7
152. Manaus area, Amazonas, Brazil	4	6
153. Iauarête, Amazonas, Brazil	0	1
154. Porto Velho Region, Rondônia, Brazil	1	3
155. Acre State, Brazil	1	2
156. Belém area, Pará, Brazil	6	10
157. Tome-Açu, Pará, Brazil	2	0
158. Left side of Rio Tocantins, Tucuruí, Pará, Brazil	8	4
159. Right side of Rio Tocantins, Tucuruí, Pará, Brazil	0	8
160. Marabá, Pará, Brazil	0	1
161. Itaituba, Pará, Brazil	0	3
162. Curral Grande, Monte Alegre, Santarém, Pará, Brazil	11	7
163. Mouth of Rio Trombetas, Pará, Brazil	0	1
164. Barra do Corda region, Maranhão, Brazil	3	0
165. Oiapoque, Amapá, Brazil	1	0
166. Serra do Navio, Amapá, Brazil	0	1
167. Guaraí, Tocantins, Brazil	0	1
200. Southern Goiás and Brasília region, Brazil	7	6
201. Presidente Prudente region, São Paulo, Brazil	4	5
202. Itaipú Dam, Foz do Iguaçu, Paraná, Brazil	6	5
203. Ilha Solteira Dam, Mato Grosso do Sul, Brazil	8	7
204. Eastern and north-eastern Mato Grosso, Brazil	4	2
205. Ilha do Bananal, Tocantins, Brazil	0	1
300. São Bento, coastal Maranhão, Brazil	8	8
301. Eastern part of Marajó Island, Pará, Brazil	3	3
302. Breves, western part of Marajó, Pará, Brazil	2	3
303. Macapá area, Amapá, Brazil	1	0
400. Conceição do Araguaia, Pará, Brazil	3	1
401. Cachimbo, Pará, Brazil	1	0
402. Almas, Natividade, Tocantins	2	1
403. Uruçui, Piauí, Brazil	1	1
404. Avelino Lopes, Piauí, Brazil	0	1
405. Formoso do Araguaia, Tocantins, Brazil	0	1
501. Southern Bahia and northern Espirito Santo, Brazil	2	1
601. Ibiraçu (formerly Pau Gigante), Espirito Santo, Brazil	2	0
701. Portuguesa State, Venezuela	2	0

Species designations follow mostly Campbell & Lamar (1989), with modifications according to phenotype: OTUs 100–199 would be classified as *B. atrox*, OTUs 200–299 as *B. moojeni*, OTUs 300–399 as *B. marajoensis*; OTUs 400–499 are populations phenotypically intermediate between *B. moojeni* and *B. atrox*, OTU 501 represents *B. leucurus*, OTU 601 *B. pradoi* and OTU 701 *B. isabelae*.

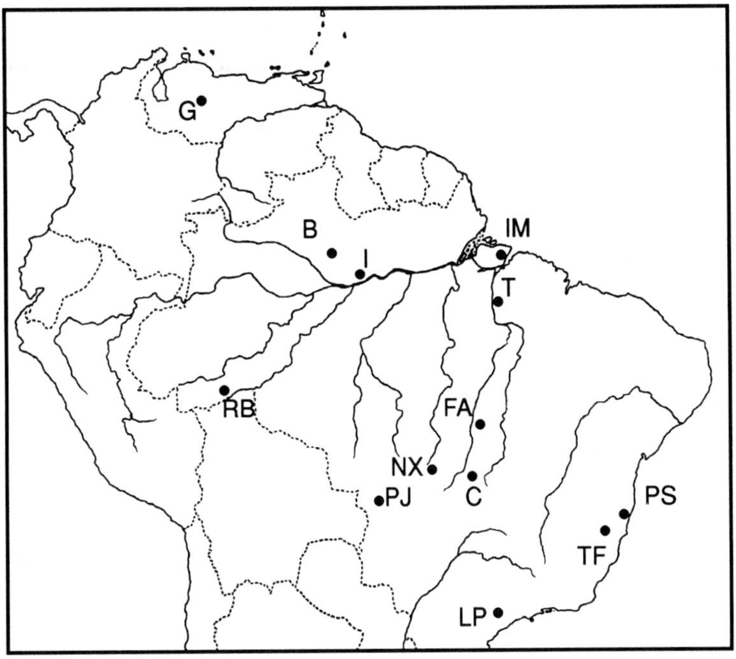

Fig. 1. Localities from which tissue samples were obtained, and nominal species to which these populations were assigned. The codes for the localities are as follows: B: UHE Balbina, Presidente Figueiredo, Amazonas, Brazil (*atrox*); C: Corumbá, Goiás, Brazil (*moojeni* — near type locality); FA: Formoso do Araguaia, Tocantins, Brazil (*atrox* × *moojeni* phenotypic intergrade); G: Guanare, Portuguesa, Venezuela (*isabelae* — type locality); I: Itacoatiara, Amazonas, Brazil (*atrox*); IM: Ilha de Marajó, Pará, Brazil (*marajoensis* — type locality); LP: Lençois Paulista, São Paulo, Brazil (*moojeni*); NX: Nova Xavantina, Mato Grosso, Brazil (*moojeni*); PJ: Porto Jofre, Mato Grosso, Brazil (*moojeni*); PS: Porto Seguro, Bahia, Brazil (*leucurus*); RB: Rio Branco, Acre, Brazil (*atrox*); T: UHE Tucurui, Pará, Brazil (*atrox*); TF: Teixeira de Freitas, Bahia, Brazil (*leucurus*).

Phylogenetic analysis

A homologous region of 520 base-pair positions was subjected to phylogenetic analysis. Four different algorithms were used in order to assess the robustness of the phylogenetic hypotheses generated here: maximum parsimony analysis (heuristic search method — PAUP 3.0 — Swofford 1989), Fitch–Margoliash with and without the assumption of a molecular clock (FITCH and KITSCH programs on PHYLIP 3.5 — Felsenstein 1993), and the Maximum Likelihood method (DNAML — PHYLIP 3.5 — Felsenstein 1993). The outgroup sequences for tree rooting were from *Bothrops alternatus* and *B. jararacussu*, both of which were shown to be suitable outgroups for the *B. atrox* complex by Salomão *et al.* (this volume pp. 89–98). All analyses were run both without character weighting and with transversions weighted twice as heavily as transitions.

Testing hypotheses with Mantel tests

Partial Mantel tests were used to test for significant associations between observed phenotypic variation in the *Bothrops atrox* group and potential causes of geographic variation. Five potential causes were investigated in this study: rainfall regime, temperature regime, vegetation type, geographic distance and hypothesized Pleistocene refugia.

Rainfall and temperature regimes and vegetation types are known to be associated with geographic variation in various reptiles (Thorpe & Brown 1989; Castellano, Malhotra & Thorpe 1994). Geographic proximity provides a measure of the opportunities for gene flow between localities, as well as accounting for other effects related to geography.

Pleistocene changes in vegetation cover in South America may have led to the fragmentation of the range of the populations of *Bothrops* and hence to differentiation. In particular, the fragmentation of the Amazonian forest into small refugia during the Pleistocene has been hypothesized to be a major cause of speciation or population differentiation among Amazonian faunal elements (Haffer 1969; Vanzolini & Williams 1970). The refuge scenario used in this paper is based on a synthesis of the ideas of Ab'Sáber (1977), Brown (1979) and Brown & Ab'Sáber (1979), as presented in Haffer (1987).

For each specimen, temperature regimes (represented as a combined score derived from the number of months with a mean temperature under 25 °C and the number of months with mean temperature under 20 °C), rainfall regimes (represented as a combined score derived from total annual rainfall, the number of months with under 100 mm and the number of months with under 25 mm of precipitation), and vegetation type of the locality of collection were converted into a between-specimen distance matrix by means of principal co-ordinates analysis. Geographic distance was calculated on the basis of latitude and longitude of the locality of each specimen. In the case of the hypothesized Pleistocene refugia, each specimen was assigned to the refugium closest to its collecting locality. The assumption is that specimens descended from the same refugium population should be morphologically more similar to each other than specimens descended from different refugium populations.

Partial Mantel tests were run on individual specimens using 1000 randomizations. One test was run on the generalized phenotype of all specimens (expressed as a taxonomic distance matrix obtained by a principal co-ordinates analysis), and on nine individual characters or constellations of functionally and statistically related characters (such as number of dorsal scale rows at various points along the body).

Results

Multivariate morphometrics

The CVA of male specimens (Fig. 2a) indicates a clear separation along the first canonical variate between populations conventionally assigned to *Bothrops moojeni* and the remaining populations. However, there are several phenotypically intermediate specimens from regions where the ranges of *B. atrox* and *B. moojeni* meet, in the Brazilian states of Tocantins, Piauí and south-western Pará. The populations normally assigned to *B. marajoensis* group with *B. atrox*, as do those assigned to *B. leucurus*, *B. pradoi* and *B. isabelae*. In the females (Fig. 2b), the pattern is similar, except that some of the populations assigned to *B. marajoensis* are more phenotypically similar to *B. moojeni*.

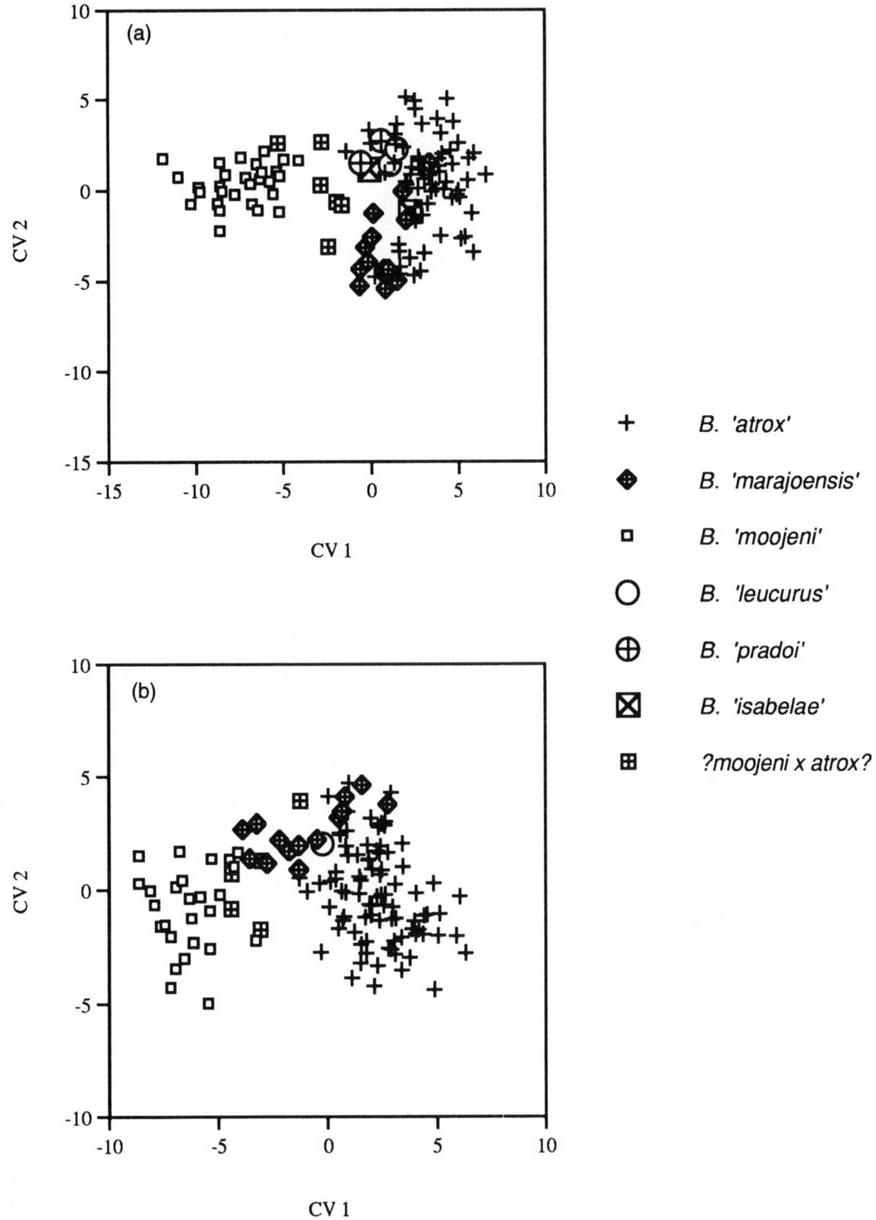

Fig. 2. Ordination of the male (a) and female (b) specimens of the *Bothrops atrox* complex along the first two canonical variates. Species names represent the nominal species to which these specimens would conventionally be assigned.

Cytochrome *b* sequence analysis

A total of 19 ingroup haplotypes were identified and used in the phylogenetic analyses. Of the 520 base-pair positions analysed, 111 (21.3%) were variable overall, and 69 (13.3%) within the ingroup. The pairwise sequence divergence among the ingroup haplotypes ranged from 0.2% to 6.6%, mean 2.87% (0.2–4.5%, mean 2.56%, if *caribbaeus* is excluded) of all positions. The transversion-to-transition ratio among ingroup haplotypes ranged from 0 to 0.77. Owing to the low level of pairwise divergences, saturation of transition sites does not appear to be a problem.

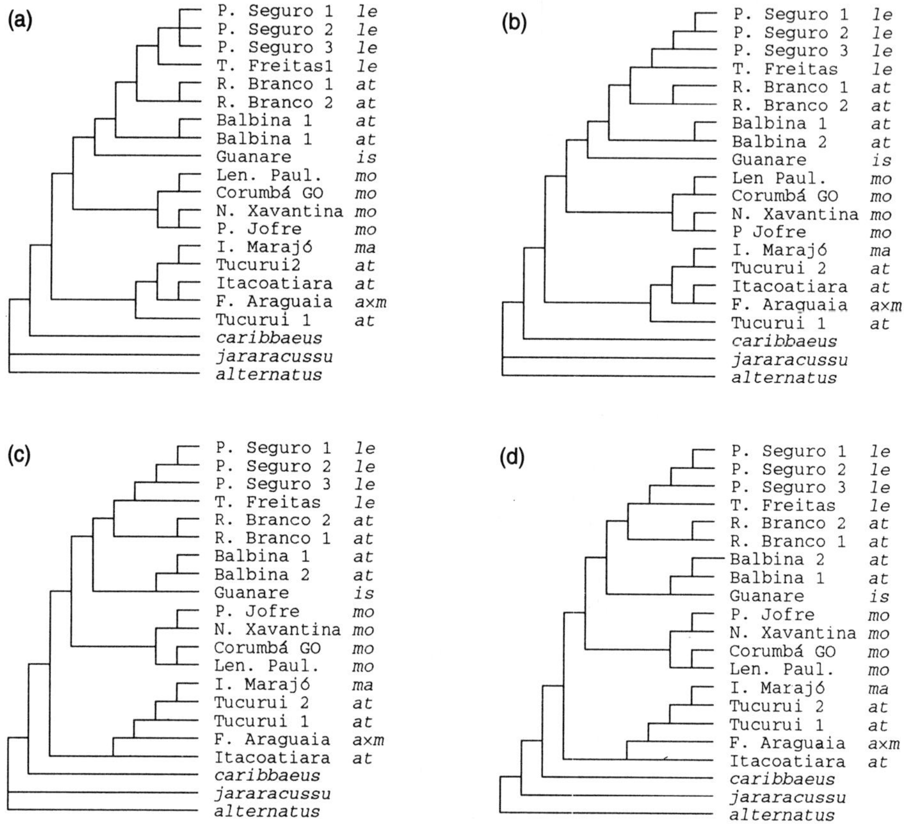

Fig. 3. Phylogenetic trees obtained through the use of four algorithms: (a) maximum parsimony analysis; (b) maximum likelihood analysis; (c) Fitch–Margoliash algorithm; (d) Fitch–Margoliash algorithm with molecular clock assumption. These trees show the results of an analysis using equal weighting for all base-pair substitutions. Double weighting of transversions results in identical trees, except for the branching order of the three Porto Seguro haplotypes. Abbreviations in italics denote the nominal species to which these populations are normally assigned. (*at* = *B. atrox*; *is* = *B. isabelae*; *le* = *B. leucurus*; *ma* = *B. marajoensis*; *mo* = *B. moojeni*; *a* × *m*: specimen from zone of morphological intergradation between *B.* '*atrox*' and *B.* '*moojeni*')

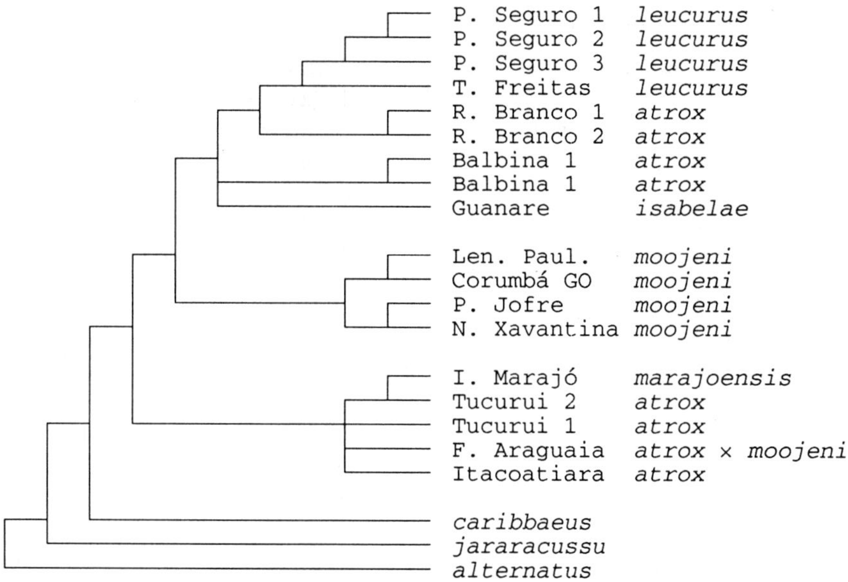

Fig. 4. Semistrict consensus of the trees generated by the four algorithms used in this study.

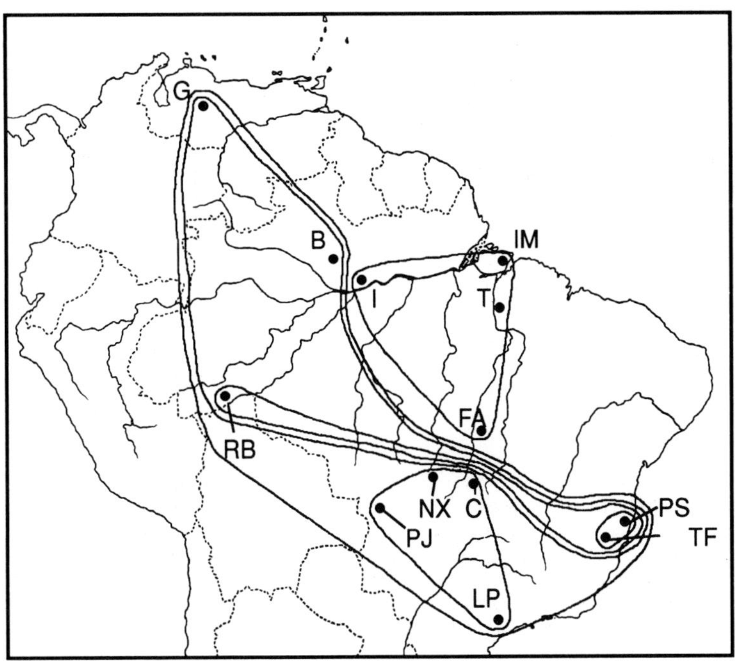

Fig. 5. Geographic distribution of the clades revealed by this study. Locality abbreviations are as in Fig. 1.

The trees resulting from analyses of unweighted substitutions are essentially identical to those resulting from analyses with transversions weighted twice as heavily as transitions (the only differences concern the branching order of the three haplotypes from Porto Seguro). The trees resulting from the analysis of unweighted data are shown in Fig. 3, and a semistrict consensus tree of the trees obtained with the four methods is shown in Fig. 4. In all trees, the South American populations of the *B. atrox* complex constitute a monophyletic group with *B. caribbaeus* as a sister taxon.

Within the *B. atrox* group, the basal dichotomy separates a group of populations from the lower Amazon Basin and the Tocantins–Araguaia drainage from the remaining South American populations. This remainder is again split into two further clades: the first of these comprises the populations conventionally referred to as *B. moojeni*; the second comprises a set of populations from the east coast of Brazil, and several populations from the upper Amazon Basin and the Orinoco drainage. The haplotypes from the eastern coast of Brazil form a monophyletic group. The clades revealed by these analyses are mapped in Fig. 5.

Mantel tests

The Mantel tests of association between geographic variation in the generalized phenotype of the South American *B. atrox* group clade show that the generalized phenotypes of both sexes are most strongly associated with the vegetational zone inhabited by the relevant population (Table 2). In the case of the females, vegetation is the only hypothesized cause significantly associated with the observed pattern of variation. In the males, generalized phenotype is also associated with Pleistocene refugia and geographic distance. Similarly, in both sexes, vegetation was associated with the highest number of single characters or character suites.

Discussion

In this analysis of the systematics of the *Bothrops atrox* complex, the pattern of geographic variation revealed by multivariate morphometrics and the population phylogeny revealed by mtDNA sequence analysis are largely incongruent, and both are incongruent with the conventional taxonomic arrangement of this group.

The multivariate analysis revealed that only the southern populations currently assigned to the nominal species *B. moojeni* are clearly morphologically distinct from the remaining populations. These populations occupy a zone of relatively open cerrado vegetation, with a highly seasonal climate, whereas most of the remaining populations originate from moister, less seasonal regions, usually covered by more or less permanently humid forest. The remaining populations, which mostly originate from rainforest habitats, are all relatively similar to each other.

The phylogenetic analysis of the cytochrome *b* sequence data reveals a population phylogeny that is largely incongruent with both the pattern of morphological variation discovered through multivariate morphometrics, and with the conventional taxonomy of these forms. The only populations that are morphologically distinct from all others, the populations commonly called *B. moojeni*, from the seasonally dry areas of central-western Brazil, do constitute a monophyletic group.

However, the remaining populations, with the '*atrox*' phenotype, fall into two different clades, among which the '*moojeni*' clade is nested. It would appear that the phenotype of the '*moojeni*' populations is a derived adaptation to the different vegetation, whereas the '*atrox*' phenotype is a plesiomorphy of the *B. atrox* group.

Table 2. Probabilities for the null hypothesis of no association between generalized phenotype and individual morphological characters and hypothesized potential causes of the observed pattern of variation

	Geographic distance	Rainfall	Temperature	Refugia	Vegetation
General phenotype	♂♂**0.002** ♀♀0.032	0.017 0.976	0.707 0.084	**0.002** 0.017	**≤0.001** **≤0.001**
Ventral number	♂♂0.652 ♀♀0.829	**0.009** 0.018	0.014 0.717	**≤0.001** **0.009**	**≤0.001** **≤0.001**
Subcaudal number	♂♂**≤0.001** ♀♀**≤0.001**	**0.005** 0.931	**0.003** 0.033	0.134 0.333	0.090 0.011
Dorsal scale rows	♂♂0.035 ♀♀**0.003**	**0.006** 0.125	0.872 0.237	0.194 0.588	0.319 0.983
Caudal scale reductions	♂♂**≤0.001** ♀♀**0.003**	0.018 0.125	0.789 0.237	0.466 0.588	**0.003** 0.983
Supralabials in postocular stripe	♂♂0.992 ♀♀0.186	0.143 **≤0.001**	**0.008** **≤0.001**	0.167 0.055	**≤0.001** **≤0.001**
Percentage of last supralabial in postocular stripe	♂♂0.327 ♀♀0.457	0.022 0.915	**≤0.001** 0.090	0.083 **0.006**	**≤0.001** 0.106
Ventral pigmentation	♂♂0.994 ♀♀0.045	**≤0.001** 0.800	0.042 0.01	**≤0.001** 0.037	**0.008** 0.013
Separation of upper & lower patterns	♂♂**0.002** ♀♀0.035	0.522 0.919	0.012 0.012	0.087 0.871	**≤0.001** **≤0.001**
Intercanthals	♂♂0.487 ♀♀0.939	0.411 0.471	0.025 **≤0.001**	0.566 0.931	0.076 **0.009**

P values in **bold** type are significant at the 0.05 level after Bonferroni correction for the number of hypotheses tested for each character.

With respect to formal classification, the results presented here do not allow any definitive conclusions, except to note that the conventional species as currently defined (Campbell & Lamar 1989) do not generally correspond either to phenetically distinct and homogeneous units, or to evolutionary lineages. *Bothrops atrox*, as conventionally understood, is polyphyletic according to the models of Farris (1974) and Oosterbroek (1987). The haplotypes of the two '*B. leucurus*' populations included in this study form a clade, but are nested within a clade containing several other populations conventionally assigned to *B. atrox*. The population of '*B. marajoensis*' is nested within the clade from lower Amazonia, and the population of *B. isabelae* is nested within a clade containing populations commonly assigned to *B. atrox* and *B. leucurus*. It is clear from this that studies which take the nominal species of the *B. atrox* complex as given units cannot provide any useful information about the sister-group relationships within the *B. atrox* complex. More information on the distribution of these haplotypes is needed, and material from the type localities of all species will be required before a revision of the nomenclature of these forms can begin. We urge other researchers working with this complex, and especially toxinological and medical researchers, to obtain and publish a maximum of information on the locality of origin of their material, to enable their results to be reconciled with future developments of the classification of the group.

The Mantel tests showed a particularly significant association between vegetation formation and external morphology of the complex as a whole, reflecting largely the different vegetational formations occupied by the morphologically differentiated '*B. moojeni*' populations. Although relatively little published information on the natural behaviour of these snakes is available, most pitvipers, including large *Bothrops*, are thought to be primarily 'sit-and-wait' predators, relying on their camouflage to remain unnoticed by approaching prey animals (Greene 1992; Sazima 1992). We therefore hypothesize that the morphological differences of the divergent '*B. moojeni*' populations represent adaptations to the vegetational differences resulting from a drier, more seasonal, climate within its range. The fact that the morphology of this complex appears to represent primarily an adaptation to ecological conditions rather than a reflection of phylogeny in this group is likely to be one of the principal causes of the general failure to arrive at a stable and robust classification for these animals (see also Markezich & Taphorn 1993).

The data presented here are insufficient to indicate whether the clades revealed in this analysis represent separate, independently evolving species, or whether the mitochondrial DNA gene trees presented here are contained within a single, highly variable species. The resolution of this problem requires more sequence information from a much greater number of localities than is currently available, in order to identify the distribution of different mtDNA haplotype lineages among the populations of this complex. Furthermore, a close investigation of contact zones between the clades identified here by means of nuclear as well as mitochondrial markers is likely to be necessary in order to understand the patterns of genetic exchange between these lineages.

Acknowledgements

We thank the staff of the Museu Paraense Emílio Goeldi, Jean-Marc Touzet (Fundación Herpetológica Gustavo Orcés, Quito) and Antonio Freire L. (Instituto Nacional de Higiene y Medicina Tropical 'L. Izquieta Pérez', Guayaquil) for their valuable help in obtaining live *Bothrops* specimens for our DNA analyses. We also thank the curators and staff of the following museums and collections for the loan of specimens or access to their collections: Academy of Natural Sciences (Philadelphia), American Museum of Natural History (New York), British Museum (Natural History) (London), California Academy of Sciences (San Francisco), Field Museum of Natural History (Chicago), Fundación Herpetológica Gustavo Orcés (Quito), Instituto Butantan (São Paulo), Instituto Nacional de Higiene y Medicina Tropical 'L. Izquieta Pérez' (Guayaquil), Museu Paraense Emílio Goeldi (Belém), National Museum of Natural History (Washington D.C.), Naturhistorisches Museum Basel (Basel), Pontificia Universidad Católica del Ecuador (Quito), Staatliches Museum für Naturkunde (Stuttgart), University of Florida (Gainesville), University of Kansas Museum of Natural History (Lawrence), and Zoologisches Museum A. Koenig (Bonn). The program PHYLIP was run using the Seqnet Facility of the Central Laboratory of the Research Councils (Daresbury Laboratory). This study was supported by the Science and Technology for Development Program of the European Community (contract no. TS3-CT91-0024), the Wellcome Trust (Research Career Development Fellowship to WW), Fundação Banco do Brasil and the British Council (fellowship to MGS).

References

Ab'Sáber, A. N. (1977). Espaços ocupados pela expansão dos climas secos na América do Sul, por ocasião dos periodos glaciais quaternários. *Paleoclimas (Inst. Geogr., Univ. S. Paulo)* 3: 1–19.

Anderson, S., Bankier, A. T., Barrell, B. G., de Bruijn, M. H. L., Coulson, A. R., Drouin, J., Eperon, I. C., Nierlich, D. P., Roe, B. A., Sanger, F., Schreier, P. H., Smith, A. J. H., Staden, R. & Young, I. G. (1981). Sequence and organization of the human mitochondrial genome. *Nature, Lond.* 290: 457–465.

Bolaños, R. (1984). *Serpientes, venenos y ofidismo en Centroamérica.* Editorial Universidad de Costa Rica, San José.

Brown, K. S., Jr. (1979). *Ecologia geográfica e evolução nas florestas Neotropicais.* Universidade Estadual de Campinas, São Paulo.

Brown, K. S., Jr. & Ab'Sáber, A. N. (1979). Ice-age refuges and evolution in the Neotropics: correlation of paleoclimatological, gemorphological and pedological data with modern biological endemism. *Paleoclimas (Inst. Geogr., Univ. S. Paulo)* 5: 1–30.

Campbell, J. A. & Lamar, W. W. (1989). *The venomous reptiles of Latin America.* Cornell University Press, Ithaca, New York.

Cardoso, J. L. C. (1992). Introdução ao estudo dos acidentes por animais peçonhentos. In *Plantas venenosas e animais peçonhentos:* 139–142. (Ed. Schvartsman, S.). Sarvier, São Paulo.

Castellano, S., Malhotra, A. & Thorpe, R. S. (1994). Within-island geographic variation of the dangerous Taiwanese snake, *Trimeresurus stejnegeri*, in relation to ecology. *Biol. J. Linn. Soc.* 52: 365–375.

Cunha, O. R. da & Nascimento, F. P. do (1993). Ofídios da Amazônia. As cobras da região leste do Pará. *Bolm Mus. para. Emílio Goeldi (Ser. Zool.)* 9: 1–191.

Farris, S. J. (1974). Formal definitions of paraphyly and polyphyly. *Syst. Zool.* 23: 548–554.

Felsenstein, J. (1993). *Phylogeny inference package (PHYLIP), version 3.5c.* Department of Genetics, University of Washington, Seattle.

Greene, H. W. (1992). The ecological and behavioral context for pitviper evolution. In *Biology of the pitvipers*: 107–117. (Eds Campbell, J. A. & Brodie Jr., E. D.). Selva, Tyler, Texas.

Haad, J. S. (1980/81). Accidentes humanos por las serpientes de los géneros *Bothrops* y *Lachesis. Mems Inst. Butantan* 44/45: 403–423.

Haffer, J. (1969). Speciation in Amazonian forest birds. *Science* **165**: 131–137.

Haffer, J. (1987). Quaternary history of tropical America. In *Biogeography and Quaternary history in tropical America*: 1–18. (Eds Whitmore, T. C. & Prance, G. T.). Clarendon Press, Oxford.

Hoge, A. R. (1947). Uma nova espécie de *Trimeresurus*. *Mems Inst. Butantan* **20**: 193–202.

Hoge, A. R. (1965). Preliminary account on Neotropical Crotalinae (Serpentes, Viperidae). *Mems Inst. Butantan* **32**: 109–184.

Kocher, T. D., Thomas, W. K., Meyer, A., Edwards, S. V., Pääbo, S., Villablanca, F. X. & Wilson, A. C. (1989). Dynamics of mitochondrial DNA evolution in animals: amplification and sequencing with conserved primers. *Proc. natn. Acad. Sci. USA* **86**: 6196–6200.

Kouyoumdjian, J. A. & Polizelli, C. (1988). Acidentes ofídicos causados por *Bothrops moojeni*: relato de 37 casos. *Revta Inst. Med. trop. S. Paulo* **30**: 424–432.

Markezich, A. L. & Taphorn, D. C. (1993). A variational analysis of populations of *Bothrops* (Serpentes: Viperidae) from western Venezuela. *J. Herpet.* **27**: 248–254.

Moritz, C., Schneider, C. J. & Wake, D. B. (1992). Evolutionary relationships within the *Ensatina eschscholtzii* complex confirm the ring species interpretation. *Syst. Biol.* **41**: 273–291.

Nishioka, S. A. & Silveira, P. V. P. (1992). A clinical and epidemiologic study of 292 cases of lance-headed viper bite in a Brazilian teaching hospital. *Am. J. trop. Med. Hyg.* **46**: 805–810.

Oosterbroek, P. (1987). More appropriate definitions of paraphyly and polyphyly, with a comment on the Farris 1974 model. *Syst. Zool.* **36**: 103–108.

Otero, R., Tobon, G. S., Fernando Gómez, L., Osorio, R., Valderrama, R., Hoyos, D., Urreta, J. E., Molina, S. & Arboleda, J. J. (1992). Accidente ofídico en Antioquia y Chocó. Aspectos clínicos y epidemiológicos (marzo de 1989–febrero de 1990). *Acta med. Colombiana* **17**: 229–249.

Pesantes, O. C. (1989). *Relações entre algumas espécies do gênero* Bothrops, *pela eletroforese do plasma e morfologia do hemipênis (Serpentes: Viperidae)*. MSc dissertation: Universidade Estadual Paulista Julio de Mesquita Filho, Rio Claro, São Paulo.

Sandner Montilla, F. (1979). Una nueva especie del género *Bothrops* (Serpentes, Crotalidae, Bothropinae) de la región de Guanare, Estado Portuguesa, Venezuela. *Mems cient. Ofidiol.* No. 4: 1–19.

Sandner Montilla, F. (1981). Una nueva subespecie de *Bothrops lanceolatus* (Lacepède, 1789) Familia Crotalidae. *Mems cient. Ofidiol.* No. 6: 1–15.

Sanger, F., Nicklen, S. & Coulson, A. R. (1977). DNA sequencing with chain-terminating inhibitors. *Proc. natn. Acad. Sci. USA* **74**: 5463–5467.

Sazima, I. (1992). Natural history of the jararaca pitviper, *Bothrops jararaca*, in southeastem Brazil. In *Biology of the pitvipers*: 199–216. (Eds Campbell, J. A. & Brodie Jr., E. D.). Selva, Tyler, Texas.

Swofford, D. L. (1989). *PAUP — phylogenetic analysis using parsimony, version 3.0.* Illinois Natural History Survey, Champaign, Illnois.

Thorpe, R. S. & Brown, R. P. (1989). Microgeographic variation in the colour pattern of the lizard *Gallotia galloti* within the island of Tenerife: distribution, pattern and hypothesis testing. *Biol. J. Lim. Soc.* **38**: 303–322.

Thorpe, R. S., Brown, R. P., Day, M., Malhotra, A., McGregor, D. P. & Wüster, W. (1994). Testing ecological and phylogenetic hypotheses in microevolutionary studies: an overview. In *Phylogenetics and ecology*: 189–206. (Eds Eggleton, P. & Vane-Wright, R.). Academic Press, London.

Thorpe, R. S., Malhotra, A., Black, H., Daltry, J. C. & Wüster, W. (1995). Relating geographic pattern to phylogenetic process. *Phil. Trans. R. Soc. (B)* **349**: 61–68.

Vanzolini, P. E. & Williams, E. E. (1970). South American anoles: the geographic differentiation and evolution of the *Anolis chrysolepis* species group (Sauria, Iguanidae). *Arqs Zool. S. Paulo* **19**: 1–298.

Werman, S. D. (1992). Phylogenetic relationships of Central and South American pitvipers of the genus *Bothrops (sensu lato)*: cladistic analyses of biochemical and anatomical characters. In *Biology of the pitvipers*: 21–40. (Eds Campbell, J. A. & Brodie Jr., E. D.). Selva, Tyler, Texas.

Wüster, W., Otsuka, S., Malhotra, A. & Thorpe, R. S. (1992). Population systematics of Russell's viper: a multivariate study. *Biol. J. Linn. Soc.* **47**: 97–113.

Wüster, W., Otsuka, S., Thorpe, R. S. & Malhotra, A. (1992). Morphological variation in Russell's viper in Burma and Thailand. *Herpet. J.* **2**: 99–101.

Wüster, W. & Thorpe, R. S. (1989). Population affinities of the Asiatic cobra (*Naja naja*) species complex in south-east Asia: reliability and random resampling. *Biol. J. Linn. Soc.* **36**: 391–409.

Wüster, W. & Thorpe, R. S. (1990). Systematics and biogeography of the Asiatic cobra (*Naja naja*) species complex in the Philippine Islands. In *Vertebrates in the tropics*: 333–344. (Eds Peters, G. & Hutterer, R.). Museum A. Koenig, Bonn.

Wüster, W. & Thorpe, R. S. (1991). Asiatic cobras: systematics and snakebite. *Experientia* **47**: 205–209.

Wüster, W. & Thorpe, R. S. (1992). Asiatic cobras: population systematics of the *Naja naja* species complex (Serpentes: Elapidae) in India and Central Asia. *Herpetologica* **48**: 69–85.

Wüster, W. & Thorpe, R. S. (1994). *Naja siamensis*, a cryptic species of venomous snake revealed by mtDNA sequencing. *Experientia* **50**: 75–79.

Wüster, W., Thorpe, R. S., Cox, M. J., Jintakune, P. & Nabhitabhata, J. (1995). Population systematics of the snake genus *Naja* (Reptilia: Serpentes: Elapidae) in Indochina: multivariate morphometrics and comparative mitochondrial DNA sequencing (cytochrome oxidase I). *J. evol. Biol.* **8**: 413–510.

Wüster, W., Thorpe, R. S., Puorto, G. & BBBSP (In press). Systematics of the *Bothrops atrox* complex (Reptilia: Serpentes: Viperidae) in Brazil: a multivariate analysis. *Herpetologica.*

9 New perspectives on the evolution of south-east Asian pitvipers (genus *Trimeresurus*) from molecular studies

ANITA MALHOTRA and ROGER S. THORPE

Synopsis

The genus *Trimeresurus* contains about 36 species, several of which are remarkably similar in superficial appearance, and it is unclear whether this similarity relates to phylogenetic relationships or to ecological convergence. Whatever the cause, it has resulted in considerable confusion which hinders the understanding of the evolution of the genus by biologists as well the advancement of snakebite treatment by medical workers. Our ongoing study of this group (part of a larger study of the systematics of medically significant snakes) focuses on a core group of species (*T. albolabris*, *T. popeorum*, *T. stejnegeri*, *T. erythrurus*, *T. macrops*) known to be or potentially sympatric in Thailand and adjacent areas and involves the use of multivariate morphometry and mitochondrial DNA analysis to study both inter- and intra-specific relationships. We propose a preliminary phylogenetic hypothesis for a set of species based on PCR-RFLP of the cytochrome *b* gene of mitochondrial DNA and discuss this in relation to its ability to shed light on morphological evolution in the genus. We then illustrate geographic variation in morphology of *T. albolabris*, a species with a large range, and relate this to current subspecies and to the phylogenetic history of populations as revealed by cytochrome *b* sequencing. Mantel tests are used to evaluate various causal relationships.

Introduction

The genus *Trimeresurus* contains a large number of species, found in most of southern Asia from the Indian subcontinent to southern China and Japan, Indochina, the Philippines and the Indo-Malayan archipelago (excluding Sulawesi). The most recent checklist (Golay *et al.* 1993) lists 36 species under *Trimeresurus*; another four formerly considered part of this genus are now listed under the genus *Ovophis* (Burger 1971; H. M. Smith & Chiszar 1988), and the species *wagleri* in a monotypic genus, *Tropidolaemus* (Hoge & Romano-Hoge 1978/79).

The genus *Trimeresurus* (*sensu stricto*) still contains a diversity of forms which can be divided, on the basis of superficial resemblance, into several groups. There are ground-living, egg-laying forms (e.g. *T. mucrosquamatus*, *T. flavoviridis*, *T. jerdoni*).

Symp. zool. Soc. Lond. (1997) No. 70: 115–128
ISBN 0 19 854986 5

The majority of species of *Trimeresurus*, however, are arboreal. Some of these are brown (e.g. *T. puniceus*) and some are patterned (e.g. *T. trigonocephalus*, *T. kanburiensis*, *T. sumatranus*). Quite a few species, however, can be described as 'green pitvipers' (e.g. *T. albolabris*, *T. popeorum*, *T. erythrurus*, *T. stejnegeri*, *T. tibetanus*, *T. macrops*, *T. gramineus*, *T. medoensis*, *T. flavomaculatus*, *T. hageni*). Many of these were originally included in a single species, *T. gramineus* (Shaw 1802). Later the work of Stejneger (1927) and Pope (Pope & Pope 1933) revealed that this contained several species, the status and relationships of which have since been the subject of much debate and rearrangement (e.g. M. A. Smith 1943; Hoge & Romano-Hoge 1978/79; Regenass & Kramer 1981). Identification of the green species remains fraught with problems, as illustrated by the number of herpetological books which contain pictures of misidentified snakes (e.g. Engelmann & Obst 1981), including some guides to snakes of particular countries or regions (e.g. Nootpand 1971).

Apart from its general biological interest, there are likely to be medical benefits arising from a thorough revision of the genus. While less life-threatening than some of the elapids and true vipers found in the same region, *Trimeresurus* species are frequently the most important cause of snake-bite as they are common and frequently come into contact with man (Romer 1963; Kuo & Wu 1972; Viravan *et al.* 1992). The problems associated with identification have led to a situation where most of the medical literature on aspects of snakebite and venom involving these species is virtually meaningless, as the actual species involved are unknown (Mitrakul 1973; Mitrakul & Impun 1973; Visudiphan *et al.* 1989; but see Hutton *et al.* 1990).

In this paper, we will briefly review some of the characters which have been used for identification purposes and some of the problems which contribute to the general confusion surrounding the 'green' species. We then give a progress report on our work on this genus to date, illustrating our approach to resolving some of these problems.

Morphological characters of use in identification of 'green' *Trimeresurus*

Various characters have been considered important in distinguishing between the species. Two major ones, which have been given considerable emphasis by previous workers in defining species affinities, are the type of hemipenis (Pope & Pope 1933; Maslin 1942) and the condition of the first supralabial scale (M. A. Smith 1943).

1. Hemipenis type. There are two markedly different types of hemipenes. One is short and covered by spines in the region of the fork, with the tips being calyculate (pictured in Mao, Yin & Guo 1984). This resembles that of other pitvipers and true vipers. The second type is long and slender and completely lacks spines; again it is calyculate for much of the forked region (pictured in Vogel 1991). These are not subtle differences, and can be easily detected without even everting the hemipenes, by the pronounced bulge in the top of the tail produced by the spiny type compared to a more gradually tapering tail in those with the slender type of hemipenis. In spite of the weight given to this character in determining species groups, descriptions of new species are often published with no mention of the hemipenis (e.g. *T. medoensis*, Zhao & Jiang 1977) and literature reports are often inconsistent.

2. Fusion of the first labial and the nasal scales. In some species (*T. albolabris*, *T. erythrurus*), the first supralabial scale is partly or completely fused to the nasal scale. In others (e.g. *T. stejnegeri*, *T. popeorum*), they are completely separated by a suture. Scoring this character can be problematic as fused scales are often creased at the point at which the suture normally occurs, therefore close examination may be necessary in order to distinguish the states.
3. Size of internasals and their contact or separation by intervening scales. *T. albolabris* normally has internasals in contact, while similar species (e.g. *T. erythrurus*) normally have one or more scales between the internasals.
4. Presence or absence of one or more scales between the nasal and the shield bordering the pit anteriorly.
5. Keeling of scales. Some species have smooth or feebly keeled scales while others have extremely keeled scales (e.g. *T. erythrurus*). Keeling of head scales may be of particular diagnostic utility.
6. Size of the eye. *T. macrops* (meaning 'big eye') was confused with *T. albolabris* until recently (Regenass & Kramer 1981) in spite of the coloration and head shape differences between the two. This confusion had particular medical significance since *T. macrops* is the second most common species in the heavily populated Bangkok area (the most common being *T. albolabris*) (Mahasandana & Juntakune 1990).
7. Number of scales around the body. In the green species, this varies between 25 (*T. erythrurus*) and 12 (*T. macrolepis*), but is usually 21 in the species most difficult to distinguish (e.g. *T. popeorum*, *T. stejnegeri*, *T. albolabris*, *T. macrops*).
8. The size and arrangement of gular scales. These may be arranged in regular pairs (e.g. *T. albolabris*) or may be more irregular (*T. stejnegeri*).

As already indicated, in spite of the fact that several of these characters can be used to separate species, misidentification of species (even by professional herpetologists) continues to be frequent.

Problems of identification of *Trimeresurus*

Within-species variability

A mixture of ontogenetic variation, sexual dimorphism and geographic variation is present in almost every character. An example of geographic variation especially relevant to identification in the field is colour pattern in a widespread species, *T. albolabris*. In a typical *albolabris* specimen from central Thailand, the upper lip is sharply demarcated from the upper part of the head, being the same yellow colour as the ventral scales. Sexual dimorphism is rather subtle in this species, males having a distinct white stripe down the side of the body which tends to be indistinct or absent in females. In the eastern Indonesian islands (e.g. Flores and Komodo), the stripe is absent in both sexes, and the lip is not sharply demarcated from the rest of the head. Since these islands are on the edge of the range of this species, it perhaps is not unexpected that they look different. However, only a few hundred kilometres north-east of Bangkok, *albolabris* are also a much more uniform green all over and have an orange eye rather than a yellow eye. A few hundred kilometres in the other direction, in the region of Chiang Mai in north Thailand, specimens that resemble *albolabris* in scalation characters (e.g. internasals in contact and fused first labial) nevertheless look quite different, having a deep-red eye and rather prominently keeled temporal and body scales. These specimens will be referred to as *Trimeresurus* sp. in the rest of the paper.

Another example of geographic variation is given by *T. stejnegeri*. Males of this species from Taiwan have a prominent lateral white stripe which continues onto the head, edged with a narrow red stripe (below the white stripe on the body, above it on the head). Females lack the lateral red pigmentation, but the eyes are red in both sexes. They are small snakes, typically reaching 0.5 m. In *T. stejnegeri* from north-east Thailand (Loei Province) the red stripe in males is much more pronounced and the eye is also deep red. Females, however, have only the white stripe and have yellow eyes. In this area this species reaches a large size, with females commonly exceeding 1 m in length. Ontogenetic variation is also present, notably in the colour of the eye in juvenile males, which are yellow, although they have a broad red lateral stripe from birth.

This variation poses problems for identification, in some cases producing pronounced confusion. In Loei Province, *T. stejnegeri* are, according to current range maps (Regenass & Kramer 1981), sympatric with *T. popeorum*. These two species present perhaps the most difficult challenge in identification as their external morphology seems to be almost perfectly convergent. The males, however, have different hemipenes, with *stejnegeri* having the short spiny type and *popeorum* the long slender type. While this means that it should be possible to correctly identify males of the two species, females are virtually indistinguishable by any currently known criteria. From time to time, characters are proposed which purport to do this, a recent example being the continuity of the reddish tail colour (continuous vs. broken up into patches). However, a recent examination of a series of 40 *T. stejnegeri* from Loei showed that this character varies ontogenetically within the species, being broken up in young animals and solid in older ones.

Inaccurate range descriptions

Snakes are cryptic animals and, in much of the area in which this genus occurs, access is politically and physically difficult. These facts make it highly likely that current range descriptions are inaccurate, yet identifications sometimes seem to be based on the species known to occur in the area. For example, a classic paper on guild structure in reptiles (Inger & Colwell 1977), carried out in Nakhon Ratchasima Province (north-east of Bangkok), lists all *Trimeresurus* specimens caught as *T. popeorum*. However, only a cursory examination of the preserved specimens is necessary to confirm that the species is actually *T. stejnegeri*, the hemipenes having been carefully everted and clearly of the spiny type. The simplest explanation that can be proposed for such a basic error is that, since it was known that *stejnegeri* did not occur so far south in Thailand, the hemipenis had not been checked. In fact, it is clear that this species occurs considerably further south than Nakhon Ratchasima, a confirmed specimen now being known from as far south in peninsular Thailand as Krabi (P. Juntakune, pers. comm.). Since this confusion came to light, we have examined almost every available male '*popeorum*' from north-east Thailand and in every case it has proved to be *T. stejnegeri*, which raises the possibility that *popeorum* does not occur in the Korat plateau at all.

Discovery of new species

Many currently accepted species will almost certainly eventually be found to be polytypic (e.g. *T. flavomaculatus* in the Philippines). Entirely new species are regularly being described (e.g. *T. medoensis*, *T. tibetanus*, *T. xiangchengensis*) although not all of them may be validated as new material surfaces (e.g. new specimens of *T. venustus* have recently been found which call into question the validity of its separation from *T. kanburiensis*). Since *Trimeresurus* are very popular with hobbyists, animals are sometimes imported into Europe that do not obviously fit into any of the currently described species. In many cases even the country of origin of these animals is unknown, making it difficult to judge whether the differences relate to intraspecific geographic variation or to interspecific differentiation.

A molecular phylogenetic approach: can new techniques help?

As we have already described, some of the 'green' species are so similar in external morphology that they may be impossible to tell apart except by hemipenis morphology (Pope & Pope 1933). This last fact suggests that they are not closely related sister species, implying a remarkable degree of evolutionary convergence. This also suggests that studying morphology alone may not be appropriate in determining the evolutionary relationships within the group. The use of molecular markers has been shown to be a powerful means of recovering phylogenetic histories for many organisms (see references in Avise 1994). While not being entirely free from selective effects, molecular markers will almost certainly show less homoplasy than morphology in this case. Nevertheless, considerable morphological variation may be present within currently recognized species, so character analysis is still appropriate to define the taxonomic units to be represented in the molecular phylogenetic tree.

Methods

Because of the problems of identification already alluded to, the samples used in this study were collected by a few people who could be relied on to make an accurate identification, and are supported by photographs or voucher specimens. The technique used to obtain the phylogeny presented here is enhanced restriction fragment length polymorphism (RFLP), also known as PCR–RFLP (Quinn 1992; Simon, McIntosh & Deniega 1993). The advantages of this technique over conventional RFLP of whole mitochondrial genomes is that it involves cutting of polymerase chain reaction (PCR)-amplified fragments, hence small amounts of tissue (blood or small section of tip of tail or even shed skin) provide sufficient DNA. Also, since the fragment is smaller (a 767 bp fragment of cytochrome *b* was used in this study), more sensitive four-cut enzymes can be used while retaining the ability to easily map the restriction sites. Its advantage over sequencing is that it is a faster, easier and cheaper method of screening large numbers of samples. However, loss of restriction sites can result from mutations occurring at any one of four positions (in the case of a four-base cutter) and involving three possible changes at each position. The chance of non-homologous site loss is therefore considerably higher than that of non-homologous site gain and this must be taken into account in the analysis. Also, as only a subset of the polymorphisms present will be detected, the technique is less sensitive than sequencing. Although it is likely to provide enough information at the interspecific level, and in some cases sufficient intraspecific variation for a phylogenetic analysis may be detectable by PCR–RFLP alone (e.g. Daltry, Wüster & Thorpe, this volume pp. 155–171), in general levels of differentiation are likely to be much lower at the within-species level. We therefore chose to sequence the fragment to provide more detailed information.

Whole genomic DNA extracts from muscle or liver tissue were PCR–amplified using primers L14841 (Kocher *et al.* 1989) and MVZ16 (Moritz, Schneider & Wake 1992). Thermal cycle parameters were 1 min at 94 °C, 1 min at 45 °C and 1 min at 72 °C for the first five cycles, followed by 94 °C for 1 min, 50 °C for 1.5 min and 72 °C for 1.5 min for an additional 30 cycles and a final 5 min extension step at 72 °C. A negative control was always included in order to exclude the possibility of contamination. Purified product was either sequenced (see Salomão *et al.*, this volume pp. 89–98, for protocol details) or digested with eight restriction enzymes (*Aci*1, *Cfo*1, *Dde*1, *Hae*3, *Nla*3, *Mse*1, *Rsa*1, *Taq*1). Digestion products were separated on a 6% acrylamide gel alongside a size marker, and visualized by staining with ethidium bromide.

Restriction site analysis was carried out using the programme REAP (McElroy *et al.* 1992) and genetic distances from this were input into PHYLIP 3.5 (Felsenstein 1993) for phylogenetic analysis. Sequence data (confirmed by sequencing from both primers) could be aligned easily by eye (since the cytochrome *b* gene is a protein coding region), and subjected to maximum parsimony analysis using PAUP 3.3 (Swofford 1993) with default settings. All sites were weighted equally.

Results

Phylogenetic tree for *Trimeresurus*: a preliminary hypothesis

At present 10 species are represented in the phylogeny. A Fitch–Margoliash tree (Fig. 1) was constructed from information on 42 restriction sites (from seven four-cut enzymes and one five-cut), using *Bothrops moojeni* and two specimens of *Tropidolaemus wagleri* as outgroups. The branch lengths are not shown, as it is the topology that is of primary interest at this stage. The tree is quite pectinate, but some interesting patterns emerge nevertheless. *T. mucrosquamatus* is close to the base of the tree, as might be expected from its ground-living, egg-laying habits. *T. albolabris* forms a monophyletic group, with a second cohesive group being formed by *T. purpureomaculatus*, *T. erythrurus*, *T. cantori* and the red-eyed *Trimeresurus* sp. from Chiang Mai. The species *purpureomaculatus* as currently recognized is clearly polyphyletic, with *erythrurus* intervening between the mainland subspecies (*T. p. purpureomaculatus*) and the Andaman island subspecies (*T. p. andersoni*).

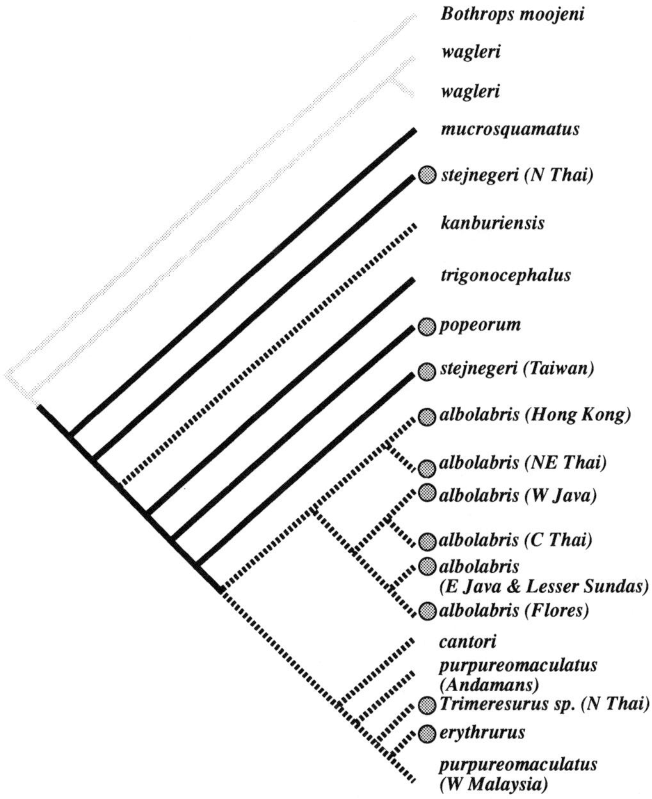

Fig. 1. Phylogenetic tree for some species of *Trimeresurus*. Outgroups are *Bothrops moojeni* and *Tropidolaemus wagleri*. The dashed lines represent the presence of the long slender hemipenis type and the fusion of the first labial and nasal scale. The grey circles indicate the green species. See text for details.

Another interesting feature is that the two *stejnegeri* specimens (from Taiwan and north-east Thailand) are quite separated in the tree, which indicates that they may be different species. It would be useful to compare them with *T. stejnegeri* from Fujian (the type locality) to determine their affinities.

Phylogenetic utility of morphological characters

How do the characters described above, that have been used to determine species affinities in the past, map onto the tree?

1. Fusion of first labial and nasal. Species within the *albolabris/purpureomaculatus* clade all possess this character (Fig. 1) but it is not restricted to this clade, also being found in *T. kanburiensis* (although it appears to be variable in this species). This could be the result of an independent evolution or loss of the character, or it might result from the hypothesis being wrong regarding the relationships of this species. It is interesting to speculate that if *T. kanburiensis* actually belongs to the *albolabris/purpureomaculatus* lineage, only a single evolutionary event would be necessary to explain the distribution of this character.

2. Hemipenis. The *albolabris/purpureomaculatus* lineage is also characterized by the possession of the long, slender hemipenis type. The speculation that *kanburiensis* may be part of this lineage is reinforced by the observation that it shares this hemipenis type. Only one other species represented on the tree has this hemipenis type, *T. popeorum*. It is possible that the *popeorum* hemipenis has evolved independently as it differs from all the rest in lacking finger-like papillae in the region of the fork, below the calyculate region.
3. Green coloration. To emphasize the point about morphological convergence, it can be seen from Fig. 1 that the green species (indicated by grey circles at the tips of the tree) have evolved independently.

Within-species evolution in *Trimeresurus albolabris*

Since significant substructure was observed within the *albolabris* clade (Fig. 1), this was subsequently examined in more detail, with particular attention to the correspondence between morphological variation and phylogenetic relationships revealed by mitochondrial DNA analysis.

Figure 2 illustrates the ranges of the three currently recognized subspecies (Regenass & Kramer 1981). The presence of *albolabris* on Sumatra and Borneo is in some doubt, depending on the existence of a few specimens with imprecise locality data (which may possibly be in error). Tissue samples have been obtained from most parts of the species range, with the exception of the subspecies *septentrionalis* from Nepal. For this within-species study, sequence was obtained for a 660 bp region of the cytochrome *b* gene (mitochondrial DNA).

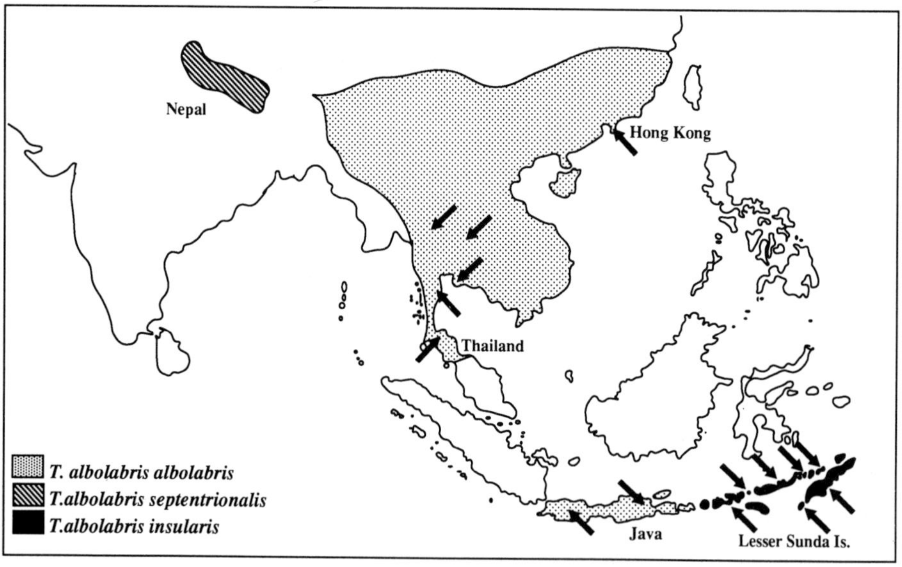

Fig. 2. Distribution of the currently recognized subspecies of *T. albolabris*. Black arrows indicate populations sampled for DNA analysis. Most parts of the illustrated range were represented in the morphological analysis.

Molecular phylogeny of *T. albolabris*

Figure 3 illustrates the majority-rule consensus tree (parsimony) with the bootstrap values (out of 100 bootstraps) given above the nodes. Three main lineages are discernible, the first comprising all specimens from the Lesser Sunda islands, together with the east Javan specimens. This clade is very well supported by the bootstrap, appearing in all 100 bootstrapped trees. The second lineage consists of all specimens from southern Thailand (Trang and Krabi) and the third consists of all other samples (north-east Thailand, central Thailand and Hong Kong and west Java). The relationships between these three clades are unresolved in the consensus tree. Although the affinities of the west Javan population appear to make no biogeographic sense on the basis of current geography, in the Pleistocene when sea levels were lower these islands formed part of the Sunda shelf (Heaney 1991). It is possible that Java was colonized from Indochina, bypassing south Thailand (Fig. 4). Many species do show boundaries between southern Thailand and northern west Malaysia, e.g. *Calloselasma rhodostoma* (Daltry *et al.*, this volume pp. 155–171), so there may have been some kind of a biogeographic barrier here. However, why east Java should be so divergent from west Java is inexplicable, unless it represents a westward recolonization of Java from the Lesser Sundas.

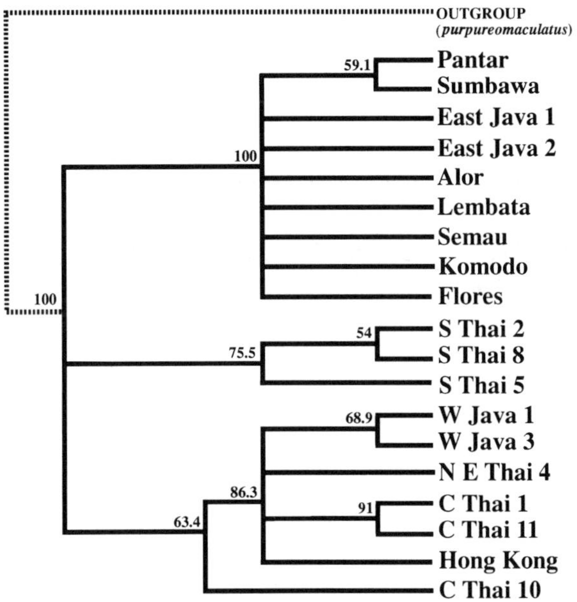

Fig. 3. Haplotype tree (50% majority-rule consensus tree from parsimony analysis) for *T. albolabris* from a 660 bp sequence of cytochrome *b*. Bootstrap values (out of 100 bootstraps) are given adjacent to the nodes. Haplotype numbers are the author's reference numbers.

Geographic variation in morphology

The phylogenetic lineages described above do not relate to the subspecific designations, since the subspecies *insularis* does not include any Javan populations. However, subspecies described by conventional taxonomic methods frequently fail to portray the real patterns of variation (Thorpe 1987). Consequently, we reanalysed morphological variation in this species.

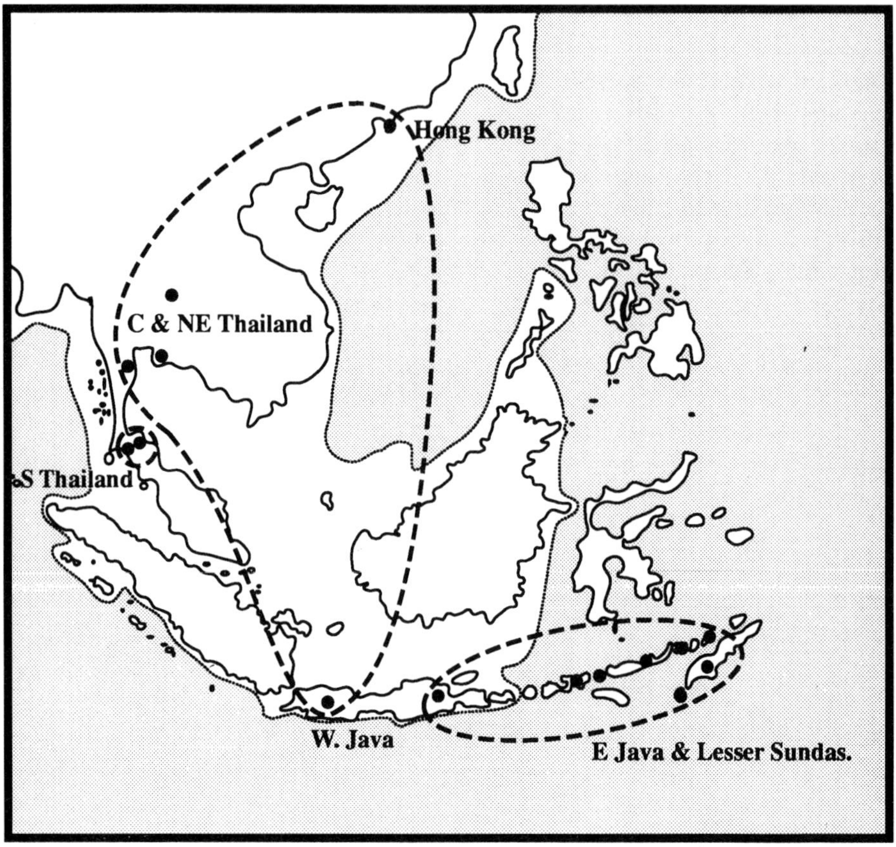

Fig. 4. Map indicating sea levels in the Pleistocene (stippled area) in relation to current sea levels (thin black line) and the three lineages represented on the phylogenetic tree (thick dashed line).

A multivariate (canonical variate) analysis on scalation characters showed some evidence of geographic variation. If the canonical variate scores are contoured on a map of Asia (Fig. 5), the closeness of the contours represents the gradient of morphological change between populations. The first canonical variate (CV1) distinguishes the Lesser Sundas from the rest, while the second canonical variate separates the north-eastern populations (Hong Kong, North Vietnam, China) from all other populations. Note that the east and west Javan populations are virtually indistinguishable and the Nepalese population is also not distinguishable from those further east in Thailand. A similar pattern of geographic variation is obtained for both sexes (but is not illustrated for males). A single female specimen from Borneo was fitted onto the axes, and appears different to any other population (Fig. 5), indicating that the locality may not be in error as suggested earlier.

The morphological analysis has shown, yet again, that the conventional subspecies are probably invalid. The groups of morphologically similar populations, however,

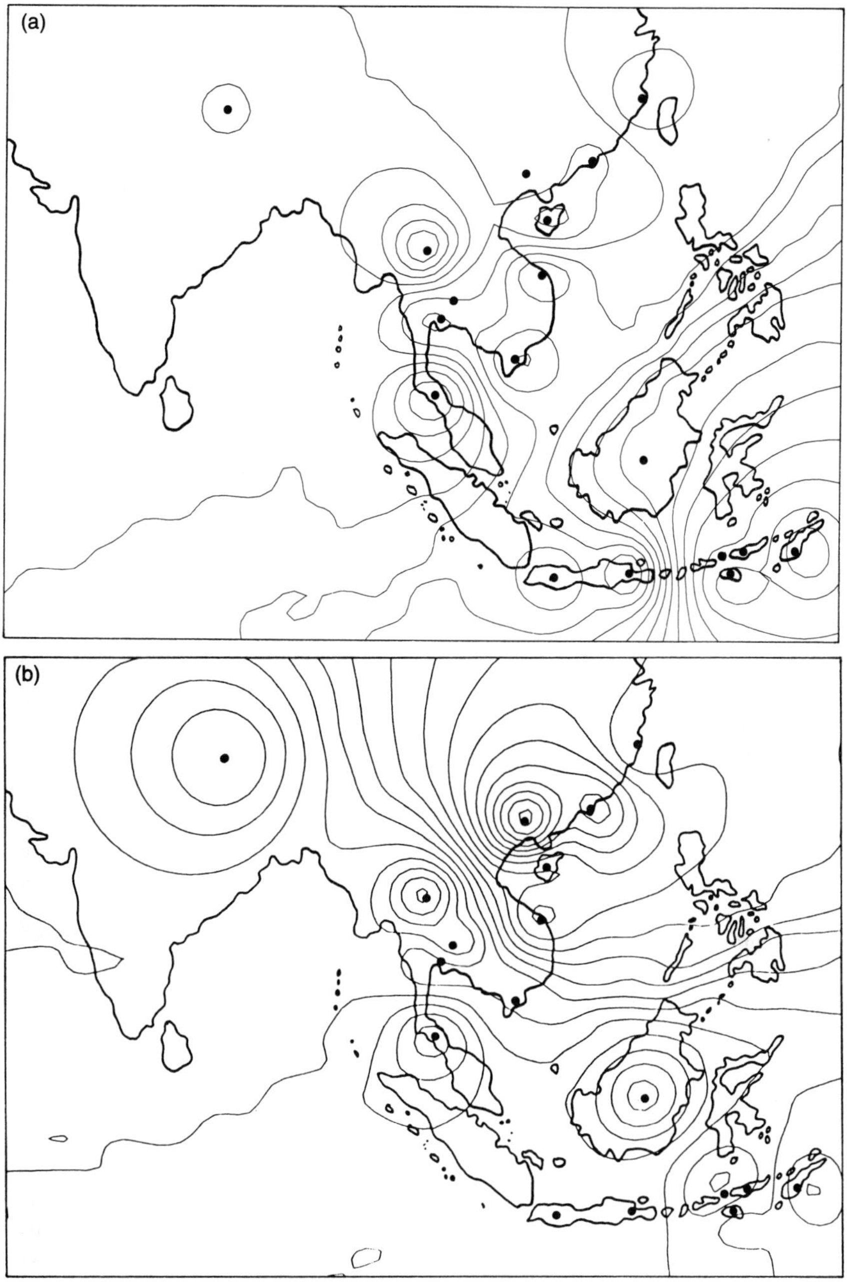

Fig. 5. Canonical variate scores contoured over a map of Asia illustrating geographic variation in scalation characters in females. (a) CV1; (b) CV2. See text for explanation.

do not correspond to lineages on the phylogenetic tree. The Hong Kong specimen, for example, is not in a separate lineage to the Thai samples as implied by the ordination based on scalation characters.

Causation

What might be causing the observed discrepancy between morphology and phylogeny? An obvious candidate is natural selection for environmental conditions, since the species is found in a wide range of environmental regimes, from seasonal subtropical areas (e.g. southern China), relatively aseasonal wet areas (e.g. southern Thailand, west Java), and arid areas (the Lesser Sundas). To test this, a matrix correlation test or Mantel test (Thorpe *et al.* 1994) was run using scalation as the dependent matrix and climate (derived from four variables obtained from climatic maps of Asia: Anon. 1981, 1992) as the independent matrix. The result was highly significant ($P < 0.001$), indicating that taxonomic units based on morphological similarity may have little to do with historical evolutionary units.

Discussion

The results presented in this paper represent a progress report rather than a completed piece of work, and are only the first step towards resolving some of the challenges posed by this genus. It seems likely that the molecular approach adopted here will prove very rewarding, but there is a long way to go before all 36 (or more) species are represented on the tree. Many of the problems of resolving the systematics of this genus result from the difficulty of finding stable morphological characters. As far as it is possible to evaluate morphological characters against the preliminary molecular phylogeny presented here, it would seem that both the type of hemipenis and the fusion of the first labial scale with the nasal may be valuable for defining species affinities, although this requires confirmation by further work. However, at the intraspecific level, the molecular phylogeny of *T. albolabris* does not correspond with morphological differentiation, making it difficult to name subspecies on any consistent criterion. The inclusion of other morphological character systems (e.g. colour pattern and body shape) in the analysis is likely to exacerbate this problem, since they are unlikely to be congruent with each other.

The results presented here also suggest the existence of several new taxa. The *Trimeresurus* sp. from Chiang Mai is probably an entirely new species. The Taiwanese population of *T. stejnegeri* also appears to be distinct from mainland populations, as are the mainland and Andaman Islands populations of *T. purpureomaculatus*. However, we feel it would be inappropriate to confuse the situation further by naming new species until further work confirms these findings.

Acknowledgements

We are grateful to the large numbers of people who have assisted us in the field and/or supplied us with tissue samples for analysis. These include (in no particular order of importance) Jenny Daltry, Wolfgang Wüster, Merel J. Cox, Romulus Whitaker, Indraneil Das, Michihisa Toriba, Anslem de Silva, Professor Sangkot Marzuki, Dr Aucky Hinting, Rick Hodges, Vincen Khartono, Pak Harwono and his staff, Professor David Warrell, Wen-hao Chou, Tanya Chan-ard, Jarujin Nabhitabhata, James D. Lazell. We also acknowledge the assistance of the curators of the following museums who in some cases allowed us to borrow large numbers of their *Trimeresurus* holdings: American Museum of Natural History; California Academy of Sciences, San Francisco;

Field Museum of Natural History, Chicago; Museum of Comparative Zoology, Harvard; Museum d'Histoire Naturelle, Geneva; Naturhistorisches Museum, Basel; Naturhistorisches Museum, Wien; University of Florida, Florida State Museum, Gainesville; National Museum of Natural History, Smithsonian Institution.

We have also examined specimens in the private collections of Khun Piboon Juntakune (Queen Savoabha Memorial Institute, Bangkok) and the Phu Luang Wildlife Sanctuary, Thailand. This study was funded by the Leverhulme Trust, with additional support from the Royal Society, the Percy Sladen Trust, the Bonhote Trust and the Carnegie Trust.

References

Anon. (1981). *Climatic atlas of Asia. I. Maps of mean temperature and precipitation.* WMO, UNESCO, Goscomgidromet USSR, UNEP. Moscow, USSR.

Anon. (1992). *'The Times' atlas of the world — comprehensive edition.* (9th edn). Bartholomew and Times Books, London.

Avise, J. C. (1994). *Molecular markers, natural history and evolution.* Chapman & Hall, New York.

Burger, W. L. (1971). *Genera of pitvipers (Serpentes: Crotalidae).* PhD thesis: University of Kansas.

Engelmann, W.-E. & Obst, F. J. (1981). *Snakes: biology, behavior and relationship to man.* Croom Helm, London & Canberra.

Felsenstein, J. (1993). *PHYLIP manual version 3.5.* University of Washington, Seattle.

Golay, P., Smith, H. M., Broadley, D. G., Dixon, J. R., McCarthy, C., Rage, J. C., Schatti, B. & Toriba, M. (1993). *Endoglyphs and other venomous snakes of the world: a checklist.* Azemiops S. A., Herpetological Data Centre, Geneva.

Heaney, L. R. (1991). A synopsis of climatic and vegetational change in South-east Asia. *Climatic Change* **19**: 53–61.

Hoge, A. R. & Romano-Hoge, S.A.R.W.L. (1978/79) [1981]. Poisonous snakes of the world. Part I. Check list of the pit vipers, Viperoidea, Viperidae, Crotalinae. *Mems Inst. Butantan* **42/43**: 179–310.

Hutton, R. A., Looareesuwan, S., Ho, M., Silamut, K., Chanthavanich, P., Karbwang, J., Supanaranond, W., Vejcho, S., Viravan, C., Phillips, R. E. & Warrell, D. A. (1990). Arboreal green pit vipers (Genus *Trimeresurus*) of south-east Asia: bites by *T. albolabris* and *T. macrops* in Thailand and a review of the literature. *Trans. R. Soc. trop. Med. Hyg.* **84**: 866–874.

Inger, R. F. & Colwell, R. K. (1977). Organization of contiguous communities of amphibians and reptiles in Thailand. *Ecol. Monogr.* **47**: 229–253.

Kocher, T. D., Thomas, W. K., Meyer, A., Edwards, S. V., Pääbo, S., Villablanca, F. X. & Wilson, A. C. (1989). Dynamics of mitochondrial DNA evolution in animals: amplification and sequencing with conserved primers. *Proc. natn. Acad. Sci. USA* **86**: 6196–6200.

Kuo, T. P. & Wu, C. S. (1972). Clinico-pathological studies on snakebites in Taiwan. *Snake* **4**: 1–22.

McElroy, D., Moran, P. Bermingham, E. & Kornfield, I. (1992). REAP: an integrated environment for the manipulation and phylogenetic analysis of restriction data. *J. Hered.* **83**: 157–158.

Mahasandana, S. & Juntakune, P. (1990). The species of green pit viper in Bangkok. *S. E. Asian J. trop. Med. publ. Hlth* **21**: 225–230.

Mao, S. H., Yin, F. Y. & Guo, Y. W. (1984). The hemipenes of common Taiwanese venomous snakes. *Herpetologica* **40**: 406–410.

Maslin, P. (1942). Evidence for the separation of the crotalid genera *Trimeresurus* and *Bothrops*, with a key to the genus *Trimeresurus*. *Copeia* **1942**: 18–24.

Mitrakul, C. (1973). Effects of green pit viper (*Trimeresurus erythrurus* and *Trimeresurus popeorum*) venoms on blood coagulation, platelets and fibrinolytic enzyme systems: studied in vivo and in vitro. *Am. J. clin. Path.* **60**: 654–662.

Mitrakul, C & Impun, C. (1973). The haemorrhagic phenomena associated with green pit viper (*Trimeresurus erythrurus* and *Trimeresurus popeorum*) bites in children. A report of studies to elucidate their pathogenesis. *Clin. Pediat.* **12**: 215–218.

Moritz, C., Schneider, C. J. & Wake, D. B. (1992). Evolutionary relationships within the *Ensatina eschscholtzii* complex confirm the ring species interpretation. *Syst. Biol.* **41**: 273–291.

Nootpand, W. (1971). *Poisonous snakes of Thailand*. Mitrahpadung Kanphim, Bangkok.

Pope, C. H. & Pope, S. H. (1933). A study of the green pit-vipers of south eastern Asia and Malaysia, commonly identified as *Trimeresurus gramineus* (Shaw), with description of a new species from Peninsular India. *Am. Mus. Novit.* No. 620: 1–12.

Quinn, T. W. (1992). The genetic legacy of Mother Goose — phylogeographic patterns of lesser snow goose *Chen caerulescens caerulescens* maternal lineages. *Mol. Ecol.* **1**: 105–117.

Regenass, U. & Kramer, E. (1981). Zur Systematik der grunen Grubenottern der Gattung *Trimeresurus* (Serpentes, Crotalidae). *Rev. suisse Zool.* **88**: 163–205.

Romer, J. D. (1963). Notes on the white-lipped pit viper, *Trimeresurus albolabris* Gray, in Hong Kong. In *Venomous and poisonous animals and noxious plants of the Pacific Region*: 379–384. (Eds Keegan, H. L. & MacFarlane, W. V.). Pergamon Press, Oxford.

Shaw, G. (1802). *General zoology, or systematic natural history*. **3**. *Amphibia*. G. Kearsley, London.

Simon, C., McIntosh, C. & Deniega, J. (1993). Standard restriction fragment length analysis of the mitochondrial genome is not sensitive enough for phylogenetic analysis or identification of 17-year periodical cicada broods (Hemiptera: Cicadidae): the potential for a new technique. *Ann. ent. Soc. Am.* **86**: 228–238.

Smith, H. M. & Chiszar, D. (1988). Occupation of *Ovophis* Burger (Serpentes: Viperidae). *Bull. Md herpet. Soc.* **24**: 23–26.

Smith, M. A. (1943). *The fauna of British India. Reptilia and Amphibia* **3**: *Serpentes*. Taylor and Francis, London.

Stejneger, L. (1927). The green pit viper, *Trimeresurus gramineus*, in China. *Proc. U. S. natn. Mus.* **72**: 1–10.

Swofford, D. L. (1993). *PAUP: phylogenetic analysis using parsimony, Version 3.1*. Illinois Natural History Survey, Champaign, Illinois.

Thorpe, R. S. (1987). Geographic variation: a synthesis of cause, data, pattern and congruence in relation to subspecies, multivariate analysis and phylogenesis. *Boll. Zool.* **54**: 3–11.

Thorpe, R. S., Brown, R. P., Day, M. L., Malhotra, A., McGregor, D. P. & Wüster, W. (1994). Testing ecological and phylogenetic hypotheses in microevolutionary studies: an overview. In *Phylogenetics and ecology*: 189–206. (Eds Eggleston, P. & Vane-Wright R.). Academic Press, London.

Viravan, C., Looareesuwan, S., Kosakarn, W., Wuthiekanun, V., McCarthy, C. J., Stimson, A. F., Bunnag, D., Harinasuta, T. & Warrell, D. G. (1992). A national hospital-based survey of snakes responsible for bites in Thailand. *Trans. R. Soc. trop. Med. Hyg.* **86**: 100–106.

Visudhiphan, S., Tonmukayakul, A., Tumliang, S., Dumavibhat, B. & Piankijagum, A. (1989). Dark green pit viper (*Trimeresurus popeorum*) bite: clinical and serial coagulation profiles in 51 cases. *Am. J. trop. Med. Hyg.* **41**: 570–575.

Vogel, G. (1991). Eine neue *Trimeresurus*-Art aus Thailand, *Trimeresurus venustus* sp. nov. (Reptilia: Serpentes: Crotalidae). *Sauria* **13**: 23–28.

Zhao, E.-M. & Jiang, Y-M. (1977). A survey of reptiles in the Xizang autonomous region, with faunal analysis and descriptions of new forms. *Acta zool. sin.* **23**: 64–71.

10 Reproductive strategies and sperm competition in the adder, *Vipera berus*

CLAES ANDRÉN, GÖRAN NILSON, MATS HÖGGREN and HÅKAN TEGELSTRÖM

Synopsis

Reproduction in temperate snakes shows large variation. In males the reproductive cycle is annual, while females have a biennial or longer cycle depending on species and environmental conditions. Ovulation (fertilization) is temporally separated from copulation, allowing sperm competition in polyandric species. This system can, however, be violated by male-mediated renal sex secretion reducing sperm transfer from later matings by contraction of posterior uteri, or by a copulatory plug. Matings can be restricted to one spring or autumn period or can be biannual. Short-term sperm storage between mating and ovulation is the normal case, but sperm can be stored over winter in species with autumn matings, or several years in species lacking aggregation periods or with low population density. In the adder (*Vipera berus*), precopulatory behaviour in spring includes violent male–male combats and mate guarding, and sex ratio is male-biased because the female reproductive cycle is biennial. During recent years quite different opinions have been expressed concerning adder reproductive behaviour. Main questions have involved the presence or importance of long-term sperm storage, a copulatory plug, multiple paternity in wild populations, male and female choice and whether promiscuity influences offspring viability. We performed DNA fingerprinting of 12 litters of adders in order to find out whether multiple paternity can be determined without collecting fathers and in order to investigate the significance of multiple paternity in wild adders. For each litter we tested the binomial probability that a single father would transmit the observed number of paternal bands in each offspring. In 83% of the litters only one male had sired the offspring, while in the remaining litters more than one father was involved. We conclude that although females may copulate more than once with a single male, multiple paternity is a rare event in the population studied.

Symp. zool. Soc. Lond. (1997) No. 70: 129–141
ISBN 0 19 854986 5

Introduction

Reproductive strategies in snakes

Reproductive tactics in snakes vary both among and within taxa. Length of the male reproductive cycle in temperate regions is normally one year (Volsoe 1944; Fitch 1970; Shine 1977a; Nilson 1980; Saint Girons 1982; Aldridge, Greenhaw & Plummer 1990). In females it can be one, two or several years depending on the mode of reproduction (i.e. viviparity or oviparity), body size, quality and quantity of food, altitude and latitude affecting the length of the activity period and annual climatic variation (Fitch 1970; Prestt 1971; Shine 1977b; Aldridge 1979; Seigel & Ford 1987). An important reason for the prolonged female reproductive cycle in temperate regions is the time needed for feeding and accumulating energy to replenish the fatbody reserves which later can be transformed to eggs (Aldridge 1979; Nilson 1981b; Seigel & Fitch 1985). During periods when food is scarce or environmental conditions are extreme, reproduction may not be possible at all (Bull & Shine 1979; Andrén 1982a; Andrén & Nilson 1983; Macartney & Gregory 1988; Brown 1991).

Timing of reproduction is controlled by environmental and genetic factors, but the period of female attractiveness determines more directly the start and termination of the mating period (Shine 1977a; Garstka & Crews 1982; King & Duvall 1990; Schuett 1992). In most species, females attract males during a short oestrous period by releasing a combination of volatile long-distance air-borne pheromones and more heavy chemicals which stick to the body and leave odour trails on the ground. These sex pheromones are transported with the blood and reach the surface of the skin on the back and sides of the female (Noble 1937; Garstka & Crews 1981; Ford 1982; Andrén 1986). The male sexual behaviour is mediated by sex attractant pheromones present on the female skin and unless this chemical information is transferred via the male tongue tips to his vomeronasal organs courtship and mating will not occur (Noble 1937; Kubie, Vagvolgyi & Halpern 1978; Andrén 1982b; Mason, Fales *et al.* 1989; Mason, Jones *et al.* 1990).

In some species the oestrous period can be prolonged and mating may occur twice a year, during an autumn and a spring mating period (Mendonca 1987; Schuett 1992). Sex pheromones are also important in the communication between males (Noble 1937; Kubie *et al.* 1978; Andrén 1982b; Mason, Fales *et al.* 1989; Mason, Jones *et al.* 1990), and a male with experimentally blocked tongue tips will not receive chemical information and will therefore not participate in reproductive activities (Andrén 1982b). Precopulatory male–male competitive behaviour can be complex and often includes violent combats (Bernström 1943; Carpenter & Ferguson 1977; Gillingham 1980; Andrén 1986; Madsen 1988). This is the case in adders and the sensational fights may attract predators as well as expending excessive energy (Andrén 1985).

In snakes, mating is temporally separated from fertilization, which takes place during ovulation. For species with only spring mating, the period between copulation and fertilization is usually a few weeks (Klauber 1972; Saint Girons 1979; Nilson 1981b). In other species, with autumn or biannual mating periods, the ovulation takes place about 10 months after the first mating period (Fitch 1960; King & Duvall

1990; Brown 1991; Schuett 1992). In yet other species, mating and fertilization may be separated by several years and this extreme long-term sperm storage may be an insurance for fertilization in subsequent years when the probability of meeting a new male is low (Schuett 1992). In some species mating is a necessary trigger for ovulation to take place (Volsoe 1944; Nilson 1981b; Luiselli 1993).

Sperm storage occurs both in males and females but can vary considerably in length, depending on species and reproductive mode. Storage and maintenance of sperm in the male accessory organs occurs, but to what extent the renal sex segment can supply stored sperm with nutrient, or affect its motility, is unknown (Tsui & Licht 1974; Saint Girons 1975; Nilson 1981b; Halpert, Gartska & Crews 1982; Rissman & Crews 1989; Schuett 1992). In females, sperm can be stored for very long periods in receptacles in the infundibulum and also in other parts of the oviduct (Fox 1956, 1977; Nilson & Andrén 1982; Saint Girons 1982; Schuett & Gillingham 1986; Gist & Jones 1987, 1989).

A male that mates with different females will increase his reproductive success if a greater number of copulations results in a greater number of offspring sired by him. Male promiscuity has been observed in several temperate snake species (Andrén 1986; Madsen *et al.* 1992, 1993; Duvall, Arnold & Schuett 1992; Duvall, Schuett & Arnold 1993). As the female reproductive cycle is usually longer than the male, the effective reproductive population is biased in favour of males, increasing the potential for male–male competition. Besides precopulatory combats and establishment of mating hierarchies (Carpenter & Ferguson 1977; Gillingham, Carpenter & Murphy 1983; Andrén 1986), different methods to prevent sperm competition occur including a copulatory plug (Devine 1975, 1977, 1984; Ross & Crews 1977, 1978), male-induced posterior oviduct constriction (Nilson & Andrén 1982), or chemically induced deterioration of the previous males' sperm (Halpert *et al.* 1982). Some species exhibit postcopulatory mate guarding (Andrén 1986; Madsen 1988).

Why do female snakes mate with several different males? The idea that a female may choose the father(s) of her offspring is controversial and female paternity control can occur at different stages of the breeding cycle (Birkhead & Möller 1993). Multiple matings by females may represent a conflict of interest between the sexes (Parker 1970, 1984). As a male ejaculate normally is more than sufficient to fertilize all ova (Parker 1970, 1984; Madsen *et al.* 1992; Duvall, Schuett *et al.* 1993), why should a female expose herself during repeated matings, as has been observed in at least a few snake populations (Barry, Weatherhead & Phillip 1992; Madsen *et al.* 1992; Duvall, Schuett *et al.* 1993), and increase the risk of predation (Ross & Crews 1977)? It has been suggested that in highly inbred populations offspring viability will be increased by multiple matings (Loman, Madsen & Håkansson 1988; Madsen *et al.* 1992; but see Luiselli 1993; Capula & Luiselli 1994, in press; and later in this paper).

The variation in reproductive modes gives rise to the potential for differences in female capacity to store sperm and it also allows the possibility of sperm competition, both within and between seasons. Males may have several means of reducing sperm competition and a female may, at least in theory, be able to counteract these efforts if it serves her interest to let her eggs be fertilized by a special male or by more than one male.

Adder reproductive biology

The general biology of the adder has long been known from field investigations and laboratory work. Pioneering works include those by Bernström (1943, 1958), Volsoe (1944), Saint Girons (1952), Viitanen (1967) and Prestt (1971). The general description of adder reproductive biology presented below is based on long-term field studies and outdoor enclosure experiments (Nilson 1980; Andrén 1981; Nilson 1981a,b; Andrén 1982a,b; Nilson & Andrén 1982; Andrén & Nilson 1983; Andrén 1985, 1986).

Males leave hibernation three weeks before females, at the end of March. They bask intensively to speed up the maturing process of spermatozoa. The reproductive period is initiated by a synchronized male sloughing in mid or late April, which causes a drastic change in behaviour and activity and the development of a pronounced sexual dichromatism. Female sloughing occurs later and is not synchronized. Females in oestrus are sedentary and usually solitary and attract males by pheromones at a distance of at least 500 m. Non-reproductive females leave the hibernation area before the mating period and disperse into feeding areas.

During the few weeks of the mating period, males are very vagile. Operational sex ratio is male-biased and several males often try to court each oestrous female. Early in the mating period intense courting and violent combats between defending and intruding males are common. After a few days, at most a week, all females in oestrus are guarded by superior males. Copulations are usually not observed until then. The same pattern has been observed in outdoor enclosures where large males won most combats and copulated more frequently than smaller males. After copulation, the male guards the female, for from a few hours up to a day. After one, or a few, copulations with the same male, the female usually leaves the arena. Female behaviour changes after the first copulation and they expose themselves less.

Female reproductive costs are high in the sense of mass loss, which is on average 55% of the preparturition mass. After parturition in August, the female is emaciated and needs at least one year to replenish fatbody capacity to produce a new litter. During longer periods of food shortage, both male and female skip the opportunity to mate and the reproductive cycles are prolonged in both sexes.

Copulatory plug function and sperm competition

In outdoor enclosure experiments we found that males transfer a renal sex secretion together with the sperm. Within a few hours after the sperm and renal sex secretion had entered the female cloaca, a sphincter muscle in the posterior part of the uteri started to contract. This resulted in a reduced and compressed folded lumen and the posterior uteri became stiff and hard. This effect lasted for the whole mating period of about two weeks (Andrén & Nilson 1987) and did not, as suggested by Stille, Madsen & Niklasson (1986), relax after a short period to allow a new mating. We also examined 45 females before, during and after copulation and found that newly mated females had long, slightly convoluted and hard posterior uteri, while non-reproductive females and those that had not yet mated had weakly contracted or flaccid uteri. Oestrous females artificially supplied with male renal sex secretion showed a significant decrease in acceptance of male courtship (Nilson & Andrén 1982).

Stille, Madsen *et al.* (1986) performed indoor terrarium breeding experiments with adders, and determined paternity from the distribution of two blood allozyme alleles among the offspring. Bipaternity was found in both single- and double-mated females. The authors concluded that, in single-mated females, long-term sperm storage explained the mixed paternity in the offspring. In double-mated females with mixed offspring, sperm storage was not considered as a possible explanation. Instead double mating within the same season was considered as the only explanation. In all double-mated females there was a first-male effect and double mating did not always result in shared paternity. This effect was more pronounced in a complementary experiment (Stille, Niklasson & Madsen 1987). The first male either was the only father or fathered more than 50% of the litters. The authors concluded that the female adders store sperm between seasons. As female adders have a biennial reproductive cycle and mate only in spring, long-term sperm storage must be over a period of at least two years.

The reproductive biology of the adder has also been studied in the Italian Alps (e.g. Capula, Luiselli & Anitaldi 1992; Luiselli 1992). Luiselli (1993) performed field experiments to test whether the hypothesis about long-term sperm storage was correct and whether within-season multiple matings were frequent in free-ranging adders. In the first study 47 potentially reproductive females were caught before the onset of mating and kept isolated from males until the end of the reproductive period. Therefore, no female had the opportunity to copulate, and no female ovulated, or gave birth to any young. Luiselli (1993) concluded that spring mating seems to be a necessary trigger for ovulation to take place in the adder, as suggested also by Volsoe (1944) and Nilson (1981b). In the second study (Luiselli 1993), the occurrence of within-season multiple matings was investigated in a free-ranging population of adders. During three successive years altogether 22 females were carefully watched for the number of matings. He found that 18 females (88%) copulated with only one male and only once. Four females (12%) copulated more than once, but usually with the same male. Thus, matings with different males in this particular free-ranging population did occur, but were a rare phenomenon.

The breeding experiments by Stille, Madsen *et al.* (1986) to determine multiple paternity in adders were based on the distribution of two blood allozyme alleles among offspring. The design assumed both multiple paternity and long-term sperm storage. However, as long-term sperm storage is not likely to exist in the adder (see below) the results are difficult to evaluate. In a recent study (Höggren & Tegelström 1995), breeding experiments were performed in order to investigate whether multiple matings result in multiple paternity. DNA fingerprinting of mothers, their offspring and possible fathers showed that multiple matings gave rise to multiple paternity in all broods investigated. Associated males in the experiment were the genetical fathers of all offspring from females brought to the enclosure before onset of the mating period. Thus, long-term sperm storage from earlier mating seasons could not be confirmed. During these experimental conditions the copulatory plug described by

Nilson & Andrén (1982) did not prevent multiple paternity following multiple matings within a short period of time.

In a later study (M. Höggren in prep.), controlled breeding experiments followed by DNA fingerprinting were used to determine paternity among all offspring of 15 litters. None of the young of 11 females that had mated in a previous season was sired by sperm from an earlier year. Obligatory over-winter sperm storage found in species with autumn matings, as in many rattlesnakes (Schuett 1992), is not relevant to the adder which has only spring mating and a female biennial reproductive cycle. This and earlier empirical data, together with results from Nilson & Andrén (1982) and Luiselli (1993), support the hypothesis that long-term sperm storage does not occur in the adder.

In a small and inbred population of adders Madsen *et al.* (1992) found that females mated several times and often with different males. The females copulated on average 3.7 times. The authors hypothesized that females intentionally mated with many different males to increase offspring viability. Multiple mating should increase the potential for intra-uterine competition among sperm from different males and Madsen *et al.* (1992) suggested that offspring viability depended on a female's choice of mates, as well as the number of times she copulated. The authors also believed that this could be a general female reproductive strategy involving other populations of adders. However, from breeding experiments Madsen and coworkers found a first-male effect (Stille, Madsen *et al.* 1986; Stille, Niklasson *et al.* 1987, but see Madsen & Shine 1994) which may indicate a conflict between the sexes.

In order to test the hypothesis that multiple matings increase offspring viability, Capula & Luiselli (1994) made an enclosure experiment where 10 female adders were mated with a single male and another 10 females were mated with three to eight different males. No significant difference in offspring viability was detected between the two groups of litters. A similar offspring viability test with the Aesculapian snake (*Elaphe longissima*) (Capula & Luiselli in press) gave the same results, i.e. multiple matings with different males did not increase offspring viability.

Paternity in free-ranging adders

DNA techniques now allow molecular ecologists to challenge many exciting questions in reproductive biology (Burke 1989; Avise 1994). We performed DNA fingerprinting of adders from an insular population at the Swedish west coast, in order to estimate relatedness among breeding individuals. We aimed to find out whether multiple paternity can be determined without collecting the fathers and to investigate the significance of multiple paternity in wild adders.

Methods

Study area and population

The study area is a 54-ha island, Stora Hästholmen, 20 km north-west of Gothenburg at the Swedish west coast. Low and flat parts are vegetated by grass and herbs and in wet areas the grass is highly tussocked forming a network of tunnels covered by the previous year's withered grass. At higher elevations there is a mosaic of bare rock and patches of heather and small bushes. South-facing slopes with overgrown stone piles serve as hibernacula in winter. The predominant prey species is the field vole (*Microtus agrestis*), but occasional field mice (*Apodemus flavicollis*), common shrews (*Sorex araneus*) and moor frogs (*Rana arvalis*) may be taken as food. Possible predators are the hooded crow (*Corvus corone*), the raven (*Corvus corax*), the rough-legged buzzard (*Buteo lagopus*), the European mink (*Lutreola lutreola*) and the fox (*Vulpes vulpes*).

The adder population size varies dramatically on the island in relation to prey availability and has been estimated to range from fewer than 10 to about 200 adults. The material included in this study was collected during 1991 and 1992 at an estimated population size of 50 adults, equivalent to one adult per hectare. Sex ratio in adults was close to 1 : 1, but the effective population taking part in reproduction was male-biased (about 2 : 1). Shortly before parturition 12 gravid females were collected and brought to the laboratory and kept in small individual terraria until the birth of young. Blood samples were taken from all females and their offspring and stored at -70°C for later analyses.

Genetic analyses

Laboratory procedures and data analyses largely follow methods as described by Höggren & Tegelström (1995), with the extension of two criteria for detecting multiple paternity in the absence of putative fathers. First, the number of maternal and paternal bands were compared in each litter. The assumption is that if the observed total number of paternal bands within a litter significantly exceeds the total number of maternal bands, more than one father is involved. Second, we compared the average number of paternal bands expected to be transmitted from a single father. As a vast majority of the observed bands are heterozygous and segregate independently (Höggren & Tegelström 1995), we expect each young to receive, on average, 50% of its father's bands. If the number of paternal bands per young is significantly lower than 50% of the paternal bands in that particular brood, at least two fathers are expected to be involved.

In order to estimate the overall similarity in our population we calculated the proportion of shared bands between the adult females (Similarity index, $S = 2N_{AB}/(N_A + N_B)$ where N_A and N_B are the numbers of bands present in individuals A and B respectively, and N_{AB} is the number of shared bands). The mean similarity index (45 pairwise comparisons) was 0.28 ± 0.15 (range = 0.00–0.63), i.e. two randomly chosen individuals share, on average, 28% of their DNA bands. As the actual father's complete DNA profile was not available in our analyses, fragments shared between the mother and father(s) were automatically interpreted as maternal. However, the similarity estimate (28%) allows us to compensate for the biased observed ratio of maternal- and paternal-derived bands in the offspring. For each mother/litter we treated 28% of the observed 'maternal' bands as non-informative bands actually shared between the parents. Accordingly, we obtained an adjusted ratio of maternal- and paternal-specific bands.

Results

The 12 females gave birth to a total of 107 young ($\bar{x}$ = 8.9, range 6–14). Females possessed, on average, 8.8 heterozygous bands (range 6–15), while the corresponding figure for males, assuming one father for each litter, was estimated at 9.0. Mean number of bands in the offspring was 8.4. The applied combination of restriction enzyme (Alu I) and DNA probe $(TG)_n$ (Fig.1) has proved to give an optimal resolution in the adder (Höggren & Tegelström 1995).

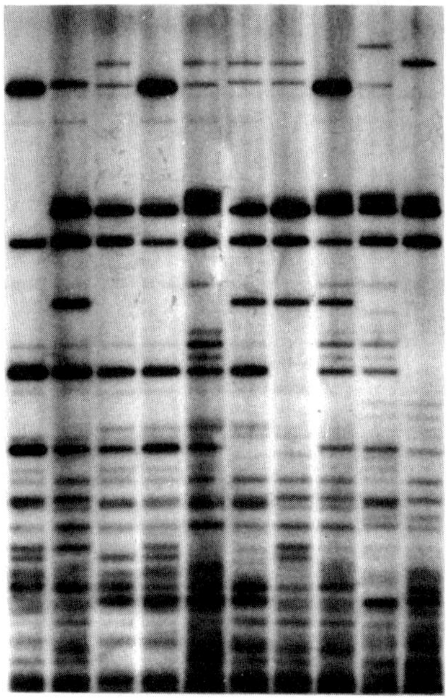

Fig. 1. Example of a DNA fingerprint of a female adder (F) with nine offspring sired in the wild by a single male. Fragments were generated from Alu-I-digested genomic DNA hybridized with a synthetic $(TG)_n$ polydinucleotide.

For each litter we tested the binomial probability that a single father would transmit the observed number of paternal bands in each offspring, assuming independent segregation of the heterozygous bands. In two of the 12 litters, observed band numbers deviated significantly (99% confidence limit) from the expected numbers. Here, each offspring exhibited significantly fewer paternal bands than would have been expected with regard to the total number of paternal bands in those two litters. The observations strongly support the possibility that more than one male had sired each of these litters. The 10 remaining litters did not deviate from expected figures, which means that multiple paternity could not be confirmed in those cases. Thus, multiple paternity was demonstrated in 17% of the litters examined. As only one multi-locus DNA probe was used, at this stage, we did not achieve enough resolution to establish the exact number of fathers that sired the mixed broods.

Discussion

We have demonstrated elsewhere, using different approaches, that long-term sperm storage, between seasons, is not likely to occur in the adder (Nilson 1981b; M. Höggren & H. Tegelström in prep.) and hence we conclude that paternity revealed in the present study was the result of within-season mating(s). In the present free-ranging adder population, being relatively dense and male-biased, mixed paternity has been shown to exist, but it does not seem to be a prevailing phenomenon. This is also in agreement with our outdoor enclosure and field observations of mating behaviour (Nilson & Andrén 1982; Andrén 1986). Although females might copulate more than once with a single male, multiple matings with different partners were rare.

From the DNA fingerprint results we cannot exclude the possibility that the mated female moved to another place and mated with a new male. However, if this was the case, the second mating was not successful, maybe as a result of the copulatory plug function described earlier. In outdoor enclosure experiments on sperm storage and paternity, repeated matings with different males also gave rise to mixed paternity. The length of time required for the copulatory plug to become effective needs to be investigated and related to the time interval between matings and to observed post-copulatory mate-guarding in males (Andrén 1986; Madsen 1988). It would also be interesting to investigate the length of time in the mating season that the copulatory plug remains effective. Copulatory plugs produced by males have been observed also in species of *Nerodia* and *Thamnophis*, but they do not seem to be effective, as multiple paternity is common in wild populations in several of these species (Schwartz, McCracken & Burghardt 1989; Barry *et al.* 1992).

Even though the reproductive biology has been more intensively studied in the adder than in any other snake, there are still many problems to be solved. Our results on long-term sperm storage and female promiscuity are largely in agreement with studies made in the Italian Alps (Luiselli 1993; Capula & Luiselli 1994), but are quite incongruent with observations made in an inbred south Swedish population, where female adders mate several times and often with different males (Madsen *et al.* 1992). However, preliminary fingerprint data indicate that the occurrence and frequency of mixed paternity may vary between different adder populations (M. Höggren *et al.* in prep.), and at this stage it is not possible to draw general conclusions on the significance of female multiple matings and multiple paternity in wild adders. Possible variance between populations may reflect spatial differences in genetic composition, density and operational ratios on the mating scene.

Acknowledgements

The Royal Society of Arts and Sciences in Gothenburg, the Swedish Natural Science Research Council, the Nilsson-Ehle Foundation and the Erik Philip-Sörensen Foundation supported the research.

References

Aldridge, R. D. (1979). Female reproductive cycles of the snakes *Arizona elegans* and *Crotalus viridis*. *Herpetologica* **35**: 256–261.

Aldridge, R. D., Greenhaw, J. J. & Plummer, M. V. (1990). The male reproductive cycle of the rough green snake (*Opheodrys aestivus*). *Amphibia-Reptilia* **11**: 165–172.

Andrén, C. (1981). *Behaviour and population dynamics in the adder* Vipera berus (*L.*). PhD thesis: Gothenburg University.

Andrén, C. (1982a). Effect of prey density on reproduction, foraging and other activities in the adder, *Vipera berus*. *Amphibia-Reptilia* **3**: 81–96.

Andrén, C. (1982b). The role of the vomeronasal organs in the reproductive behavior of the adder *Vipera berus*. *Copeia* **1982**: 148–157.

Andrén, C. (1985). Risk of predation in male and female adders, *Vipera berus* (Linné). *Amphibia-Reptilia* **6**: 203–206.

Andrén, C. (1986). Courtship, mating and agonistic behaviour in a free-living population of adders, *Vipera berus* (L.). *Amphibia-Reptilia* **7**: 353–383.

Andrén, C. & Nilson, G. (1983). Reproductive tactics in an island population of adders, *Vipera berus* (L.), with a fluctuating food resource. *Amphibia-Reptilia* **4**: 63–79.

Andrén, C. & Nilson, G. (1987). The copulatory plug of the adder, *Vipera berus*: does it keep sperm in or out? *Oikos* **49**: 230–232.

Avise, J. C. (1994). *Molecular markers, natural history and evolution*. Chapman & Hall, New York.

Barry, F. E., Weatherhead, P. J. & Phillip, D. P. (1992). Multiple paternity in a wild population of northern water snakes *Nerodia sipedon*. *Behav. Ecol. Sociobiol.* **30**: 193–199.

Bernström, J. (1943). Till kännedom om huggormen *Vipera berus berus* (Linné). *Göteborgs K. Vetensk. o. VitterhSamh. Handl.* (6) **2B**(10): 3–34. (In Swedish.)

Bernström, J. (1958). Huggormarna vid Tidö. In *Natur i Västmanland*: 193–206. (Eds Waldén, I. B. & Curry-Lindahl, K.). Almqvist & Wiksell, Uppsala. (In Swedish.)

Birkhead, T. R. & Möller, A. P. (1993). Female control of paternity. *Trends Ecol. Evol.* **8**: 100–104.

Brown, W. S. (1991). Female reproductive ecology in a northern population of the timber rattlesnake, *Crotalus horridus*. *Herpetologica* **47**: 101–115.

Bull, J. J. & Shine, R. (1979). Iteroparous animals that skip opportunities for reproduction. *Am. Nat.* **114**: 296–303.

Burke, T. (1989). DNA fingerprinting and other methods for the study of mating success. *Trends Ecol. Evol.* **4**: 139–144.

Capula, M. & Luiselli, L. (1994). Can female adders multiply? *Nature, Lond.* **369**: 528.

Capula, M. & Luiselli, L. (In press). Does promiscuity influence offspring viability in a population of snakes (*Elaphe longissima*). *Anim. Behav.*

Capula, M., Luiselli, L. & Anibaldi, C. (1992). Complementary study on the reproductive biology in female adder, *Vipera berus* (L.), from eastern Italian Alps. *Vie Milieu* **42**: 327–336.

Carpenter, C. C. & Ferguson, G. W. (1977). Variation and evolution of stereotyped behavior in reptiles. In *Biology of the Reptilia* **7**: 335–554. (Eds Gans, C. & Tinkle, D. W.). Academic Press, London.

Devine, M. C. (1975). Copulatory plugs in snakes: enforced chastity. *Science* **184**: 844–845.

Devine, M. C. (1977). Copulatory plugs, restricted mating opportunities and reproductive competition among male garter snakes. *Nature, Lond.* **267**: 345–346.

Devine, M. C. (1984). Potential for sperm competition in reptiles: behavioral and physiological consequences. In *Sperm competition and the evolution of animal mating systems*: 509–521. (Ed. Smith, R. L.). Academic Press, Orlando, San Diego etc.

Duvall, D., Arnold, S. J. & Schuett, G. W. (1992). Pitviper mating systems: ecological potential, sexual selection and microevolution. In *Biology of the pitvipers*: 321–336. (Eds Campbell, J. A. & Brodie, E. D.). Selva, Tyler, Texas.

Duvall, D., Schuett, G. W. & Arnold, S. J. (1993). Ecology and evolution of snake mating systems. In *Snakes, ecology and behaviour*: 165–200. (Eds Seigel, R. A. & Collins, J. T.). McGraw-Hill, New York.

Fitch, H. S. (1960). Autecology of the copperhead. *Univ. Kans. Publs Mus. nat. Hist.* **13**: 85–288.

Fitch, H. S. (1970). Reproductive cycles of lizards and snakes. *Misc. Publs. Mus. nat. Hist. Univ. Kans.* **52**: 1–247.

Ford, N. B. (1982). Species specificity of sex pheromone trails of sympatric and allopatric garter snakes (*Thamnophis*). *Copeia* **1982**: 10–13.

Fox, W. (1956). Seminal receptacles of snakes. *Anat. Rec.* **124**: 519–540.

Fox, W. (1977). The urinogenital systems of reptiles. In *Biology of the Reptilia* **6**: 1–157. (Eds Gans, C. & Parsons, T. S.). Academic Press, London & New York.

Garstka, W. & Crews, D. (1981). Female sex pheromone in the skin and circulation of a garter snake. *Science* **214**: 681–683.

Garstka, W. & Crews, D. (1982). Female control of male reproductive function in a Mexican snake. *Science* **217**: 1159–1160.

Gillingham, J. C. (1980). Communication and combat behavior of the black rat snake (*Elaphe obsoleta*). *Herpetologica* **36**: 120–127.

Gillingham, J. C., Carpenter, C. C. & Murphy, J. B. (1983). Courtship, male combat and dominance in the western diamondback rattlesnake, *Crotalus atrox*. *J. Herpet* **17**: 265–270.

Gist, D. H. & Jones, J. M. (1987). Storage of sperm in the reptilian oviduct. *Scanning Microsc.* **1**: 1839–1849.

Gist, D. H. & Jones, J. M. (1989). Sperm storage within the oviduct of turtles. *J. Morph.* **199**: 379–384.

Halpert, A. P., Garstka, W. R. & Crews, D. (1982). Sperm transport and storage and its relation to the annual sexual cycle of the female red-sided garter snake, *Thamnophis sirtalis parietalis*. *J. Morph.* **174**: 149–159.

Höggren, M. & Tegelström, H. (1995). DNA fingerprinting shows within-season multiple paternity in the adder (*Vipera berus*). *Copeia* **1995**: 271–277.

King, M. B. & Duvall, D. (1990). Prairie rattlesnake seasonal migrations: episodes of movements, vernal foraging and sex differences. *Anim. Behav.* **39**: 924–935.

Klauber, L. M. (1972). *Rattlesnakes: their habits, life histories, and influence on mankind* (2nd edn) **2**. University of California Press, Los Angeles and Berkeley.

Kubie, J. L., Vagvolgyi, A. & Halpern, M. (1978). Roles of the vomeronasal and olfactory systems in courtship behaviour of male garter snakes. *J. comp. physiol. Psychol.* **92**: 627–641.

Loman, J., Madsen, T. & Håkansson, T. (1988). Increased fitness from multiple matings, and genetic heterogeneity: a model of a possible mechanism. *Oikos* **52**: 69–72.

Luiselli, L. (1992). Reproductive success in melanistic adders: a new hypothesis and some considerations on Andrén and Nilson's (1981) suggestions. *Oikos* **64**: 601–604.

Luiselli, L. (1993). Are sperm storage and within-season multiple mating important components of the adder reproductive biology? *Acta oecol.* **14**: 705–710.

Madsen, T. (1988). Reproductive success, mortality and sexual size dimorphism in the adder, *Vipera berus*. *Holarct. Ecol.* **11**: 77–80.

Madsen, T. & Shine, R. (1994). Costs of reproduction influence the evolution of sexual size dimorphism in snakes. *Evolution* **48**: 1389–1397.

Madsen, T., Shine, R., Loman, J. & Håkansson, T. (1992). Why do female adders copulate so frequently? *Nature, Lond.* **355**: 440–441.

Madsen, T., Shine, R., Loman, J. & Håkansson, T. (1993). Determinants of mating success in male adders, *Vipera berus*. *Anim. Behav.* **45**: 491–499.

Mason, R. T., Fales, H. M., Jones, T. H., Pannell, L. K., Chinn, J. W. & Crews, D. (1989). Sex pheromones in snakes. *Science* **245**: 290–293.

Mason, R. T., Jones, T. H., Fales, H. M., Pannell, L. K. & Crews, D. (1990). Characterization, synthesis, and behavioral responses to sex attractiveness pheromones of red-sided garter snakes (*Thamnophis sirtalis parietalis*). *J. chem. Ecol.* **16**: 2353–2369.

Macartney, J. M. & Gregory, P. T. (1988). Reproductive biology of female rattlesnakes (*Crotalus viridis*) in British Columbia. *Copeia* **1988**: 47–57.

Mendonca, M. T. (1987). Timing of reproductive behaviour in male musk turtles, *Sternotherus odoratus*: effects of photoperiod, temperature and testosterone. *Anim. Behav.* **35**: 1002–1014.

Nilson, G. (1980). Male reproductive cycle of the European adder, *Vipera berus*, and its relation to annual activity periods. *Copeia* **1980**: 729–737.

Nilson, G. (1981a). *Endocrine and ecological interrelations in reproduction of the adder* Vipera berus *(L.)*. PhD thesis: University of Göteborg.

Nilson, G. (1981b). Ovarian cycle and reproductive dynamics in the female adder, *Vipera berus* (Reptilia: Viperidae). *Amphibia-Reptilia* **2**: 63–82.

Nilson, G. & Andrén, C. (1982). Function of renal sex secretion and male hierarchy in the adder, *Vipera berus*, during reproduction. *Horm. Behav.* **16**: 404–413.

Noble, G. K. (1937). The sense organs involved in the courtship of *Storeria, Thamnophis*, and other snakes. *Bull. Am. Mus. nat. Hist.* 73: 673–725.

Parker, G. A. (1970). Sperm competition and its evolutionary consequences in the insects. *Biol. Rev.* **45**: 525–567.

Parker, G. A. (1984). Sperm competition and the evolution of animal mating systems. In *Sperm competition and the evolution of animal mating systems*: 1–60. (Ed. Smith, R. L.). Academic Press, Orlando, San Diego etc.

Prestt, I. (1971). An ecological study of the viper *Vipera berus* in southern Britain. *J. Zool., Lond.* **164**: 373–418.

Rissman, E. F. & Crews, D. (1989). Effect of castration on epididymal sperm storage in male musk shrews (*Suncus murinus*) and mice (*Mus musculus*). *J. Reprod. Fert.* **86**: 219–222.

Ross, P. & Crews, D. (1977). Influence of the seminal plug on mating behaviour in the garter snake. *Nature, Lond.* **267**: 344–345.

Ross, P. & Crews, D. (1978). Stimuli influencing mating behavior in the garter snake, *Thamnophis radix*. *Behav. Ecol. Sociobiol.* **4**: 133–142.

Saint Girons, H. (1952). Ecologie et éthologie des vipères de France. *Ann. Sci. nat. (Zool.)* (11) **14**: 263–341.

Saint Girons, H. (1975). Sperm survival and transport in the female genital tract of reptiles. In *The biology of spermatozoa*: 105–113. (Eds Hafez, E. S. E. & Thibault, C. G.). S. Karger, Basel.

Saint Girons, H. (1979). Les cycles alimentaires des vipères européennes dans des conditions seminaturelles. *Annls Biol. anim. Biochim. Biophys.* **19**: 125–134.

Saint Girons, H. (1982). Reproductive cycles of male snakes and their relationships with climate and female reproductive cycles. *Herpetologica* **38**: 5–16.

Schuett, G. W. (1992). Is long-term sperm storage an important component of the reproductive biology of temperate pitvipers? In *Biology of the pitvipers*: 169–184. (Eds Campbell, J. A. & Brodie, E. D.). Selva, Tyler, Texas.

Schuett, G. W. & Gillingham, J. C. (1986). Sperm storage and multiple paternity in the copperhead *Agkistrodon contortrix*. *Copeia* **1986**: 807–811.

Schwartz, J. M., McCracken, G. F. & Burghardt, G. M. (1989). Multiple paternity in wild populations of the garter snake, *Thamnophis sirtalis*. *Behav. Ecol. Sociobiol* **25**: 269–273.

Seigel, R. A. & Fitch, H. S. (1985). Annual variation in reproduction in snakes in a fluctuating environment. *J. Anim. Ecol.* **54**: 497–505.

Seigel, R. A. & Ford, N. B. (1987). Reproductive ecology. In *Snakes: ecology and evolutionary biology*: 210–252. (Eds Seigel, R. A., Collins, J. T. & Novak, S. S.). Macmillan Publishing Co., New York.

Shine, R. (1977a). Reproduction in Australian elapid snakes. I. Testicular cycles and mating seasons. *Aust. J. Zool.* **25**: 643–657.

Shine, R. (1977b). Reproduction in Australian elapid snakes. II. Female reproductive cycles. *Aust. J. Zool.* **25**: 655–666.

Stille, B., Madsen, T. & Niklasson, M. (1986). Multiple paternity in the adder, *Vipera berus*. *Oikos* **47**: 173–175.

Stille, B., Niklasson, M. & Madsen, T. (1987). Within season multiple paternity in the adder, *Vipera berus*, a reply. *Oikos* **49**: 232–233.

Tsui, H. W. & Licht, P. (1974). Pituitary independence of sperm storage in male snakes. *Gen. comp. Endocr.* **22**: 277–279.

Viitanen, P. (1967). Hibernation and seasonal movements of the the viper, *Vipera berus berus* (L.), in southern Finland. *Ann. zool. fenn.* **4**: 472–546.

Volsoe, H. (1944). Structure and seasonal variation of the male reproductive organs of *Vipera berus* (L.). *Spolia zool. Mus. haun.* **5**: 1–157.

11 Growth and survival of *Vipera berus* in a variable environment

ANDERS FORSMAN

Synopsis

Data from a capture–mark–recapture study (1986–1994) of a population of adders *Vipera berus* inhabiting a group of islands off the Swedish east coast in the Baltic Sea are used to examine whether changing food availability and weather conditions influence growth and survival of large relative to small individuals. Mean air temperature and population density of the field vole *Microtus agrestis*, the primary prey for adders, were used as measures of weather conditions and food availability, respectively.

The relationship between growth rate and body size varied significantly among years, but was parallel in males and females. Large and small snakes responded differently to changes in vole densities; growth rates decreased at a slower rate with increasing body size when voles were abundant than when voles were scarce. There was no effect of changing weather conditions on adder growth rates.

Logistic regression analyses of recapture data revealed that survival probability of adders increased with increasing body size in some years and decreased with increasing body size in other years. The survival advantage of large body size tended to increase with increasing food availability, but was not affected by changing weather conditions. Within any given year, the relationship between survival and body size was similar in the two sexes.

Years characterized by rapid individual growth rates were associated with a subsequent reduced survival probability of large relative to small snakes, a pattern that should prevent consistent shifts in body size through time. Although an observed lack of variation in mean body size of female adders across years corroborates this prediction, mean body size of male adders increased significantly during the nine years of the study.

Introduction

Several previous studies of snakes suggest that many important life-history components, such as reproductive expenditure, growth rate and survival, vary among populations, as well as through time within populations, in parallel with food availability (e.g Andrén 1982; Andrén & Nilson 1983; Platt 1984; Ford & Seigel 1989; Forsman & Lindell 1991, in prep.; Madsen & Shine 1993). Presumably, ambient temperature variation can also be an important source of variation, because snakes

Symp. zool. Soc. Lond. (1997) No. 70: 143–154
ISBN 0 19 854986 5

rely upon external heat sources to increase their own body temperature, and because both behavioural and physiological capacities are temperature-dependent (e.g. Huey 1982; Peterson, Gibson & Dorcas 1993). Consequently, adverse weather conditions may counter the positive effects of high prey densities by restricting activity levels and body temperatures of snakes (Andrews 1982). However, little is known about the relative importance of food and temperature in determining performance such as growth and survival in snakes, and how different environmental conditions influence the relative performance of different phenotypes. For instance, it is possible that periods with high resource levels may disproportionately favour larger individuals because they can handle more food per unit time, relative to their energetic requirement, than smaller individuals can (Sibly & Calow 1986). Also, adverse weather conditions may have a more severe impact on small than on large individuals, owing to thermoregulatory constraints (e.g. Slip & Shine 1988; Willmer 1991).

Here, data from a nine-year capture–mark–recapture study of a population of the adder *Vipera berus* (L.) are used to examine what effects temporal variation in food availability and environmental temperature may have on growth and survival of large relative to small individuals.

Materials and methods

Life history

Vipera berus is a small, viviparous, venomous snake species widely distributed through Europe and northern Asia. On islands in the Baltic, where this study was conducted, adders enter hibernation in late September or October and emerge in late March or April. During the mating period, beginning about four weeks after emergence, males move large distances and fight with other males for access to receptive females (e.g. Andrén 1986; Madsen 1988). Large males often win these combats and as a result obtain more matings than small males do (Madsen 1988). Females reproduce at the most every second year and give birth in late August to between five and 20 young, depending on maternal body size. Snout–vent length (henceforth SVL) of newly born young is 150–200 mm. Reproduction commences at three or four years of age when SVL is 400–500 mm (Prestt 1971; Madsen 1988). Growth continues throughout life as a decreasing function of body size, and females grow faster and attain a larger body size than males (Prestt 1971; Madsen 1988; Forsman 1991a). Average adult body sizes vary among populations, but maximum individual SVL seldom exceeds 750 mm (Forsman 1991a).

Study site and data gathering

Growth and survival were studied in a population of adders inhabiting a group of islands, Ängskär, off the Swedish east coast in the Baltic Sea (59°32′N 19°18′E, see Forsman 1991a,b, or 1995 for a map of the study area). A detailed description of the study area and methodology has been given elsewhere (Forsman & Lindell 1991; Forsman 1993a,b, 1995). Field work was conducted from April 1986 through August

1994. I visited seven different islands (ranging in size from 3 to 8 ha) several times between mid April and mid September each year on days with favourable weather conditions and captured snakes by hand while walking slowly across the islands. For each snake I recorded date of capture and sex, and measured SVL to the nearest 5 mm. Measurement repeatability of SVL is very high (0.997, $F_{14,30} = 1120$, $P < 0.0001$: Forsman 1993b). All snakes were individually marked by branding on their ventral scales and released at the place of capture after marking.

The primary and only mammalian prey species available for adders on these islands is the field vole *Microtus agrestis* (Forsman 1991b). The adders experience a variable food availability because these island populations of field voles fluctuate in numbers through time (Pokki 1981; Forsman & Lindell in prep.). Although adders occasionally feed also on frogs (*Rana* spp.) and nestling birds, data on stomach contents suggest that these constitute at the most 10% of the energy intake (L. E. Lindell in prep.). To obtain an estimate of relative prey availability voles were trapped on five days in September each year from 1989 to 1993 on the same islands as those where snakes were captured (see Forsman & Lindell (1991) for details). Mean annual vole density (calculated as the mean of number of individual voles captured in five days per hectare on the different islands) remained relatively low from 1989 through 1993 and ranged from 0.86 to 3.00 individuals/ha.

To obtain a relative measure of weather conditions during the growth season of adders, mean daily air temperature for the period 1 May through 31 August for each of the eight years was calculated from weather data for Svenska Högarna (an island located approximately 20 km south-east of the study area) supplied by The Swedish Meteorological and Hydrological Institute. Mean daily air temperature for this period ranged from 10.2 to 14.0°C among years and was highly negatively correlated with both rainfall and proportion of cloudy days.

Analysing growth

Individual growth rates, in millimetres per day, were calculated as the increment in SVL between two capture dates divided by the time elapsed between captures for all snakes that were captured twice in the same growth season (set to 1 June–14 September for males and 16 May–14 September for females: Forsman 1993b). No individual contributed more than one growth observation per year. Snout–vent increments smaller than 5 mm were regarded as measurement error, and were not included in the analyses. To test whether growth rates varied significantly among years, the general linear model procedure in SAS was used (SAS Institute Inc. 1988). Growth rate was regarded as the dependent variable, SVL at first capture in a growth season as a covariate, and sex and year as independent class variables. A model including all possible interaction terms was run first and then a reduced model without non-significant interaction terms.

Analysing survival

Analyses of recapture data from this population using computer software SURGE (Pradel & Lebreton 1991) have yielded estimates of annual survival probabilities of adders ranging from 70 to 88% (Forsman 1995; Forsman & Lindell in prep.). These analyses also revealed an increased survival probability with increasing vole density (Forsman & Lindell in prep.). Here the question is rather whether food availability and weather conditions influence the survival probability of large relative to small individuals.

Snakes captured in a given year were classified as survivors or non-survivors depending on whether or not they were recaptured in one or more subsequent years. Individuals not recaptured were assumed to be dead. Although some individuals may have survived and remained in the area without being recaptured, this will only bias the results if large and small snakes were not equally likely to be caught. However, the sign of the difference in body size between recaptured and non-recaptured individuals shifted between years (see below), and this cannot be accounted for simply by size-dependent catchability. Although mortality cannot be distinguished from dispersal, previously published evidence suggests that the tendency to emigrate and body size are not associated in this population (Forsman 1993b).

Statistical tests for a relationship between SVL and survival were carried out using logistic regression analysis (Hosmer & Lemeshow 1989), available under the procedure CATMOD in SAS (SAS Institute Inc. 1988). SAS provides a likelihood ratio statistics as a goodness-of-fit for the model, and maximum likelihood estimators (β, a slope coefficient that represents the change in the logit for a change of one unit in the independent variable) and standard errors of the parameters. The difference in deviance for a model with and without the variable of interest was used to assess the significance of the independent variables (Hosmer & Lemeshow 1989). Each year's data were analysed separately starting with a model including SVL, sex and the interaction between the two as independent variables. Sex was included to control for the body-size difference between male and female snakes. The interaction between sex and body size was included to determine whether the relationship between SVL and probability of survival was different in males and females. If the interaction was not significant, a simpler model without interaction was fitted to the data.

Whether changes in weather conditions and food availability affect the relative performance of large and small individuals was examined by testing for a correlation across years between: (1) the linear regression coefficient of growth rate on body size and mean temperature and vole density, respectively; and (2) the slope coefficient of survival on body size (obtained from logistic regression analysis) and mean temperature and vole density, respectively.

Results

Growth

The analyses of growth rates were based on a total of 202 observations (males 144; females 58). Mean SVL at first capture across individuals and years was 498 mm (SD = 68, range 280–645 mm). The relationship between growth rate and body size for each of the eight years is shown in Table 1. Growth rate decreased with increasing

SVL at the same rate in males and females, but females grew significantly faster than similar-sized males ($F_{1,185}$ = 30.87, $P < 0.0001$). A significant year by SVL interaction ($F_{7,185}$ = 4.36, $P < 0.001$) showed that the relationship between growth rate and body size varied among years (Table 1). This variability among years was parallel in males and females.

Table 1. Results from least squares linear regression of growth rate (mm/day) on snout–vent length (mm) in *Vipera berus* in eight different years

Year	*n*	Intercept	Slope	r^2	*P*
1986	11	1.96	-0.00319	0.68	0.0018
1987	16	1.00	-0.00149	0.35	0.0150
1988	32	1.73	-0.00273	0.55	0.0001
1989	42	1.58	-0.00253	0.67	0.0001
1990	33	0.99	-0.00143	0.23	0.0041
1991	22	1.03	-0.00156	0.51	0.0002
1992	27	0.87	-0.00115	0.27	0.0058
1993	19	0.61	-0.00075	0.22	0.0446

The results revealed no effect of prey availability on the yearly variation in overall growth rates of adders when statistically controlling for the effects of sex and body size ($F_{1,139}$ = 0.11, P = 0.74). Closer inspection, however, shows that large and small individuals responded differently to changes in food availability. The slopes from the separate regressions of growth rate on SVL increased with increasing vole densities across years (Fig. 1). This result shows that growth rate decreased at a slower rate with increasing body size when voles were abundant than when voles were scarce. Note that these analyses were restricted to the period 1989–1993 because vole densities prior to 1989 were unknown.

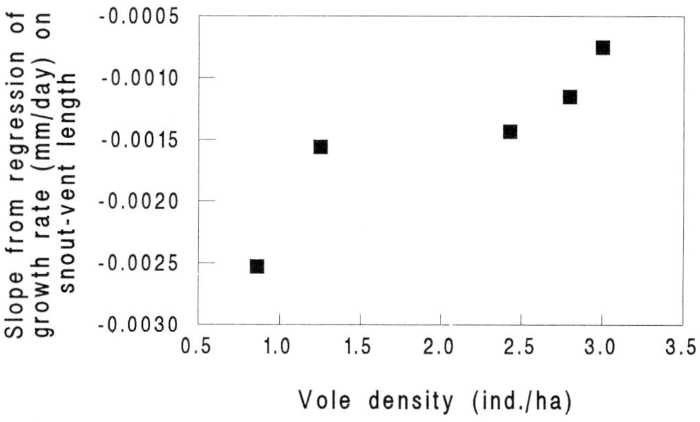

Fig. 1. Relationship across years between slopes from the least squares linear regression of growth rate (mm/day) on snout–vent length (mm) and vole density. r = 0.90, n = 5 years, $P < 0.05$.

The variation in growth rates among years was not associated with changes in weather conditions. Although overall growth rates tended to increase with increasing mean daily air temperature across years (slope = 0.18, SE = 0.011) when statistically controlling for the effects of sex and body size, this effect of ambient temperature was not significant ($F_{1,198}$ = 2.65, P = 0.10). Moreover, temperature accounted for no more than 1% of the total variance in growth rates, whereas year and the interaction between year and SVL accounted for 10%. The yearly regression coefficients of growth rate on body size were not correlated with mean ambient temperatures (r = -0.08, n = 8, P = 0.48), suggesting that changing weather conditions had the same (if any) effect on growth rates of large and small individuals.

Table 2. Results of logistic regression analyses of survival on snout–vent length and sex in *Vipera berus*

Year	n	Snout–vent length				Sex	
		β	SE	G	$P <$	G	$P <$
1986	23	-0.0036	0.0057	0.41	0.75	0.46	0.50
1987	53	-0.0156	0.0064	7.76	0.01	0.81	0.50
1988	77	0.0053	0.0031	2.95	0.10	0.02	0.90
1989	78	-0.0048	0.0030	2.74	0.10	0.10	0.75
1990	121	0.0021	0.0017	1.58	0.25	3.52	0.10
1991	81	0.0037	0.0027	1.86	0.25	2.24	0.10
1992	69	0.0074	0.0028	8.54	0.005	2.92	0.10
1993	99	0.0076	0.0045	3.31	0.10	17.18	0.001

β denotes the maximum likelihood estimate of survival on snout–vent length and SE the associated standard error.
Logistic regression analysis revealed a significant heterogeneity of β among years (G = 24.85, d.f. = 7, P < 0.001).

Survival

Overall, logistic regression analyses of recapture data revealed that the relationship between survival and body size varied significantly among years, but was parallel in males and females. The SVL-by-year interaction was highly significant (G = 24.85, d.f. = 7, P < 0.001), demonstrating heterogeneity of slope coefficients (β) among years. Separate analyses of the eight different years revealed that not only did the slope coefficients differ, but they also shifted sign between years, i.e. survivorship increased with increasing body size in some years and decreased in other years (Table 2). This was not a simple consequence of yearly differences in body size because there was no correlation across years between mean SVLs of adders (pooled sexes) and slope coefficients (r = -0.09, n = 8, P = 0.83). The interaction between SVL and sex was far from significant in all years (P > 0.25), showing that the relationship between survival and body size was independent of sex.

Survival probability of large relative to small individuals appears to have been influenced by food abundance but not by weather conditions. Analysis of data from 1989 through 1993 revealed an increase in the slope coefficient from the regression of survival on body size with increasing vole density (Fig. 2), suggesting that the survival advantage of large body size increased with increasing food availability. The slope coefficients increased also with increasing ambient air temperature across years (r = 0.74, n = 8, P < 0.05), suggesting that large individuals enjoyed a survival advantage relative to small individuals in warm but not in cold years. However, closer inspection revealed that this correlation was due to a single outlier (1987), and the association disappeared when this outlier was omitted (r = 0.06, n = 7, P = 0.89).

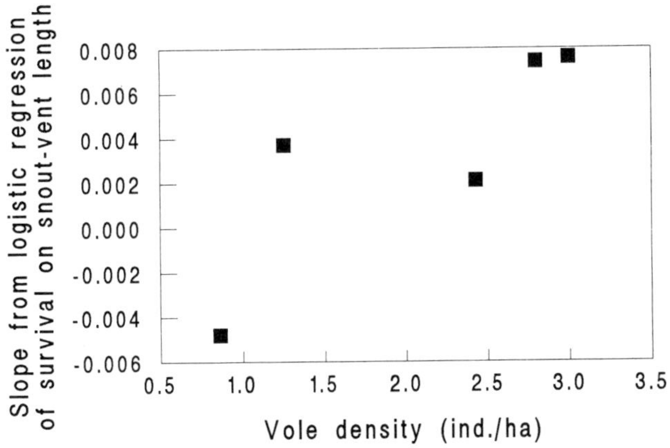

Fig. 2. Survival of large relative to small adders in relation to vole density. r = 0.83, n = 5 years, P = 0.08.

Time lags

There is evidence for a trade-off between growth and energy storage in adders (Forsman & Lindell 1991). Individuals that increase length fast for their body size pay a cost in terms of lower body mass relative to length the following spring. This may translate into an increased subsequent mortality due to starvation in the event of food shortage. Given this, one would predict that mean annual survival should be lower in years following seasons of rapid growth than in years following seasons of slower growth. In an attempt to test this prediction I tested for a relationship between estimates of average annual survival probabilities (from Forsman 1995 and Forsman & Lindell in prep.) and the growth rate of an average-sized individual the previous year (i.e. the growth rate of a 498 mm SVL individual expected from the least squares linear regression of growth rate on SVL, see Table 1 for equations). The result did not support the hypothesis; although average survival tended to decrease after years with rapid growth, the correlation was far from significant (r = -0.41, n = 7, P = 0.36). One possible explanation for this negative result is that high resource levels may translate into both rapid growth and large energy reserves. However, the survival advantage of large body size decreased after years with rapid growth, as indicated by a negative correlation across years between the slope coefficient from the logistic regression of survival on body size and growth rate of an average-sized individual the previous year (Fig. 3). In other words, years characterized by rapid individual growth rates were associated with a subsequent reduction in the survival probability of large relative to small snakes. This should prevent consistent increments or decrements of mean body size through time, a prediction that is supported by data on body size of female adders from 1986 to 1994. Thus, SVLs of females did not vary significantly among years (range 478–590 mm, $F_{8,241} = 1.51$, $P = 0.15$), and there was no consistent increase or decrease of average body size through time ($r = 0.39$, $n = 9$ years, $P = 0.3$, Fig. 4). By contrast, the prediction is not supported by data for males; SVLs of males varied significantly among years (range 462–533 mm; $F_{8,372} = 2.72$, $P < 0.01$), and average male body size consistently increased through time ($r = 0.72$, $n = 9$ years, $P < 0.05$, Fig. 4). The shift in body size of male adders was evident even when data for each animal in each year were treated as independent ($r = 0.18$, $n = 381$, $P < 0.001$). Applying this last-mentioned approach did not alter the result for females ($r = 0.04$, $n = 253$, $P > 0.5$).

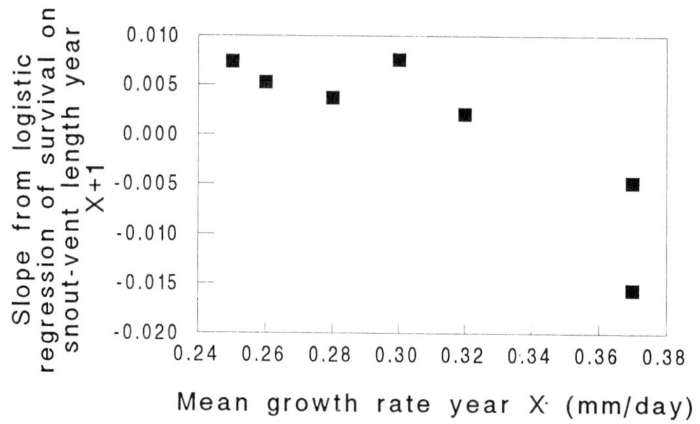

Fig. 3. Survival of large relative to small adders in relation to growth rate the previous year of an average-sized individual (SVL = 498 mm). See text for details. $r = -0.84$, $n = 7$ years, $P < 0.05$.

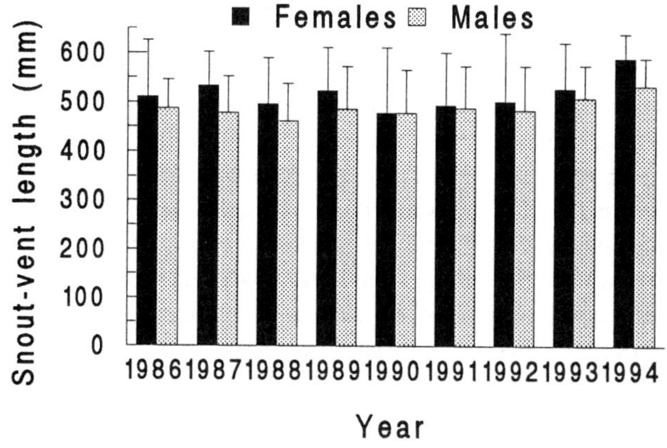

Fig. 4. Temporal variation in mean body sizes of male and female adders. Bars denote standard deviations.

Discussion

Food availability appears to be an important determinant of both survival and growth in adders. Previous analyses have shown that annual survival probabilities increase with increasing availability of prey (Forsman & Lindell in prep.). The present results further suggest that food availability may affect also the relative survival probabilities of large and small individuals. Thus, survivorship tended to increase with increasing body size in years with high vole densities and decrease with increasing body size when voles were scarce. Higher survival of small compared to large individuals during food shortage was observed also in another island population of adders in southern Sweden (Madsen & Stille 1988; but see Andrén 1982), and may reflect an increased survival cost of reproduction with diminishing food resources. If larger (and older) individuals are unable to recover from the substantial mass losses associated with reproduction, when food is scarce, they should be more susceptible to starvation than smaller individuals that do not reproduce. The relationship between growth rate and body size also varied among years in parallel with vole density; growth rates decreased with increasing body size in all years but less so when voles were abundant than when they were scarce.

But why should high prey abundance have a greater positive effect on large than on small adders? First, the proportion of encountered prey items that are sufficiently small to be ingestible is higher for large than for small individuals because larger snakes are capable of ingesting larger prey (Forsman & Lindell 1993). Large individuals therefore should have higher foraging success. Second, large individuals can eat more food per unit time relative to their energy demand as compared to small individuals, because gut volume scales isometrically or even positively to body mass whereas metabolic rate scales to the 0.75 power of body mass (Sibly & Calow 1986). Obviously, the second advantage will disappear with declining food availability.

Given that the size increase generated by rapid growth is followed by viability selection against large body size, as was apparently the case in this population, one would not expect body size to change through time. It has been previously suggested that changes in the direction of selection may have the same long-term effect on a population as stabilizing selection (Gibbs & Grant 1987). However, although mean snout–vent lengths of female adders did not vary among years, mean body size of male adders increased significantly in the course of the nine years of study. There is no apparent explanation for this change in male body size. Madsen & Shine (1992) observed a similar rapid shift in body size in a small and isolated population of adders in southern Sweden, but in their case in both males and females. In the population studied by these authors the shift was associated with the spread of a novel allele, the bearers of which apparently enjoyed increased growth and higher reproductive success, compared to individuals without the allele. However, this explanation is not applicable to the population in the present study.

Growth and survival of adders were apparently unaffected by annual variations in weather conditions. This is an unexpected result for an ectothermic animal (e.g. Huey 1982; Peterson *et al.* 1993). However, there are several possible explanations for the lack of relationship with ambient temperature in this study: (1) Mean air temperature may be a poor measure of the thermal constraints imposed by the environment. (2) There may have been an effect of temperature which was not detected owing to a small sample size and low variance in ambient temperature. (3) Adders may have been able to maintain body temperatures, activity levels and foraging performance despite variable environmental temperatures. (4) The optimum temperature for growth may be affected by food abundance. For example, it has been shown in fish that the temperature at which growth is fastest is higher when food is abundant than when food is in short supply (Brett 1971). Finally, (5) because vole density was measured for five years only, it was not possible to control statistically for differences in vole density when testing for an effect of temperature.

Conclusions and future research

This study shows not only that temporal variations in food and weather conditions may confound studies of intraspecific geographic variation but also that data from long-term field studies may help to identify potential ecological sources of variation in ecology. It is particularly noteworthy that environmental changes may influence differently the performance of large and small individuals (see also Ferguson & Fox 1984). Although the results suggest that prey availability is more important than weather conditions in determining growth and survival of adders, further studies are needed to determine the relative importance of food, temperature and their interaction. Both variables can easily be manipulated and their effects on growth and reproduction can be measured in laboratory experiments. However, results from such studies will reflect direct effects only and may have little, if any, relevance for natural populations. Under natural conditions individuals are exposed also to indirect effects mediated through their prey, competitors and predators. If such indirect effects are to be taken into account, field studies with experimental manipulation of prey availability and subsequent monitoring of marked free-ranging individuals in isolated subdivided populations appears to be the best option available. Because of the logistical problems of manipulating the range of available temperatures, it will be necessary to repeat the experiment in different years and make use of the natural variation in environmental temperatures. Alternatively, one can compare results from experiments carried out in two or more geographically separated populations exposed to different environmental temperatures, but care must then be taken to consider intersite differences other than temperature.

The results further suggest that temporal environmental heterogeneity may have important consequences for strategies of energy allocation and life-history evolution. Given environmentally mediated yearly differences in the potential for growth and a variable relationship between survival and body size across years, the effect of individual growth on body size, future fecundity and subsequent survival will be different in different years. This may select for genotypes with a flexible strategy, adjusting their allocation of energy to the competing needs of growth, storage and reproduction according to prevailing conditions. Alternatively, fluctuating environments may favour different genotypes with different but fixed allocation strategies in different years. Although both hypotheses predict that the variance in life-history traits should be higher in populations experiencing temporally variable environments than in populations inhabiting more stable environments, they can be separated by raising individuals from the two categories of populations under controlled and stable conditions. The first hypothesis predicts no difference in variance between offspring whose ancestors originated from variable and stable environments, respectively, whereas the second predicts a higher variance among descendants of the population from the most variable environment.

Acknowledgements

I acknowledge Roger S. Thorpe, Wolfgang Wüster and Anita Malhotra for asking me to contribute to this volume, and Lars Erik Lindell, Staffan Ulfstrand, and Stefan Ås for helpful comments on an earlier draft of the manuscript. The study was financially supported by Olle & Signhild Engkvists Stiftelser and The Royal Swedish Academy of Sciences.

References

Andrén, C. (1982). Effect of prey density on reproduction, foraging and other activities in the adder, *Vipera berus*. *Amphibia–Reptilia* **3**: 81–96.

Andrén, C. (1986). Courtship, mating and agonistic behaviour in a free-living population of adders, *Vipera berus* (L.) *Amphibia–Reptilia* **7**: 353–383.

Andrén, C. & Nilson, G. (1983). Reproductive tactics in an island population of adders, *Vipera berus* (L.), with a fluctuating food resource. *Amphibia–Reptilia* **4**: 63–79.

Andrews, R. M. (1982). Patterns of growth in reptiles. In *Biology of the Reptilia* **13**: 273–320. (Eds Gans, C. & Pough, F. H.). Academic Press, London, New York, etc.

Brett, J. R. (1971). Energetic responses of salmon to temperature: a study of some thermal responses in the physiological and freshwater ecology of sockeye salmon (*Oncorhynchus nerka*). *Am. Zool.* **11**: 99–113.

Ferguson, G. W. & Fox, S. F. (1984). Annual variation of survival advantage of large juvenile side-blotched lizards, *Uta stansburiana*: its causes and evolutionary significance. *Evolution* **38**: 342–349.

Ford, N. B. & Seigel, R. A. (1989). Phenotypic plasticity in reproductive traits: evidence from a viviparous snake. *Ecology* **70**: 1768–1774.

Forsman, A. (1991a). Variation in sexual size dimorphism and maximum body size among adder populations: effects of prey size. *J. Anim. Ecol.* **60**: 253–267.

Forsman, A. (1991b). Adaptive variation in head size in *Vipera berus* L. populations. *Biol. J. Linn. Soc.* **43**: 281–296.

Forsman, A. (1993a). Growth rate in different colour morphs of the adder, *Vipera berus*, in relation to yearly weather variation. *Oikos* **66**: 279–285.

Forsman, A. (1993b). Survival in relation to body size and growth rate in the adder, *Vipera berus*. *J. Anim. Ecol.* **62**: 647–655.

Forsman, A. (1995). Opposing fitness consequences of colour pattern in male and female snakes. *J. evol. Biol.* **8**: 53–70.

Forsman, A. & Lindell, L. E. (1991). Trade-off between growth and energy storage in male *Vipera berus* (L.) under different prey densities. *Funct. Ecol.* **5**: 717–723.

Forsman, A. & Lindell, L. E. (1993). The advantage of a big head: swallowing performance in adders, *Vipera berus*. *Funct. Ecol.* 7: 183–189.

Forsman, A. & Lindell, L. E. (In preparation). *Responses of a predator to variation in prey availability: survival and emigration of adders in relation to vole density.*

Gibbs, H. L. & Grant, P. R. (1987). Oscillating selection on Darwin's finches. *Nature, Lond.* **327**: 511–513.

Hosmer, D. W. Jr. & Lemeshow, S. (1989). *Applied logistic regression.* John Wiley & Sons, New York.

Huey, R. B. (1982). Temperature, physiology, and the ecology of reptiles. In *Biology of the Reptilia* **12**: 25–91. (Eds Gans, C. & Pough, F. H.). Academic Press, London, New York etc.

Madsen, T. (1988). Reproductive success, mortality and sexual size dimorphism in the adder, *Vipera berus*. *Holarct. Ecol.* **11**: 77–80.

Madsen, T. & Stille, B. (1988). The effect of size dependent mortality on colour morphs in male adders, *Vipera berus*. *Oikos* **52**: 73–78.

Madsen, T. & Shine, R. (1992). A rapid, sexually selected shift in mean body size in a population of snakes. *Evolution* **46**: 1220–1224.

Madsen, T. & Shine, R. (1993). Phenotypic plasticity in body sizes and sexual size dimorphism in European grass snakes. *Evolution* **47**: 321–325.

Peterson, C. R., Gibson, A. R. & Dorcas, M. E. (1993). Snake thermal ecology: the causes and consequences of body-temperature variation. In *Snakes — ecology and behavior*: 241–314. (Eds Seigel, R. A. & Collins, J. T.). McGraw-Hill, New York.

Platt, D. R. (1984). Growth of bullsnakes (*Pituophis melanoleucus sayi*) on a sand prairie in south central Kansas. In *Vertebrate ecology and systematics. A tribute to Henry S. Fitch*: 41–55. (Eds Seigel, R. A., Hunt, L. E., Knight, J. L., Malaret, L. & Zuschlag, N. L.). Museum of Natural History, University of Kansas, Lawrence.

Pokki, J. (1981). Distribution, demography and dispersal of the field vole, *Microtus agrestis* (L.), in the Tvärminne Archipelago, Finland. *Acta zool. fenn.* No. 164: 1–48.

Pradel, R. & Lebreton, J. D. (1991). *User's manual for program SURGE version 4.1.* CEPE/CNRS, Montpellier-Cedex, France.

Prestt, I. (1971). An ecological study of the viper *Vipera berus* in southern Britain. *J. Zool., Lond.* **164**: 373–418.

SAS Institute Inc. (1988). *SAS/STAT user's guide, release 6.03 edition.* SAS Institute Inc., Cary, NC.

Sibly, R. M. & Calow, P. (1986). *Physiological ecology of animals. An evolutionary approach.* Blackwell, Oxford.

Slip, D. J. & Shine, R. (1988). Thermoregulation of free-ranging diamond pythons, *Morelia spilota* (Serpentes, Boidae). *Copeia* **1988**: 984–995.

Willmer, P. (1991). Thermal biology and mate acquisition in ectotherms. *Trends Ecol. Evol.* **6**: 396–399.

12 The role of ecology in determining venom variation in the Malayan pitviper, *Calloselasma rhodostoma*

JENNIFER C. DALTRY, WOLFGANG WÜSTER and ROGER S. THORPE

Synopsis

Intraspecific variation in snake venom composition is a widespread phenomenon with important implications for snakebite therapy, yet few attempts have previously been made to identify its cause. The venom from almost 100 Malayan pitvipers (*Calloselasma rhodostoma*), captured in 36 localities in Vietnam, Thailand, Malaysia and Java, was analysed by using isoelectric focusing (IEF). A clear pattern of geographic variation emerged in the IEF profiles of adult *Calloselasma*, which was compared to patterns predicted from hypothetical causes by using a partial Mantel test. This test simultaneously evaluated the importance of geographic proximity ('venom variation is influenced by gene flow'), phylogeny ('venom reflects phylogenetic relationships among populations') and an ecogenetic hypothesis ('venom varies in accordance with geographic variation in prey type'). The intraspecific phylogeny was reconstructed by PCR-enhanced RFLP analysis of the cytochrome *b* gene (mtDNA), while dietary variation was assessed from the faeces and stomach contents of hundreds of wild-caught and museum specimens. Overall venom variation was significantly associated with the relative contribution of amphibians, reptiles and endotherms (mammals, birds) to local diet ($P < 0.0001$). When 13 variable IEF bands were tested in turn, most were significantly associated with diet and only one reflected phylogeny, thereby calling into question the frequent use of venom as a 'neutral' marker of phylogeny in snake systematics. The pattern of geographic variation in IEF profiles was significantly congruent with the variation in the venom's enzymatic effects, indicating adaptive differences in the snakes' ability to subdue and digest different prey. This species also exhibits significant ontogenetic and sexual variation in venom composition and diet. Studies of the venom of captive Malayan pitvipers indicate that the dietary association is genetically based rather than induced in the short term; venom composition and prey selection may have co-evolved to optimize foraging efficiency.

Symp. zool. Soc. Lond. (1997) No. 70: 155–171
ISBN 0 19 854986 5

Introduction

Intraspecific variation in venom composition has been detected in numerous snakes (reviewed by Chippaux, Williams & White 1991). As well as being of academic interest to studies of the evolution and ecology of venomous snakes, this phenomenon is of major applied importance to snakebite therapy. For example, diagnosis of the species responsible for a bite can be confounded by variation in symptomatology, and the antivenom prepared against one venom type may be substantially less effective against another. Snake venom also represents a valuable natural source of biochemicals for research and medicine (Mebs 1978; Russell 1983), and it is often useful to determine whether the components of interest are more abundant in the venom of certain individuals than others. Yet despite its far-reaching consequences, the underlying causes of intraspecific variation in snake venom have never been clarified. Literature concerning the biochemistry/pharmacology of venoms, and the systematics/ecology of snakes have tended to travel separate paths, but all of these disciplines need to be integrated to ascertain how variation arises.

The Malayan pitviper, *Calloselasma rhodostoma*, was selected as the species through which to investigate this question. This terrestrial viperid is the leading cause of venomous snakebite morbidity throughout much of South-East Asia (Warrell 1986). Although the fatality rate of untreated bites is relatively low, many victims suffer severe necrosis, often leading to permanent deformity and long-term economic hardship (Warrell 1986). Ironically, the venom of this snake has also served a beneficial role in healthcare as the source of ancrod (Arvin™), used clinically as a defibrinating agent. There have been numerous studies of the biochemistry and pharmacology of *C. rhodostoma* venom (see Hardy 1990 for bibliography), but the potential influence of intraspecific variation has been largely ignored. However, there are indications of regional differences in the bite symptomatology of *C. rhodostoma* (Warrell 1986) and in the venom's immunological properties (Tu & Ganthavorn 1978), which suggests that venom composition varies geographically.

The causes of geographic variation can be elucidated with the aid of numerical hypothesis testing. A distance (similarity) matrix based on the observed pattern of variation in phenotype can be compared with matrices representing hypothesized causes by using Mantel tests to evaluate the probability of their association (see Manly 1985). A partial Mantel test is preferred because it evaluates rival hypotheses simultaneously, regressing out any effects of intercorrelation between them. This rigorous test has previously proven invaluable for determining the causes of geographic variation in reptiles (e.g. Thorpe 1991).

The aim of this paper is to provide a more precise evaluation of the pattern of variation in the venom of *C. rhodostoma* and to identify its most probable evolutionary cause through the use of partial Mantel tests. Although the primary focus will be geographic variation, within-population variation in venom composition will also be discussed.

Materials and methods

Venom collection and electrophoresis

Venom samples were obtained from 96 *C. rhodostoma* captured in 36 localities in Vietnam, Thailand, Malaysia and Java (Fig. 1). Venom was extracted within 12 h of capture and desiccated to prevent deterioration in storage (Tan & Tan 1987).

The samples were rehydrated with reverse osmosis water to a concentration of 10 mg/ml soluble protein, and compared by using IEF across polyacrylamide gels of pH range 3.5–9.5 (Ampholine PAGplates^{TM} purchased precast from Pharmacia LKB). Fifteen microlitres of each venom solution were focused for 80 min at up to 1500 V, and stained with Coomassie Blue.

To determine whether venom composition changes with body size, a comparison was made of the IEF profiles of specimens from the best-represented region, West Java (groups 29–36; $n = 20$), of snout–vent length (SVL) 274 to 626 mm. A 20×20 distance matrix derived from the occurrence of all varying bands was tested against SVL by a Mantel test with 10 000 randomizations.

Causal hypotheses

Three hypotheses to explain geographic variation in venom were tested:

1. Variation in venom is associated with population phylogeny

Interspecific variation in venom is widely assumed to be a function of phylogeny, and similarities in venom properties between species (or higher order taxa) are often interpreted as evidence of common ancestry (e.g. Githens & George 1931; Foote & MacMahon 1977). If geographic variation in the venom of *C. rhodostoma* evolved in accordance with its population phylogeny, populations with a more recent common ancestor would be expected to produce more similar venom than phylogenetically remote populations.

The historical relationships among populations are most reliably inferred (as free as feasible from ecogenetic selection) by comparing their mitochondrial DNA (mtDNA) (Avise 1994). A few millimetres of caudal tissue were collected from each snake after venom extraction. Biopsies were also taken from other specimens obtained in the field, to give a total of 131 samples. From the tissue, 767 bp fragments of mtDNA, chiefly the cytochrome *b* gene (709 bp), were amplified by means of the polymerase chain reaction using the primers L14841′ (Kocher *et al.* 1989) and MVZ16 (Moritz, Schneider & Wake 1992). Seven restriction endonucleases (*DdeI*, *EcoRI*, *HinfI*, *NciI*, *NlaIII*, *ScrFI*, *TaqI*) identified 19 different haplotypes among the 131 mtDNA samples. Genetic distances between the populations were computed with program DA, REAP (McElroy *et al.* 1992) and used to construct a Fitch–Margoliash phylogenetic tree (program KITSCH, PHYLIP 3.3: Felsenstein 1991) (Fig. 2). Three main lineages were resolved: (1) north-western Thailand; (2) Vietnam with the rest of Thailand; and (3) Malaysia with Java.

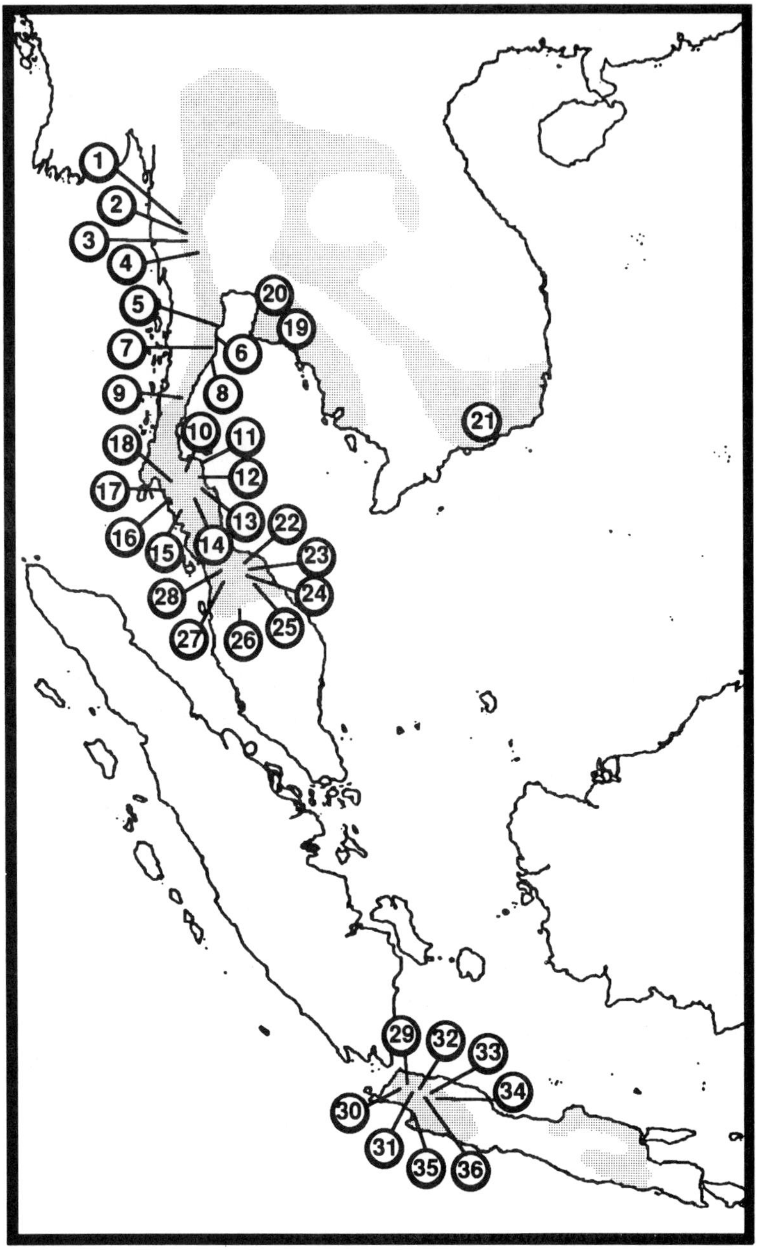

Fig. 1. Sites from which venom samples were collected: the Thai provinces of Kanchanaburi (sites 1–4); Prachuap Khiri Khan (5–8); Chumphon (9); Nakhon Si Thammarat (10–14); Krabi (15–18); Rayong (19, 20); Vung Tau province, Vietnam (21); Kedah State, Malaysia (22–28); West Java, Indonesia (29–36). Shading represents the known distribution range of *C. rhodostoma*.

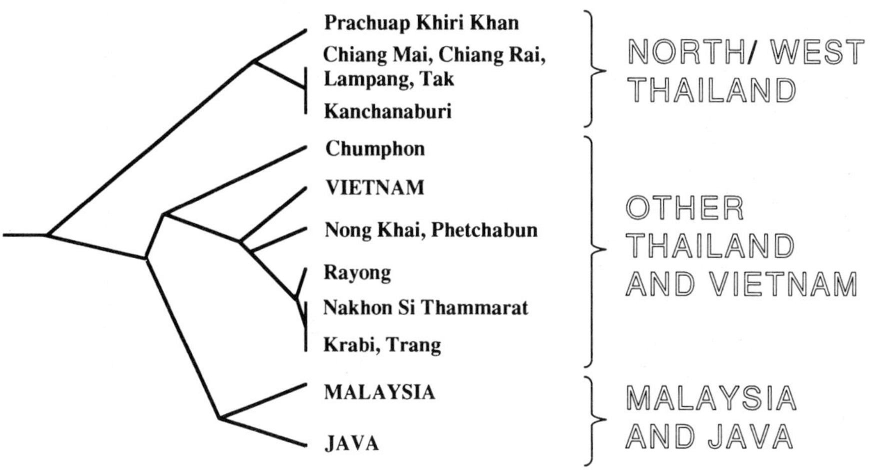

Fig. 2. Population phylogeny of *C. rhodostoma* reconstructed from RFLP analysis of the cytochrome *b* gene. Branch lengths are drawn to scale.

2. Variation in venom is associated with the geographic variation in diet

Since venom plays an important role in feeding (Thomas & Pough 1979; Russell 1983; Mackessy 1988), venom variation might be linked to dietary variation. Diet was recorded from the faeces of the specimens that provided venom, supplemented by the gut contents of museum and roadkilled specimens from the same localities. Two hundred and sixteen prey items were identified to taxonomic class, including insects, chilopods, fish, amphibians, reptiles, rodents and birds. There was significant geographic variation in the contribution of different classes to the diet of adults (Fig. 3).

3. Variation in venom is a function of the geographic distance between groups

Geographic proximity might influence venom evolution in several ways. Firstly, the opportunity for exchange of venom-coding genes tends to be higher between neighbouring demes and may enable spatially-close populations to produce more similar venom than remote conspecifers; secondly, neighbouring populations are more likely to share a similar habitat, which might promote convergent evolution of venom.

Numerical hypothesis testing

To evaluate the causes of geographic variation, the venom profiles of adult *C. rhodostoma* of >400 mm SVL were segregated into 36 groups according to capture locality (Fig. 1). (Smaller specimens were excluded to remove the ontogenetic influence on venom composition, and the gender influence was removed by excluding bands of pI 5.90 and 6.90: see Results.) Each regional group was given a series of digits denoting the mean occurrence of each band (e.g. $0 =$ band absent from all group members; $0.5 =$ present in 50% group members).

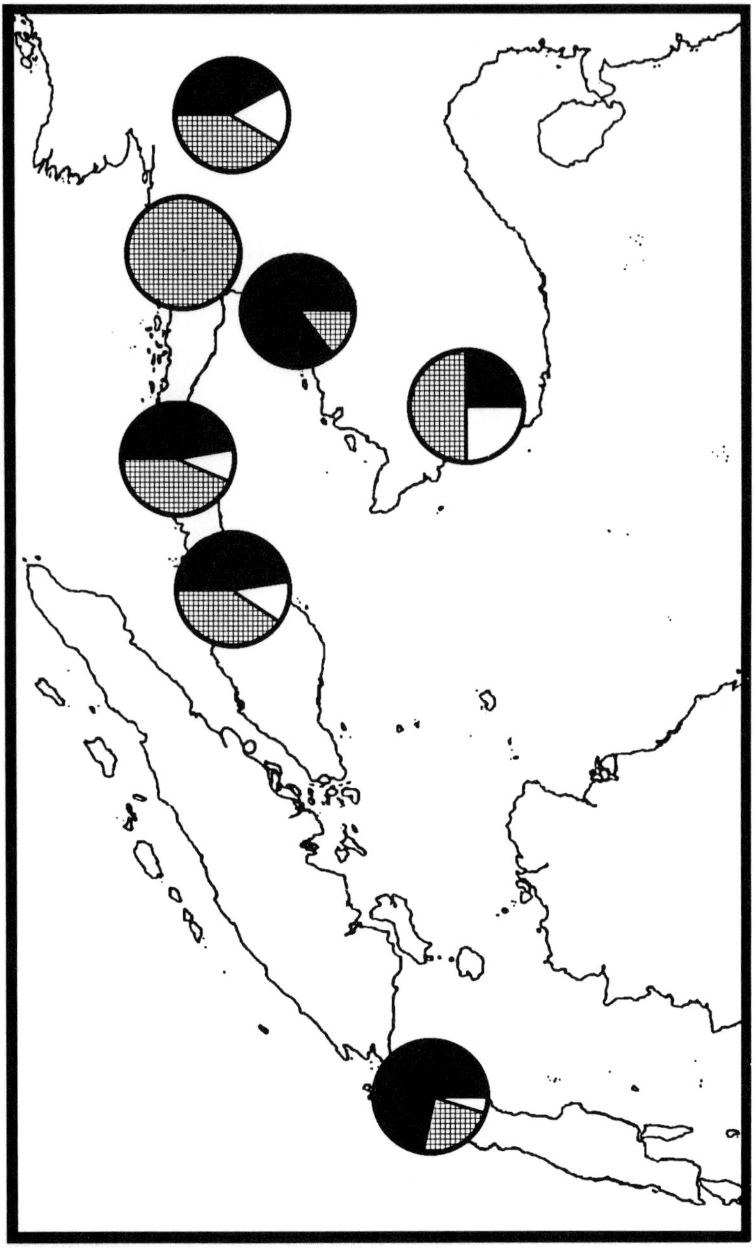

Fig. 3. Geographic variation in the diet of adult *C. rhodostoma*. The pie charts represent the contribution of amphibians (white), reptiles (cross-hatched) and endotherms (black) as a percentage number of items identified from specimens larger than 400 mm SVL.

Observed variation in venom among the groups was expressed as a 36×36 distance matrix. A partial Mantel test was used to partially correlate the observed matrix against three matrices derived from the three hypotheses above. (1) The phylogenetic relationships among the 36 groups were assessed by RFLP analysis of mtDNA, and patristic distances were computed between each pair of groups by using a Fitch–Margoliash phylogenetic tree (program KITSCH, PHYLIP 3.3: Felsenstein 1991) constructed from genetic distances calculated with program DA (REAP: McElroy *et al.* 1992). By regressing out the patristic distances between populations, the partial Mantel test makes it possible to test for ecogenetic adaptation free of phylogenetic effects (Thorpe *et al.* 1995). (2) Diet was recorded as the mean proportions of amphibians, reptiles and endotherms (birds, mammals) constituting the diet of snakes exceeding 400 mm SVL in each region (see Fig. 3). (3) Geographic proximity was computed from the latitude and longitude of the 36 localities.

In the first partial Mantel test, the dependent matrix was based upon *all* variable venom bands and thus represents the overall similarities between the groups. However, since there is no reason to suppose that all venom components evolve in the same way, subsequent tests compared the geographic variation in each band in turn with the three hypotheses of phylogeny, diet and geographic proximity.

The association between venom and diet was also tested by the comparative method of independent contrasts (Felsenstein 1985). The continuous variables representing venom and diet variation were taken as the largest principal coordinate of the venom and diet matrices respectively, and entered into Garland's Comparative Method Analysis Package (CMAP) under the gradual change model. This approach is commonly used to elucidate evolutionary cause, but is inferior to the partial Mantel test for studying variation within species because it cannot take into account gene flow between populations (estimated from geographic proximity) or other multi-dimensional hypotheses. Furthermore, the comparative method does not, on its own, consider multiple competing hypotheses free of distribution.

Results

The patterns of variation in venom composition

About 25 intense bands and numerous trace bands were resolved per isoelectrically focused venom sample. For comparison, only intense bands (the most abundant components) will be considered.

Ontogenetic variation was evident in all regions where both adults and juveniles had been sampled; for example, 11 bands varied among the venoms from West Java ($n = 20$), and occurrence was significantly associated with body size ($P = 0.0190$).

Owing to this influence of body size, only venoms from adult specimens exceeding 400 mm SVL ($n = 67$) were compared to determine the pattern of geographic variation. Across these venoms, bands representing 37 different isoelectric points were identified, of which 19 were not present in all samples. Two of these were gender-specific. Adult females from west Java, Malaysia and west and peninsular Thailand ($n = 19$) produced a dark band of pI 6.90 that was absent from all males in these regions ($n = 19$). Two males from south-east Thailand also lacked this band, but no data is available on the venom of local females. In Vietnam, both genders lacked band pI 6.90, but every female ($n = 9$) produced a dark band of pI 5.90 which was absent from Vietnam males ($n = 6$).

Each specimen was given a score of 17 digits (denoting the occurrence of the remaining 17 variable bands) which was entered into a principal component analysis (Fig. 4). Fifty-three variants were identified (specimens with identical venoms invariably came from the same locality). The first two component scores (PC1, PC2) account for 42.64% of the total variation.

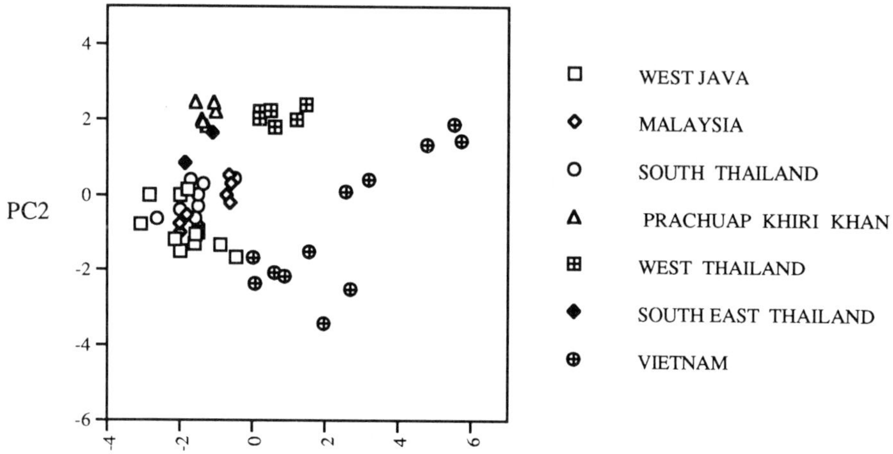

Fig. 4. Principal components analysis of isoelectrically focused electrophoretograms of venom from adult *C. rhodostoma*.

Partial Mantel tests

When venoms from 67 adult *C. rhodostoma* were grouped by capture locality (Fig. 1), 13 bands varied in occurrence among the 36 groups.

Geographic variation in overall venom composition was not significantly associated with geographic distance or patristic distance, but *was* significantly correlated to the proportions of amphibians, reptiles and mammals constituting local diet (Table 1). Gene flow and phylogenesis must therefore be dismissed as primary influences, but we cannot reject the hypothesis that venom variation is in some way linked to diet.

When the variable bands were tested in turn, almost half were significantly associated with diet, two with geographic proximity and only one with population phylogeny. Note that even when other tree-building algorithms (Fitch–Margoliash without a molecular clock assumption, neighbour-joining, and parsimony or maximum likelihood using modal haplotypes) were used to reconstruct the population phylogeny, the results were still as strongly, and in some cases more strongly, in favour of diet.

Independent contrasts method

The findings of the Mantel tests were supported by the independent contrast method. The correlation between diet and venom, free of phylogenetic effects, was highly significant ($r = 0.67$; $P < 0.01$).

Table 1. Probabilities of significance associated with partial Mantel tests comparing geographic variation in venom with causal hypotheses

	Patristic	Diet distance	Geographic proximity
Overall venom composition	0.0225	**<0.0001***	0.5358
Band pI 9.40	0.7012	0.0046	0.0576
Band pI 9.15	0.0018	0.0506	**0.0009***
Band pI 8.55	0.1680	0.1023	0.0041
Band pI 8.45	0.2217	0.3542	0.6542
Band pI 8.25	0.6280	**0.0011***	0.0083
Band pI 7.85	0.0862	0.4000	0.6182
Band pI 7.45	**0.0015***	0.7296	0.8052
Band pI 6.55	0.6700	**0.0012***	0.0321
Band pI 6.10	0.6429	0.7233	0.1113
Band pI 6.00	0.0160	0.0624	**0.0009***
Band pI 5.80	0.4829	**0.0012***	0.0542
Band pI 5.70	0.4727	**0.0002***	0.0247
Band pI 5.45	0.4944	**0.0006***	0.0044

* $P < 0.05$ after applying sequential Bonferroni correction (Rice 1989) to all tests in the table.

Discussion

The biological function of the venom of *C. rhodostoma*

Before attempting to interpret the partial Mantel test results, it is useful to consider the biochemistry and biological role of the venom of *C. rhodostoma*.

The primary roles of the venom of most snakes are to subdue prey and initiate prey digestion (Russell 1983). Snakes in general, and the Viperidae in particular, consume relatively large prey (Pough & Groves 1983); *C. rhodostoma*, for instance, commonly tackles prey equalling one third of its body weight, including rats and scolopendrid centipedes (Daltry 1995). Such formidable prey would undoubtedly be more difficult, even hazardous, to subdue without the aid of venom.

It is also crucial that such large meals are digested quickly enough to prevent the prey's gut bacteria from causing excessive putrefaction. Digestive enzymes produced in the stomach can only permeate the prey via its largely intact epidermis (which may be shielded by scales, fur or feathers), whereas enzymes in the venom are injected below the dermis, and are dispersed by diffusion and the prey's circulatory system. In accordance with their proclivity for large prey, the Viperidae have long fangs; a *C. rhodostoma* 1 m long, for example, has a penetration depth of 16 mm. Long fangs do not necessarily improve the efficiency of venom delivery (Hayes, Kaiser & Duvall 1992), but do ensure that digestive enzymes are released deep within the prey.

The venoms of various viperid snakes have been shown to have a powerful digestive effect (Thomas & Pough 1979). A major constituent of these venoms is proteolytic enzymes, which cause the breakdown of structural proteins in the muscles, organs and subcutaneous tissue. The venom of *C. rhodostoma* contains several different basic proteases and exhibits exceptionally high L-amino acid oxidase activity (Tan 1991), which is implicated in protease activation (see Mackessy 1988). Other known digestive components are reviewed by Tan (1991).

The toxicity of *C. rhodostoma* venom, on the other hand, is relatively low (see Mebs 1978). Large prey are often released after striking and it is essential that they should be rapidly immobilized by the venom before fleeing too far or injuring the snake, but it is not necessary or even advantageous to cause rapid death since the venom digestive enzymes may be usefully dispersed by the prey's circulatory system. Mice injected with small doses of *C. rhodostoma* venom rapidly lose locomotory ability, but death is protracted and accompanied by nervous spasms (Plagnol & Martin 1957), which may benefit the snake by ensuring thorough dissemination of the venom (Mackessy 1988).

Causes of geographic variation in venom

IEF revealed a strong pattern of geographic variation in *C. rhodostoma* venom (Fig. 4). Interestingly, this pattern is significantly congruent with variation in the venom's enzymatic activities (Daltry 1995); that is, venoms with similar IEF profiles have similar biological effects.

The overall pattern of variation in venom composition was significantly correlated to the diet of adults (Table 1). This might be interpreted as evidence of environmental induction. However, venoms collected from captive-bred *C. rhodostoma* closely matched those of wild specimens from the native origin of the captive stock, even though the all-mouse diet of the captives is incongruent with the more varied natural prey. Conversely, intraspecific variation is often evident even in the venom of captives maintained on the same diet (Bonilla, Faith & Minton 1973; Gregory-Dwyer *et al.* 1986). These observations, in addition to findings that hybrid snakes produce a venom exhibiting characteristics of both parents (e.g. Aird *et al.* 1989), suggest that venom variation is predominantly dictated by fixed, heritable genes.

A more plausible explanation for the association between diet and venom is evolution through natural selection. Pitvipers may specialize on certain prey, according to which are locally abundant or obtained with least 'cost', and selection will favour the individuals producing the most effective venom for subduing and ingesting these prey. The adaptive evolution of the venom might in turn promote increased dietary specialization, with the snakes focusing upon the animals that are most efficiently handled by means of their venom. (It is not yet clear whether the variation in diet is primarily a passive response to variation in local availability of prey types, or reflects inherent differences in prey 'choice'. Congenital variation in prey selection has been demonstrated in other snakes: see Arnold 1981.)

Of course, the hypothesis that venom is adapted to local diet hinges upon variation among prey species in susceptibility to venom. Indeed, venom lethalities vary between taxa by orders of magnitude (Mebs 1978; Russell 1983). Prey 'digestibility' has less frequently been compared, but the digestive role of venom may be especially valuable where the animal is relatively large, has a low surface-area : volume ratio and/or has a resistant outer coating, since these attributes retard normal digestion in the stomach. Intuitively, the highly permeable dermis of a frog presents a lesser challenge to the stomach enzymes than a reptilian skin. Squamate reptiles have a high surface-area : volume ratio, however, and may therefore be digested more easily than a bulky rodent (Thomas & Pough 1979; Mackessy 1988). Taxonomic differences in membrane structure and physiology may also affect the efficacy of the venom. Consequently, specialization upon different prey could select for different quantities and types of toxins and digestive enzymes in the venom.

Snakes frequently exhibit geographic variation in diet (e.g. Arnold 1981; Shine 1991), and further research is necessary to determine whether this is commonly correlated with geographic variation in venom composition. We would suggest, however, that a venom–diet association will not be equally strong in all species. *C. rhodostoma* is common and widespread, and the fixation of venom-coding genes by different populations under natural selection is therefore unlikely to have been significantly perturbed by the vagaries of genetic drift. In species typified by small, isolated populations, on the other hand, the influence of drift can become accentuated (Mayr 1975), thereby reducing the likelihood of similar venoms being produced by different demes even if their diet is identical. Indeed, Williams *et al.* (1988) compared the venom electrophoretograms of insular tiger snakes (*Notechis ater*) and found that within-population variation is correlated with population size, suggesting that drift has reduced genetic diversity on small islands. Unlike *C. rhodostoma*, the geographic variation in the venom was *not* significantly associated with available prey.

Ontogenetic and sexual variation

Distinct banding differences were observed between the IEF profiles of venoms from juvenile and adult *C. rhodostoma* from the same localities. Ontogenetic changes in venom electrophoretograms do not necessarily reflect changes in biological activity, but this is frequently the case (e.g. Bonilla *et al.* 1973; Aird 1985; Andrade & Abe 1993). Marked changes in the effects of the venom are commonplace in viperid species (see Chippaux *et al.* 1991), and there is mounting evidence that they are adaptively associated with changes in diet. Juvenile *Crotalus viridis*, for instance, prey chiefly upon lizards and their venom is 2.5 times more toxic to lizards than the venom of the rodentivorous adults (Mackessy 1988). Generally speaking, mammalian prey can be subdued by a less toxic venom than reptile prey, but, being bulky, are difficult to digest. This might explain why the venom of adult *C. viridis* displays proteolytic activity five times higher than that of juvenile venom (Mackessy 1988). Many other venomous snakes undergo ontogenetic shifts in diet from ectothermic to endothermic prey (Klauber 1972; Sazima 1992; Andrade & Abe 1993) and concurrent changes to a more proteolytic but less toxic venom are widely reported (Chippaux *et al.* 1991).

In most regions, small *Calloselasma rhodostoma* prey chiefly upon ectotherms whereas adults show a stronger proclivity for endotherms (Daltry 1995), so it is possible that the venom of different age-groups is specifically adapted to their contrasting foraging requirements. Of course, an alternative, non-selective explanation is that the ontogenetic changes merely reflect unavoidable developmental constraints. However, not all snakes undergo distinct changes in venom; the venoms of juvenile and adult *Oxyuranus microlepidotus* (Elapidae) exhibit similar electrophoretic patterns and biological activities (Tan, Ponnudurai & Mirtschin 1993) and *all* age groups eat small mammals (Shine 1991). Similar conservatism of both venom and diet is also exhibited by *Crotalus durissus terrificus* in south-east Brazil (Salomão, Santos & Puorto 1995).

Irrespective of body size, female *C. rhodostoma* consistently produced high levels of a venom protein(s) that was not produced by local males. Sexual variation may be associated with diet since females tend to grow larger, have proportionally larger heads, and display a stronger proclivity for larger prey taxa (Daltry 1995). Yet even small females produce this extra band. Either there are more subtle gender differences in diet which have not yet been detected (our dietary analysis compared prey classes and further data are required regarding actual species), or the sexual variation, though widely conserved, has no adaptive function. After all, by definition, the sexes have an unequal distribution of sex chromosomes and hormones which may promote divergent development of non-debilitating (but not necessarily advantageous) secondary sexual characteristics.

Differences were also frequently apparent between venoms from pitvipers collected from the same locality on a similar date, of the same gender and similar body size. This probably reflects innate individual variation in venom synthesis, since genetic variation is to be expected in the gene pool of most sexually reproducing species.

Potential implications for snakebite therapy

Identification of the species responsible is often crucial for effective snakebite therapy, but few victims can reliably identify the snake responsible. Diagnosis of *C. rhodostoma* bite is typically founded upon the symptoms presented, but this may be confused by intraspecific variability in clinical effects. Geographic differences in the symptomatology of *C. rhodostoma* bites between Malaysia and Thailand were reported by Warrell (1986) and, given the ontogenetic, sexual and individual variation in venom composition shown herein, within-locality clinical variation is also likely (see Russell 1983).

Unfortunately, even when the species has been correctly identified, the appropriate antivenoms for bites by different populations or age groups may radically differ. Gutiérrez *et al.* (1991), for example, found that antivenom prepared against adult Costa Rican *Crotalus durissus durissus* was a poor neutralizer of local juvenile venom, which differs from the adult venom in this population, whereas in Brazilian *C. d. terrificus* it does not. In the case of *Calloselasma rhodostoma*, physicians treating bites in Malaysia suspect that antivenom prepared from Thai populations is less effective than antivenom prepared from local specimens (S. Ambu pers. comm.).

Producers of *C. rhodostoma* antivenom typically obtain venoms from areas where the species is most easily collected and, since juveniles are hard to find and produce little venom, large specimens are favoured. Yet Chippaux *et al.* (1991) stated that 'producers of antivenom must . . . ensure representation of all venom types required within each antivenom'. In the case of *C. rhodostoma* venom, this approach might entail obtaining venom from specimens of both genders and various age-groups from across the geographic range in which the antivenom is to be used, which would inevitably raise the cost of antivenom production. It is therefore crucial to ascertain the extent to which the intraspecific variation in venom composition revealed in the present study reflects differences of importance to antivenom efficacy. After all, some of the components in the venoms might not be of pharmacological importance, while others may have serious clinical effects yet fail to stimulate antibody responses (Chinonavanig *et al.* 1988).

Potential implications for biochemical research and snake systematics

Venom proteins are frequently used as tools to investigate biological processes and structure (Mebs 1978; Russell 1983). Research using ancrod, for instance, has shed light on the processes of blood coagulation in humans (Denson 1969). A better understanding of intraspecific variation in quantity and types of venom components could help to pinpoint the optimal sources of the desired compound.

Chippaux *et al.* (1991) advised investigators of the biochemistry and biological activity of venom to monitor the precise origin of the sample under analysis, so that the findings can be replicated and interpreted in a biologically meaningful way. *C. rhodostoma* venom has been subjected to intense research (Hardy 1990), but almost every study has been based upon commercial sources of uncertain geographic origin, body size or gender. Perhaps inconsistencies in the venom source partly explain why experimental findings often vary widely among authors. The intravenous LD_{50} of *C. rhodostoma* venom, for instance, ranges from 2.8 to 6.2 mg/kg mice (Mebs 1978).

Intraspecific variation in venom also has serious consequences for taxonomists. Given the special importance of accurate systematics among dangerous snakes (Wüster & Thorpe 1991), the means by which phylogenetic relationships are reconstructed require serious consideration. Venoms are frequently compared to elucidate taxonomic relationships, with compositional similarities usually interpreted as evidence of common ancestry (e.g. Githens & George 1931; Chen, Wu & Zhao 1984), but findings may vary according to the specimen(s) chosen to represent each taxon; for example, the venom of juvenile *Crotalus d. durissus* from Costa Rica is more similar to adult Brazilian *C. d. terrificus* than adult *C. d. durissus* (Gutiérrez *et al.* 1991).

Even when specimens of equal size are compared, there is no guarantee that their venoms will accurately reflect their phylogenetic affinities. The venom of adult *C. d. durissus*, for example, shows a greater immunological affinity with *C. adamanteus* and *C. viridis* than with the conspecific *C. d. terrificus* (Anderson, Gutiérrez & Ownby 1993). Only one IEF band in adult *C. rhodostoma* venom reflected phylogeny (Table 1). Clearly, venom may not be a reliable phylogenetic marker, particularly at low taxonomic levels.

Is intraspecific variation in venom predictable?

Much remains to be learned of the patterns of inter- and intraspecific variation in venom composition and pharmacology of medically important snakes, but collecting and analysing venom from a broad geographic area is laborious, expensive and hazardous. Is it possible to predict where and to what extent variation is present? Regrettably, the findings of this study suggest it is not: diet was a better predictor of venom differences between populations of *C. rhodostoma* than either geographic proximity or phylogeny, but the prey of most species are too poorly known to predict venom variation.

Can venom variation instead be inferred from the snakes' appearance? Many studies have found differences in composition that *seem* to accord with the distribution of subspecies characterized by morphological features (Chippaux *et al.* 1991), but the comparisons were often tautologically made between pooled samples from the ranges of the nominal subspecies. Closer scrutiny reveals that the correlation between venom and morphology is less clear-cut; for example, the venom of *Crotalus lepidus klauberi* is usually similar to *C. l. lepidus* venom, but *C. l. klauberi* from Chihuahua and parts of Arizona and New Mexico produce a potent toxin and can be a hundred times more lethal (Rael *et al.* 1992).

Calloselasma rhodostoma also exhibits significant geographic variation in coloration, scalation and body shape, but it is impossible to identify different venom variants simply from such variation in appearance (Daltry 1995). It would seem that painstaking biochemical analysis is the only sure means of determining intraspecific variation in venom.

Acknowledgements

We thank the Zoo Negara (Malaysia) for venom, N.-H. Tan and G. Ponnudurai for enzymological assays, J. Norman for advice on molecular methods, B. J. F. Manly and T. Garland for programs, and the museums that loaned specimens. T. K. Anh and T. X. Kiem (Choray Hospital, Vietnam), S. Ambu and B. L. Lim (Institute for Medical Research, Malaysia) and C. K. Shin (Zoo Negara) provided stimulating discussions. This research was primarily funded by a SERC studentship (JCD) and the Leverhulme Trust (RST, A. Malhotra), with support from NERC (WW) and the Royal Society (RST).

References

Aird, S. D. (1985). A quantitative assessment of variation in venom constituents within and between three nominal rattlesnake subspecies. *Toxicon* 23: 1000–1004.

Aird, S. D., Thirkhill, L. J., Seebart, C. S. & Kaiser, I. I. (1989). Venoms and morphology of western diamondback/Mojave rattlesnake hybrids. *J. Herpet.* **23**: 131–141.

Anderson, S. G., Gutiérrez, J. M. & Ownby, C. L. (1993). Comparison of the immunogenicity and antigenic composition of ten Central American snake venoms. *Toxicon* **31**: 1051–1059.

Andrade, D. V. & Abe, A. S. (1993). Toxicidade do veneno de *Bothrops moojeni* e sua relação com a variação ontogenética de dieta. (Abstract). In *III congresso Latino Americano de herpetologia*: 77. Instituto de Biologia, Universidade Estadual de Campinas, Campinas, São Paulo.

Arnold, S. J. (1981). Behavioral variation in natural populations. II. The inheritance of a feeding response in crosses between geographic races of the garter snake, *Thamnophis elegans*. *Evolution* **35**: 510–515.

Avise, J. C. (1994). *Molecular markers, natural history and evolution*. Chapman and Hall, London.

Bonilla, C. A., Faith, M. R. & Minton, S. A. (1973). L-amino acid oxidase, phosphodiesterase, total protein and other properties of juvenile timber rattlesnake (*Crotalus h. horridus*) venom at different stages of growth. *Toxicon* **11**: 301–303.

Chen, Y., Wu, X. & Zhao, E. (1984). Classification of *Agkistrodon* species in China. *Toxicon* **22**: 53–61.

Chinonavanig, L., Billings, P. B., Matangkasombut, P. & Ratanabanangkoon, K. (1988). Antigenic relationships and relative immunogenicities of venom proteins from six poisonous snakes of Thailand. *Toxicon* **26**: 883–890.

Chippaux, J.-P., Williams, V. & White, J. (1991). Snake venom variability: methods of study, results and interpretation. *Toxicon* **29**: 1279–1303.

Daltry, J. C. (1995). *The evolutionary biology of the Malayan pit viper: a study of the causes of intraspecific variation*. PhD thesis, University of Aberdeen.

Denson, K. W. E. (1969). Coagulant and anticoagulant action of snake venoms. *Toxicon* **7**: 5–11.

Felsenstein, J. (1985). Phylogenies and the comparative method. *Am. Nat.* **125**: 1–15.

Felsenstein, J. (1991). *PHYLIP, version 3.3*. University Herbarium, University of California, Berkeley.

Foote, R. & MacMahon, J. A. (1977). Electrophoretic studies of rattlesnake (*Crotalus* and *Sistrurus*) venom: taxonomic implications. *Comp. Biochem. Physiol. (B)* **57**: 235–241.

Githens, T. S. & George, I. D. (1931). Comparative studies on the venoms of certain rattlesnakes. *Bull. Antivenin Inst. Am.* **5**: 31–34.

Gregory-Dwyer, V. M., Egen, N. W., Bosisio, A. B., Righetti, P. G. & Russell, F. E. (1986). An isoelectric focusing study of seasonal variation in rattlesnake venom proteins. *Toxicon* **24**: 995–1000.

Gutiérrez, J. M., dos Santos, M. C., Furtado, M. F. & Rojas, G. (1991). Biochemical and pharmacological similarities between the venoms of newborn *Crotalus durissus durissus* and adult *Crotalus durissus terrificus* rattlesnakes. *Toxicon* **29**: 1273–1277.

Hardy, D. L. (1990). Venoms and envenomation: a selected bibliography of the recent literature on *Agkistrodon* and its allies. In *Snakes of the* Agkistrodon *complex: a monographic review*: 553–571. (Eds Gloyd, H. K. & Conant, R.). Society for the Study of Amphibians and Reptiles, Oxford, Ohio.

Hayes, W. K., Kaiser, I. I. & Duvall, D. (1992). The mass of venom expended by prairie rattlesnakes when feeding on rodent prey. In *Biology of the pitvipers*: 383–388. (Eds Campbell, J. A. & Brodie, E. D.). Selva, Tyler, Texas.

Klauber, L. M. (1972). *Rattlesnakes: their habits, life histories, and influence on mankind*. (2nd edn). University of California Press, Berkeley.

Kocher, T. D., Thomas, W. K., Meyer, A., Edwards, S. V., Pääbo, S., Villablanca, F. X. & Wilson, A. C. (1989). Dynamics of mitochondrial DNA evolution in animals: amplification and sequencing with conserved primers. *Proc. natn. Acad. Sci. USA* **86**: 6196–6200.

Mackessy, S. P. (1988). Venom ontogeny in the Pacific rattlesnakes *Crotalus viridis helleri* and *C. v. oreganus*. *Copeia* **1988**: 92–101.

Manly, B. J. F. (1985). *The statistics of natural selection*. Chapman and Hall, London.

Mayr, E. (1975). *Populations, species and evolution*. (Fourth printing). Harvard University Press, Cambridge, Massachusetts.

McElroy, D., Morgan, P., Bermingham, E. & Kornfield, I. (1992). REAP: an integrated environment for the manipulation and phylogenetic analysis of restriction data. *J. Hered.* **83**: 157–158.

Mebs, D. (1978). Pharmacology of reptilian venoms. In *Biology of the Reptilia* **8**: 437–560. (Eds Gans, C. & Gans, K. A.). Academic Press, London & New York.

Moritz, C., Schneider, C. J. & Wake, D. B. (1992). Evolutionary relationships within the *Ensatina eschscholtzii* complex confirm the ring species interpretation. *Syst. Biol.* **41**: 273–291.

Plagnol, H. & Martin, P. (1957). Electrophorèse du venin d'*Ancistrodon rhodostoma* Boie. *Annls Inst. Pasteur, Paris* **92**: 525–533.

Pough, F. H. & Groves, J. D. (1983). Specializations of the body form and food habits of snakes. *Am. Zool.* **23**: 443–454.

Rael, E. D., Johnson, J. D., Molina, O. & McCrystal, H. K. (1992). Distribution of a Mojave toxin-like protein in rock rattlesnake (*Crotalus lepidus*) venom. In *Biology of the pitvipers*: 163–168. (Eds Campbell, J. A. & Brodie, E. D.). Selva, Tyler, Texas.

Rice, W. R. (1989). Analyzing tables of statistical tests. *Evolution* **43**: 223–225.

Russell, F. E. (1983). *Snake venom poisoning*. Scholium International, New York.

Salomão, M. G., Santos, S. M. A. & Puorto, G. (1995). Activity patterns of the rattlesnake *Crotalus durissus* (Viperidae, Crotalinae): feeding, reproduction and snakebite. *Stud. neotrop. Fauna Envir.* **30**: 101–106.

Sazima, I. (1992). Natural history of the jararaca pitviper, *Bothrops jararaca*, in southeastern Brazil. In *Biology of the pitvipers*: 199–216. (Eds Campbell, J. A. & Brodie, E. D.). Selva, Tyler, Texas.

Shine, R. (1991). *Australian snakes: a natural history*. Reed Books, Balgowlah, New South Wales.

Tan, N.-H. (1991). The biochemistry of venoms of some venomous snakes of Malaysia — a review. *Trop. Biomed.* **8**: 91–103.

Tan, N.-H., Ponnudurai, G. & Mirtschin, P. J. (1993). A comparative study of the biological properties of venoms from juvenile and adult inland taipan (*Oxyuranus microlepidotus*) snake venoms. *Toxicon* **31**: 363–367.

Tan, N.-H. & Tan, C. S. (1987). Thermal stability of venom enzymatic activities. In *Progress in venom and toxin research*: 188–198. (Eds Gopalkrishnakone, P. & Tan, C. K.). National University of Singapore, Singapore.

Thomas, R. G. & Pough, F. H. (1979). The effect of rattlesnake venom on digestion of prey. *Toxicon* **17**: 221–228.

Thorpe, R. S. (1991). Clines and cause: microgeographic variation in the Tenerife gecko (*Tarentola delalandii*). *Syst. Zool.* **40**: 172–187.

Thorpe, R. S., Malhotra, A., Black, H., Daltry, J. C. & Wüster, W. (1995). Relating geographic pattern to phylogenetic process. *Phil. Trans. R. Soc. (B)* **349**: 61–68.

Tu, A. T. & Ganthavorn, S. (1978). Comparison of snake venoms (Reptilia, Serpentes) from Java, Indonesia and Thailand and its significance in evolution and zoogeography. *J. Herpet.* **12**: 105–107.

Warrell, D. A. (1986). Tropical snake bite: clinical studies in South-East Asia. In *Natural toxins — animal, plant and microbial*: 25–45. (Ed. Harris, J. B.). Clarendon Press, Oxford.

Williams, V., White, J., Schwaner, T. D. & Sparrow, A. (1988). Variation in venom proteins from isolated populations of tiger snakes (*Notechis ater niger, N. scutatus*) in South Australia. *Toxicon* **26**: 1067–1075.

Wüster, W. & Thorpe, R. S. (1991). Asiatic cobras: systematics and snakebite. *Experientia* **47**: 205–209.

13 Molecular evolution of phospholipase A_2s and metalloproteinase/ disintegrins from venoms of vipers

ANA M. MOURA-DA-SILVA, R. D. G. THEAKSTON and JULIAN M. CRAMPTON

Synopsis

The evolutionary relationships of venom toxins have been studied by independently analysing venom group II phospholipases A_2 (PLA_2s) and mammalian and snake venom metalloproteinase/disintegrin/cysteine-rich (MDC) families of proteins. Phylogenetic trees grouped both venom group II PLA_2s and MDC proteins according to current taxonomic divisions and/or functional activities. Gene duplications may have occurred at different points during the evolution of these protein families. The pattern of variability of the cDNAs coding for venom toxins has also been analysed. The greatest sequence variation of venom group II PLA_2s has been observed in the region coding for the mature protein, with most of the substitutions occurring in non-synonymous positions. In MDC proteins, the snake venom metalloproteinase domain appears to be the most variable, whilst neutral selection appears to be responsible for the evolution of the mammalian proteins, the only exception being the region comprising the catalytic site of the mammalian metalloproteinase domain. These data suggest that gene duplication followed by hypervariability of the amino acids in the functional domains and positive selection may have occurred during the evolution of venom toxins, assisting the generation of diversity in these protein families.

Introduction

Snake venoms are complex mixtures of proteins with different toxic or enzymatic properties. The type of toxic activity of the venoms appears to reflect their taxonomic classification; for example, Elapidae and Hydrophiidae venoms are generally neurotoxic whilst Viperidae venoms are more complex, including a number of enzymes that may induce local tissue damage, haemorrhage and often also coagulopathy (Warrell 1983). In the Viperidae these enzymes generally belong to few gene families. The most important of these, as far as toxicity is concerned, are the proteinases and the phospholipases A_2 (PLA_2s). Proteinases acting on different substrates, or PLA_2s with different tissue specificity, have often been reported in viper venoms (Kini & Evans 1989; Bjarnson & Fox 1994).

Symp. zool. Soc. Lond. (1997) No. 70: 173–187
ISBN 0 19 854986 5

The evolution of a gene family with different functions is currently associated with rapid amino acid divergence among duplicated copies of the genes, thus increasing the functional diversity of the gene family (Ohta 1994). For example, functionally distinct members of the serine protease family are thought to have arisen through gene duplication followed by divergence of the copies by hypervariability and positive Darwinian selection (Creighton & Darby 1989). In this report, the phylogenies of PLA_2s and metalloproteinases present in venoms of vipers are presented, with the suggestion that gene duplication followed by hypervariability and positive selection may have occurred during the evolution of these gene families, leading to slightly different molecules with distinct toxic activities.

Materials and methods

The sequences used in this paper were obtained by scanning GenBank, Swiss-Prot and PIR data bases using the BLAST programme (Altschul *et al.* 1990). The following PLA_2 sequences were analysed and the respective accession numbers are shown: *Bothrops jararacussu*, $BJUPLA_2$ (X76289) and Bothropstoxin I and II (Cintra *et al.* 1993); *B. asper*, Myotoxin II (P24605) and III (P20474); *Crotalus atrox*, PLA_2 (P00624); *C. adamanteus*, PLA_2 (P00623); *C. durissus terrificus*, Crotoxin (P08878, spP24027); *C. scutulatus scutulatus*, Mojave toxin (P18998, spP23559); *Trimeresurus flavoviridis*, BP2 (P20381), and PL-X (P06860); *T. gramineus*, PLA_2-II (A44179); *T. mucrosquamatus*, PLA_2 (S15133); *Agkistrodon halys blomhoffi* (current name: *Agkistrodon blomhoffii*), basic PLA_2-I (P04417) and acidic PLA_2-II (P20249); *A. h. pallas* (probably refers to *Agkistrodon blomhoffi brevicaudus*), Agkistrotoxin (P14421); *A. piscivorus piscivorus*, APP K-49 (P04361); *Vipera ammodytes ammodytes*, Ammodytoxin A (P00626), B (P14424) and C (P11407), Ammodytin L (P17935) and I_2 (X56878), and Vipoxin (P14420); *V. russelli formosensis* (current name: *Daboia russelli siamensis*, Taiwan population), RV-4 (Q02471) and RV-7 (P31100); *V. berus berus*, PLA_2 (P31854); *Cerastes cerastes cerastes*, PLA_2 (P21789); *Bitis caudalis*, Caudoxin (P00622); *Bitis nasicornis*, PLA_2 (P00621); and *Bitis gabonica*, PLA_2 (P00620). The precursor sequences of the following MDCs were used: human collagenase 3 (X75308), and tumour suppressor gene product, MDC (D17390); mouse cyritestin (S18968), monocyte surface antigen (A60385), and interstitial collagenase MMP-1 (P33435); rat Stromelysin-2, MMP-10 (P07152), and epididymal protein, EAP (S28259); guinea pig Fertilin alpha (Z11719) and beta (Z11720); monkey epididymal protein, EAP (S28258), and tMDC-I (X76637), and tMDC-II (X77619); and snake venom Rhodostomin (S33792) from *Calloselasma rhodostoma*; Atrolysin E (A43296), B (41608), C (S41609), and D (S41610) from *Crotalus atrox*; Trigramin (A43296) from *Trimeresurus gramineus*; Jararhagin (S24789) from *Bothrops jararaca*; Trimucin (S43125) from *Trimeresurus mucrosquamatus*; a metalloproteinase (U18233) from *Agkistrodon contortrix*; Halystatin (D28870) from *Agkistrodon halys*; and EcH I (X78970), and EcH II (X78971) from *Echis pyramidum leakeyi*.

Phylogenies were constructed by means of the Clustal V programme (Higgins, Bleasby & Fuchs 1992). Progressive alignments used the multiple alignment algorithms described by Higgins & Sharp (1989), with a fixed gap penalty of 10 and the Dayhoff PAM 250 protein weight matrix (Dayhoff, Schartz & Orcutt 1978: 345). The phylogenetic trees according to distances were generated from the above alignments by a neighbour-joining method (Saitou & Nei 1987), and rooted by using the Group I PLA_2 sequence of Bungarotoxin A1 chain (P00617) from *Bungarus multicinctus* venom as the outgroup for PLA_2s and Thermolysin (M21663) from *Bacillus stearothermophilus* for MDC proteins. The degree of error was calculated for each branch by bootstrapping (Felsenstein 1985). Values below the 95% confidence limits are shown in the tree.

Results and discussion

The snake venom PLA_2 gene family

Venom PLA_2s are classified into two distinct groups according to the organization of the cysteines (Heinrikson, Krueger & Keim 1977). Elapidae and Hydrophiidae venoms contain Group I PLA_2s which resemble the primary structure of mammalian pancreatic PLA_2. In Viperidae, the venom PLA_2s possess the same cysteine organization as non-pancreatic mammalian PLA_2s, and are classified as Group II PLA_2s. Several toxic activities are associated with both Group I and Group II PLA_2s, including neurotoxic, myotoxic, haemolytic, anticoagulant and oedema-inducing activities (Kini & Iwanaga 1986). The most lethal toxic activity associated with Group II PLA_2s is the presynaptic blocking of acetylcholine release from the nerve terminals resulting in inhibition of neuromuscular transmission (Kini & Evans 1989). A class of PLA_2-like molecules with the same cysteine organization as Group II PLA_2s, but differing in a conservative Lys substitution of the active site residue, D-49, is also found in Viperidae venoms (Maraganore & Heinrikson 1986). The PLA_2 activity is very low or absent in these molecules, which are often associated with myotoxic activity (Cintra *et al.* 1993).

Phylogenies of venom group II PLA_2s imply gene duplications

Evolutionary studies of snake venom PLA_2 sequences have been carried out by Dufton & Hider (1983) and Kostetsky, Arkhipova & Vladimirova (1991). The dendrograms they obtained suggest very distinct groupings for Group I and Group II venom PLA_2s. However, these authors generally used a single sequence from each species and few data on Group II PLA_2s, making it difficult to understand the relationship among Group II PLA_2s. Davidson & Dennis (1990) recently reported a more complex phylogenetic tree using 16 Group II PLA_2 sequences from venoms of different species. In this later alignment, the authors suggest that Group II PLA_2s are randomly mixed together, irrespective of their taxonomic or geographical distribution, through great mutational flexibility and/or the release of selective pressure operating on these genes. This hypothesis, however, seems inadequate considering the importance of some of these toxins in the overall toxicity of certain venoms and the interactions with specific target cells which are necessary to develop the toxic effect.

The evolutionary tree that we have reported (Moura-da-Silva, Paine *et al.* 1995) is focused on Group II PLA_2s. The tree generated from the alignment of 29 sequences (Fig. 1) demonstrates two primary divergent groups. Sequences from the subfamily Viperinae occur in the first group with a further subdivision between the sequences within the genera *Bitis* or *Vipera*. The second major group comprises the sequences from venoms of the subfamily Crotalinae. The sole exception within these groupings is the presence of *V. a. ammodytes* ammodytoxins within the Crotalinae cluster, which is closely associated with the *Bothrops* myotoxins. There are two possible explanations for this: first, since Ammodytoxins A, B, and L are present in the venom of the same species as Ammodytin I_2 (which is located in the Viperinae group), a gene duplication may have occurred before the divergence of Crotalinae/Viperinae. One copy evolved in the subfamily Viperinae and the other is more fully represented in Crotalinae and *V. a. ammodytes*, coding for ammodytoxins. Second, a gene duplication appears to have occurred after the divergence of the Viperinae and the gene coding for ammodytoxins could represent a convergent evolution with *Bothrops* myotoxins. In this respect, both are single-chained toxins with very similar hydrophatic profiles (not shown) suggesting that the structural configurations of the *Bothrops* myotoxins would be favourable for interactions with membranes without the necessity of a chaperon chain, which is usually present together with PLA_2 chains in other neurotoxins.

Sequences which possess similarity in their structure or function also tend to be clustered together. Well-defined clusters comprising functionally related toxins were observed. For example, all K-49 PLA_2-like molecules were clustered together, representing the most diverged branch from the root (Group I PLA_2) close to *Bothrops* myotoxins. This suggests a more recent divergence of this gene. Pre-synaptic neurotoxins and the non-toxic or acidic PLA_2s are also clustered in two distinct branches of the tree (Fig. 1).

Whether gene duplications have occurred generating copies coding for each type of toxin is speculative and requires further investigation. However, a gene duplication may have occurred generating the acidic and basic chains of the pre-synaptic neurotoxins (Bouchier *et al.* 1991). Gene duplications may also have taken place leading to the generation of the K-49 PLA_2-like analogues. Venoms from five specimens of *B. jararacussu* were independently analysed to investigate the chromatographic distribution of their basic proteins (Moura-da-Silva, Desmond *et al.* 1991). At least two fractions with PLA_2 activity were found in each sample, one corresponding to Bothropstoxin I (K-49) and the other probably related to $BJUPLA_2$ (unpubl. results).

Our data do not fully agree with those of Davidson & Dennis (1990), probably because different methodologies were used to generate the alignments and the phylogenetic trees. However, our data do agree with the phylogenetic distribution of pit vipers (Brattstrom 1964) deduced by osteology. Sequences from *Agkistrodon* spp. are preferentially clustered with those from *Crotalus* spp., and the *Bothrops* spp. sequences are closer to those of *Trimeresurus*. The separation of the sequences belonging to Viperinae and Crotalinae spp. is likely to be real, and the clustering of the toxins according to function is also likely to have occurred.

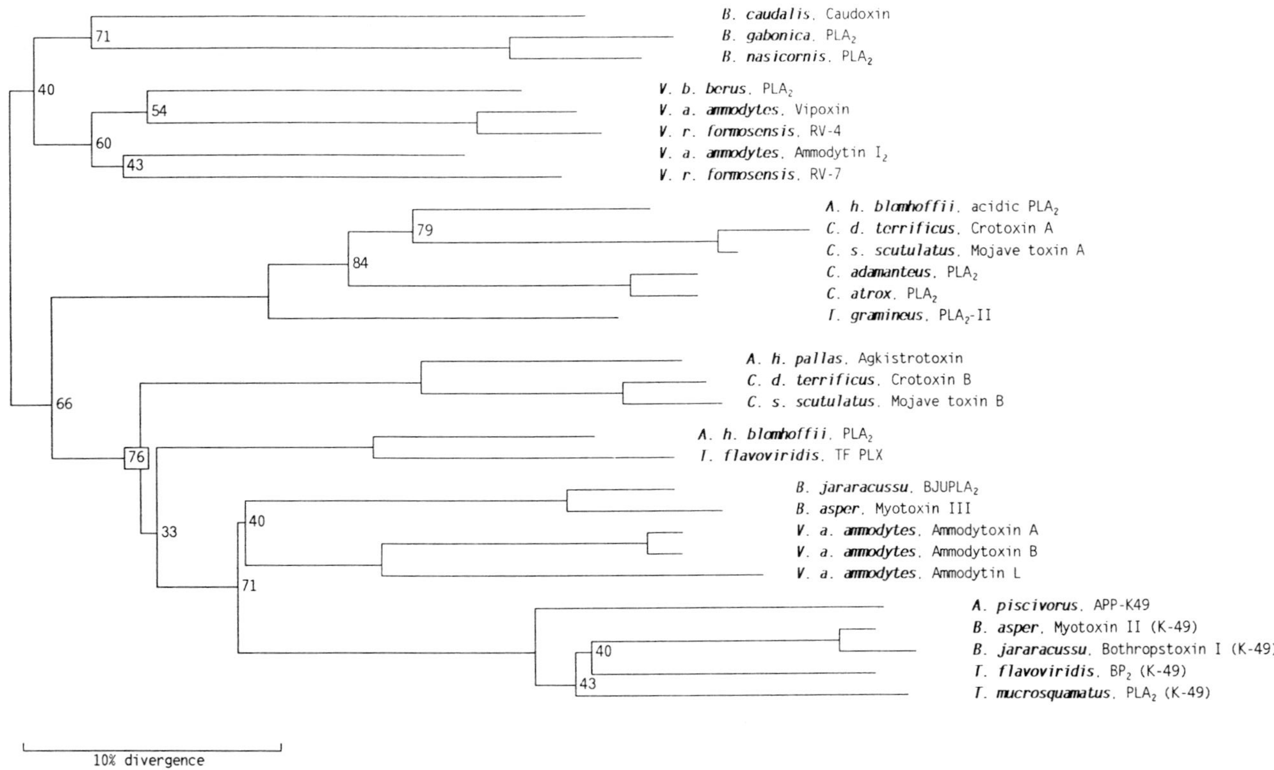

Fig. 1. Gene tree of the snake Group II PLA_2s protein sequences. All amino acid positions of the predicted mature proteins have been included in the analysis using the CLUSTAL V program. The horizontal scale is proportional to the calculated distance (percentage divergence). The numbers indicated at nodes represent the percentage confidence limits by bootstrapping. Only those values less than 95% are indicated. Scientific names in the figure are as used in the original databases. For an updated nomenclature, see p. 174.

Unusual pattern of sequence variation in Group II PLA_2s cDNAs

Figure 2 shows a schematic representation of the structural domains of Group II PLA_2 cDNAs in comparison with the $BJUPLA_2$ cDNA sequence recently cloned in our laboratory (Moura-da-Silva, Paine *et al.* 1995), and the frequency of substitutions which have occurred in each domain. The cDNAs coding for the signal peptides and 3′ and 5′ untranslated regions are highly conserved.

Variation is highest in the domains coding for the mature proteins. The pattern of substitutions among sequences from three different species of snake suggested that most of the substitutions occurred in non-synonymous sites, thus accounting for the observed amino acid substitutions (Table 1).

Table 1. Pattern of nucleotide substitution per 100 sites of cDNAs coding for the mature proteins of PLA_2s from Ammodytoxin A (*V. a. ammodytes*), Crotoxin B (*C. d. terrificus*), and PLX (*T. flavoviridis*) compared with $BJUPLA_2$ cDNA from *B. jararacussu*

Toxin	Total	Synonymous	Non-synonymous
Ammodytoxin A	22.4	7.7	14.7
Crotoxin B	20.2	2.7	17.5
PLX	19.9	4.6	15.3

Despite the variability in the coding region, the expected conservation of regions coding for the active site and Ca^{2+} binding loop is apparent. It is interesting that the catalytic activity of the molecule is not always essential for toxicity (Rosenberg 1986), which suggests that the variation of the other regions of the molecules, already correlated to their toxic activities (Kini & Iwanaga 1986), could be the domains under selection.

The high similarity of the 5′ and 3′ untranslated regions in relation to the coding regions has already been reported *within* the same species (Bouchier *et al.* 1991). Unusual conservation of the intron sequences of different PLA_2 genes from the same species of snake has also recently been reported (Nakashima *et al.* 1993; John, Smith & Kaiser 1994). Our data now extend these observations to include higher sequence conservation of 3′ and 5′ untranslated regions in relation to the coding regions *between* species.

The metalloproteinase, disintegrin, cysteine-rich (MDC) protein family

Proteolytic enzymes are the most abundant components in most viper venoms. Some of these enzymes, the haemorrhagic toxins, are responsible for the local and systemic haemorrhage observed in most cases of viperine envenoming (Kamiguti *et al.* 1991). Haemorrhagic toxins belong to a family of metalloproteinases which contains a conserved zinc-chelating sequence also present on matrix-degrading metalloproteinases (MMPs) characteristic of the astacin family (Gomis-Ruth, Kress & Bode 1993).

Disintegrins are a family of RGD-containing peptides isolated from viper venoms with high affinity for the platelet glycoprotein GP IIb/IIIa integrin receptor (Gould *et al.* 1990). They prevent platelet aggregation through inhibition of fibrinogen and von Willebrand factor binding to platelets (Huang *et al.* 1989).

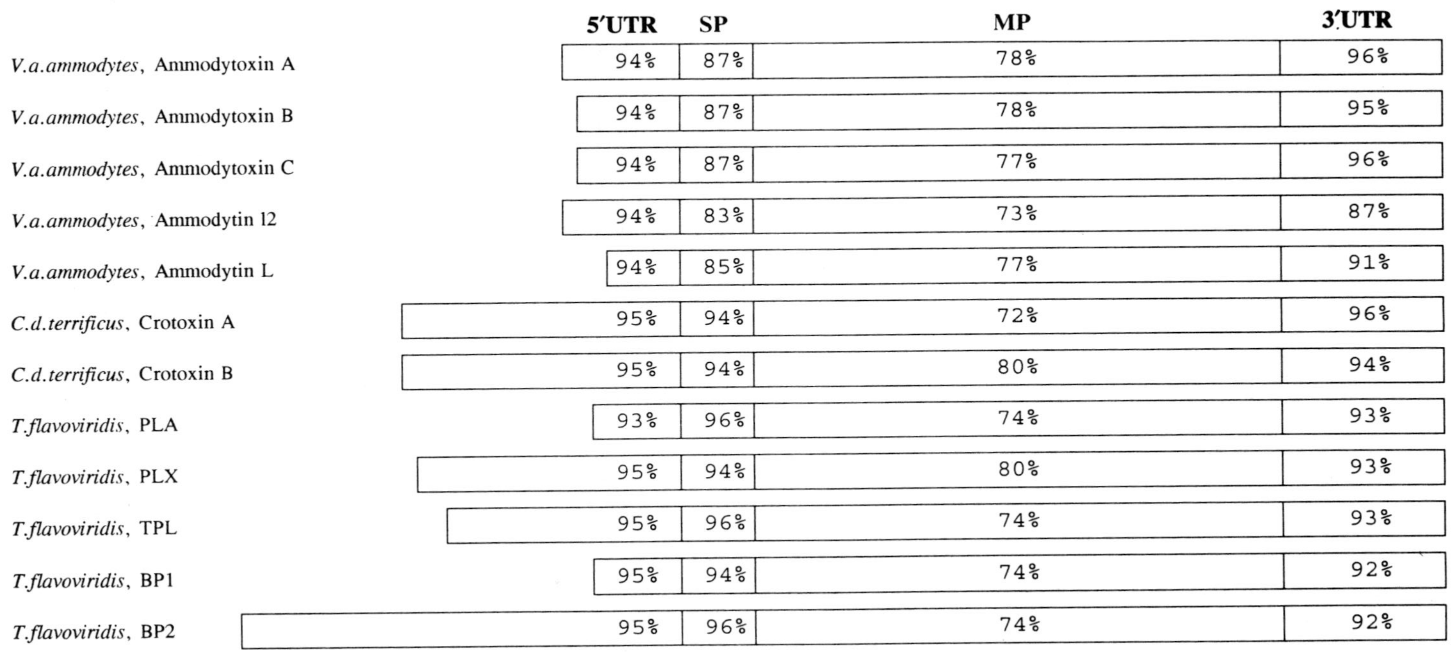

Fig. 2. Schematic representation of the similarity of snake Group II PLA_2s cDNA sequences compared with $BJUPLA_2$ cDNA. The percentage similarity of the 5′ untranslated region (5′ UTR), signal peptide coding region (SP), mature protein coding region (MP) and 3′ untranslated region (3′ UTR) is shown inside the bars. The sizes of the bars are proportional to the number of nucleotides.

Interestingly, both haemorrhagic toxins and disintegrins are coded by cDNAs that predict zymogen molecules containing a large precursor pro-peptide domain followed by a metalloproteinase domain and a carboxy-terminal disintegrin domain of varying length that replaces the haemopexin domain present in the structurally related MMPs (Paine, Desmond *et al.* 1992). The pro-peptide domain is apparently involved in the inactivation of the proteolytic activity prior to the secretion of the enzymes through a cysteine switch mechanism, as proposed for the MMPs (Woessner 1991) correlated with the sequence PKMCGV. The metalloproteinase domain has a zinc-chelating sequence, MXHGXXHXXGXXH, which is associated with the enzymatic activity of the toxins. The disintegrin domain is correlated with the binding to the platelet integrins through the RGD motif that is classically represented in venom disintegrins. Large haemorrhagins present a region of sequence similarity with the disintegrins with a substitution of the RGD motif with E/DCD at the same position, and a cysteine-rich extension of the carboxy-terminus.

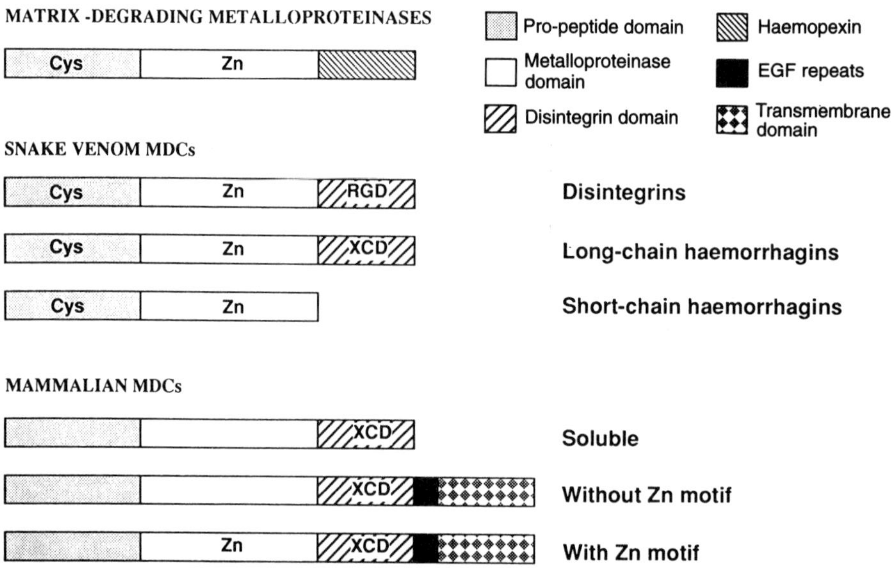

Fig. 3. Schematic representation of matrix-degrading metalloproteinases, venom and mammalian proteins belonging to the metalloproteinase/disintegrin/cysteine-rich gene family.
The presence of cysteine switch and zinc binding motifs is represented by Cys and Zn, respectively. RGD and XCD represent the potential integrin binding peptides in the disintegrin domain.

In recent years it has become apparent that a number of cysteine-rich proteins with similar structural organization occur in the mammalian male reproductive tract. The best characterized of these are Fertilin (former PH-30), a guinea pig sperm surface protein involved with sperm–egg fusion (Blobel *et al.* 1992), and the androgen-regulated epididymal apical protein-I, EAP-I (Perry *et al.* 1992). These sequences include a carboxy-terminal extension added to venom metalloproteinase sequences comprising an epidermal growth factor (EGF) repeat, a transmembrane domain and a cytoplasmatic tail (Weskamp & Blobel 1994). Representatives of this protein family were also detected in other tissues as the MS-2 antigen present on the surface of certain lineages of macrophages (Yoshida *et al.* 1990), and the protein predicted by the MDC gene, which has been associated with tumour suppression (Emi *et al.* 1993).

All the mammalian proteins present a non-RGD disintegrin domain characteristic of the large haemorrhagins. The zinc-binding motif is present only in EAPs and Fertilin alpha metalloproteinase domains, and the cysteine switch motif is absent in all the pro-domains. Figure 3 shows a schematic representation of this protein family.

Gene duplication is also suggested by the MDC family phylogeny

Evolutionary aspects of this complex gene family are still poorly understood. In our recent studies (A. M. Moura-da-Silva *et al.* in prep.), evolutionary trees have been constructed according to the alignments of the predicted proteins coded by the complete cDNA sequences of mammalian proteins, venom disintegrins and haemorrhagic toxins and some representatives of the MMPs (Fig. 4a). The two primary divergent groups comprise the MMPs and the MDC proteins respectively. The MDC cluster is apparently monophyletic and the sequences are grouped mainly according to function. The first group contains the sperm proteins related to Fertilin, and the second group includes the EAPs. The tumour suppressor gene and the monocyte surface antigen are located in the first and second groups respectively. Snake venom sequences are also distributed according to function, the first group representing the long-chained haemorrhagic toxins, the second clustering the RGD-disintegrins, and the third enclosing the short haemorrhagins/metalloproteinases.

It is possible that these results may be artificial, generated by the differences in chain length of each group of proteins. To exclude this possibility, we constructed trees using the fragments corresponding to the independent domains of each toxin. The trees corresponding to the pre + pro domain, metalloproteinase domain or disintegrin domain are shown in Fig. 4b, c and d, respectively. The same clustering characteristics are suggested by the trees generated using whole sequences or independent domains. The only exceptions are the Jararhagin and Rhodostomin sequences which diverged from the other long-chain haemorrhagins or disintegrins, respectively, in metalloproteinase and pro-domain alignments.

The apparent monophyletic distribution of MDC proteins and their independent domains suggests that both mammalian and snake venom proteins have evolved from a common ancestor gene, already assembled as the multidomain structure, either by speciation or by gene duplication. Gene duplications generating the alpha and beta chains of Fertilin and RGD and non-RGD venom disintegrins are likely to have occurred. Gene duplications may also have taken place between EAPs and Fertilin and between short-chained haemorrhagins and RGD-disintegrins. Speciation and the great variability of the metalloproteinase regions could explain the random positions of Jararhagin and Rhodostomin in the different trees. The generation of the different venom MDC proteins has been correlated with post-translational proteolytic processing or RNA splicing (Bjarnson & Fox 1994). However, if those were the only mechanisms operating, a taxonomically related distribution would be expected, at least for the trees generated with the sequences comprising the metalloproteinase

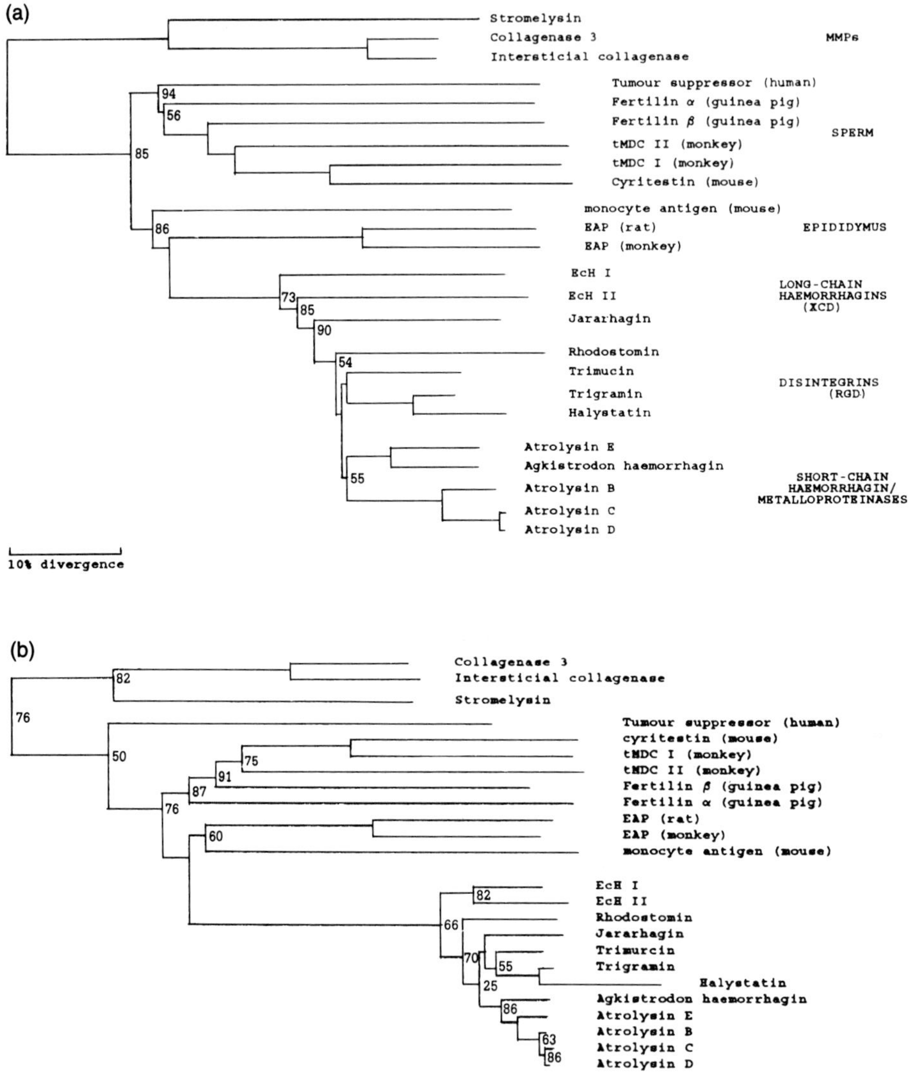

Fig. 4. Gene tree of the MDC proteins. All amino acid positions predicted by the CDNA regions coding for the whole protein (a), pro-peptide (b), metalloproteinase (c) or disintegrin (d) domains have been included in the analysis. Same representation as Fig. 1.

(c)

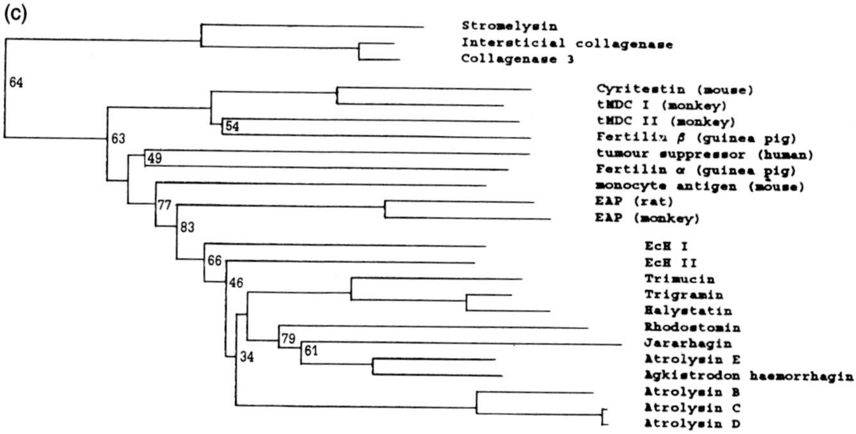
Stromelysin
Intersticial collagenase
Collagenase 3
Cyritestin (mouse)
tMDC I (monkey)
tMDC II (monkey)
Fertilin β (guinea pig)
tumour suppressor (human)
Fertilin α (guinea pig)
monocyte antigen (mouse)
EAP (rat)
EAP (monkey)
EcH I
EcH II
Trimucin
Trigramin
Halystatin
Rhodostomin
Jararhagin
Atrolysin E
Agkistrodon haemorrhagin
Atrolysin B
Atrolysin C
Atrolysin D
64
63
54
49
77
83
66
46
79
61
34

(d)

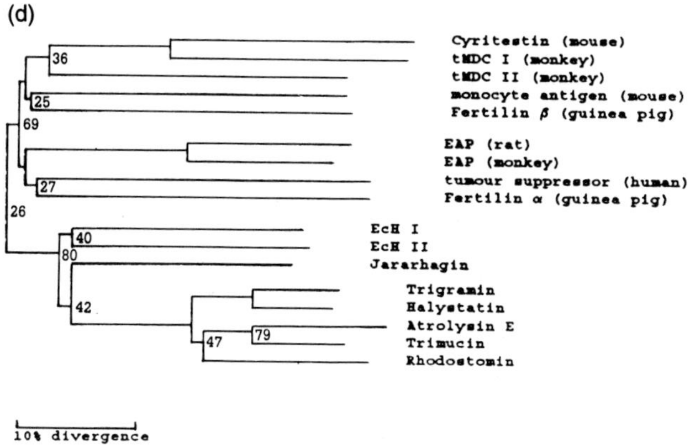
Cyritestin (mouse)
tMDC I (monkey)
tMDC II (monkey)
monocyte antigen (mouse)
Fertilin β (guinea pig)
EAP (rat)
EAP (monkey)
tumour suppressor (human)
Fertilin α (guinea pig)
EcH I
EcH II
Jararhagin
Trigramin
Halystatin
Atrolysin E
Trimucin
Rhodostomin
36
25
69
27
26
40
80
42
79
47
10% divergence

and pro-domains. The possible occurrence of different processing of RNA and translated proteins is not eliminated; however, this may have arisen on products generated by different gene copies. Further studies using alignments of genomic DNA sequences and independent exons may provide another basis for this observation once a number of genomic DNA sequences for MDC gene family become available.

The trees also suggest late divergence of venom toxins. This could imply that venom toxins have evolved from putative proteins in the reproductive tract of snakes. However, MDC proteins have not yet been identified on snake sperm cells or on the epididymis, probably because of lack of investigation.

Variability in MDC family

The similarity of cDNA sequences coding for independent domains of snake venom MDCs has already been reported (Paine, Moura-da-Silva *et al.* 1994). The most variable domain appeared to be the metalloproteinase, followed by the disintegrin domain. The fragment of cDNA coding for the pro-peptide was the most conserved, together with the 5′ and 3′ untranslated regions.

In this paper, nucleotide substitutions for each independent domain have been estimated by comparison of Fertilin alpha and monkey EAP-I as a mammalian example and Halystatin and EcH-I for snake venom (Table 2). Mammalian proteins were much more diverged than venom toxins, and this makes it difficult to compare precisely the rate of substitutions with few sequences. Nevertheless, a high variability of the snake venom metalloproteinase region has been observed. The pattern suggests that more substitutions have occurred in non-synonymous positions. Neutral selection appears to be responsible for the evolution of mammalian MDCs. Synonymous substitutions have been higher only in the regions coding for the snake pro-peptide domains and mammalian zinc chelating motifs. These regions may comprise important functional motifs under strong selective pressure.

Table 2. Pattern of nucleotide substitution per 100 sites (shortest sequence excluding gaps) in different domains of cDNAs coding for snake venom and mammal metalloproteinase/disintegrins

		PRO	MET	DIS	MET (cat.site)
Snake	Total	23	36	25	11
	Synonymous	11 (49%)	8 (22%)	6 (24%)	2 (18%)
	Non-synonymous	12	28	19	9
Mammal	Total	62	58	47	28
	Synonymous	18 (29%)	16 (27%)	14 (30%)	22 (80%)
	Non-synonymous	44	42	33	6

Conclusions

In this study, gene duplications followed by high levels of variability in regions coding for functional domains have been suggested as mechanisms that may have made a significant contribution to the evolution of venom Group II PLA_2s and MDC proteins. The different gene copies subjected to variability have probably been fixed by positive Darwinian selection, thus generating slightly different products responsible for the wide spectrum of toxicity of viper venoms. Similar mechanisms also appear to have occurred in other gene families such as the serine proteases (Ohta 1994; Creighton & Darby 1989). However, alignments of a greater number of sequences will be necessary in order to predict more precisely where gene duplications have occurred and how these relate to the toxic activity of the proteins coded for by each of the resulting gene copies.

Acknowledgements

This research was funded by the Medical Research Council and the Wellcome Trust. JMC is a Wellcome Trust Senior Research Fellow in Basic Biomedical Sciences.

References

Altschul, S. F., Gish, W., Miller, W., Myers, E. W. & Lipman, D. J. (1990). Basic local alignment search tools. *J. molec. Biol.* **215**: 403–410.

Bjarnson, J. B. & Fox, J. W. (1994). Hemorrhagic metalloproteinases from snake venoms. *Pharmac. Ther.* **62**: 325–372.

Blobel, C. P., Wolfsberg, T. G., Turck, C. W., Myles, D. G., Primakoff, P. & White, J. M. (1992). A potential fusion peptide and an integrin ligand domain in a protein active in sperm–egg fusion. *Nature, Lond.* **356**: 248–252.

Bouchier, C., Boulain, J.-C., Bon, C. & Menez, A. (1991). Analysis of cDNAs encoding the two subunits of crotoxin, a phospholipase A_2 neurotoxin from rattlesnake venom: the acidic non enzymatic subunit derives from a phospholipase A_2-like precursor. *Biochim. biophys. Acta* **1088**: 401–408.

Brattstrom, B. H. (1964). Evolution of the pit vipers. *Trans. S. Diego Soc. nat. Hist.* **13**: 185–268.

Cintra, A. C. O., Marangoni, S., Oliveira, B. & Giglio, J. R. (1993). Bothropstoxin I: amino acid sequence and function. *J. Protein Chem.* **12**: 57–64.

Creighton, T. E. & Darby, N. J. (1989). Functional evolutionary divergence of proteolytic enzymes and their inhibitors. *Trends biochem. Sci.* **14**: 319–324.

Davidson, F. F. & Dennis, E. A. (1990). Evolutionary relationships and implications for the regulation of phospholipase A_2 from snake venom to human secreted forms. *J. molec. Evol.* **31**: 228–238.

Dayhoff, M. O., Schartz, R. M. & Orcutt, B. C. (1978). *Atlas of protein sequence and structure* **5** (suppl. 3). NBRF, Washington.

Dufton, M. J. & Hider, R. C. (1983). Classification of phospholipases A_2 according to sequence: evolutionary and pharmacological implications. *Eur. J. Biochem.* **137**: 545–551.

Emi, M., Katagiri, T., Harada, Y., Saito, H., Inazawa, J., Ito, I., Kasumi, F. & Nakamura, Y. (1993). A novel metalloprotease/disintegrin-like gene at 17q21.3 is somatically rearranged in two primary breast cancers. *Nature Genet.* **5**: 151–157.

Felsenstein, J. (1985). Confidence limits on phylogenies: an approach using the bootstrap. *Evolution* **39**: 783–791.

Gomis-Ruth, F. X., Kress, L. F. & Bode, W. (1993). First structure of a snake venom metalloproteinase: a prototype for matrix metalloproteases/collagenases. *EMBO J.* **12**: 4151–4157.

Gould, R. J., Polokoff, M. A., Friedman, P. A., Huang, T.-F., Holt, J. C., Cook, J. J. & Niewiarowski, S. (1990). Disintegrins — a family of integrin inhibitory proteins from viper venoms. *Proc. Soc. exp. Biol. Med.* **195**: 168–171.

Heinrikson, R. L., Krueger, E. T. & Keim, P. S. (1977). Amino acid sequences of phospholipase A_2 from the venom of *Crotalus adamanteus*. A new classification of phospholipases A_2 based upon structural determinants. *J. biol. Chem.* **252**: 4913–4921.

Higgins, D. G., Bleasby, A. J. & Fuchs, R. (1992). Clustal V: improved software for multiple sequence alignment. *Comput. Appl. Biosci.* **8**: 189–191.

Higgins, D. G. & Sharp, P. M. (1989). Fast and sensitive multiple sequence alignments on a microcomputer. *Comput. Appl. Biosci.* **5**: 151–153.

Huang, T. F., Holt, J. C., Kirby, E. P. & Niewiarowski, S. (1989). Trigramin—primary structure and its inhibition of von Willebrand factor binding to glycoprotein IIB/IIIA complex on human platelets. *Biochemistry* **28**: 661–666.

John, T. R., Smith, L. A. & Kaiser, I. I. (1994). Genomic sequences encoding the acidic and basic subunits of mojave toxin: unusually high sequence identity in non-coding regions. *Gene* **139**: 229–234.

Kamiguti, A. S., Theakston, R. D. G., Desmond, H. P. & Hutton, R. A. (1991). Systemic haemorrhage in rats induced by a haemorrhagic fraction from *Bothrops jararaca* venom. *Toxicon* **29**: 1097–1105.

Kini, R. M. & Evans, H. J. (1989). A model to explain the pharmacological effects of snake venom phospholipases A_2s. *Toxicon* **27**: 613–635.

Kini, R. M. & Iwanaga, S. (1986). Structure–function relationships of phospholipases II: charge density distribution and the myotoxicity of presynaptically neurotoxic phospholipases. *Toxicon* **24**: 895–905.

Kostetsky, P. V., Arkhipova, S. F. & Vladimirova, R. R. (1991). Conservative and variable regions of homologous snake phospholipases A_2 sequences: prediction of the taxon specific peptides structure. *J. Protein Chem.* **10**: 593–601.

Maraganore, J. M. & Heinrikson, R. L. (1986). The lysine-49 phospholipase A_2 from the venom of *Agkistrodon piscivorus piscivorus*. *J. biol. Chem.* **261**: 4797–4804.

Moura-da-Silva, A. M., Desmond, H., Laing, G. & Theakston, R. D. G. (1991). Isolation and comparison of myotoxins isolated from venoms of different species of *Bothrops* snakes. *Toxicon* **29**: 713–723.

Moura-da-Silva, A. M., Paine, M. J. I., Diniz, M. R. V., Theakston, R. D. G. & Crampton, J. M. (1995). The molecular cloning of a phospholipase A_2 from *Bothrops jararacussu* venom: evolution of venom Group II PLA_2's may imply gene duplication. *J. molec. Evol.* **41**: 174–179.

Nakashima, K. I., Ogawa, T., Oda, N., Hattori, M., Sakaki, Y., Kihara, H. & Ohno, M. (1993). Accelerated evolution of *Trimeresurus flavoviridis* venom gland phospholipase A_2 isozymes. *Proc. natn. Acad. Sci. USA* **90**: 5964–5968.

Ohta, T. (1994). On hypervariability at the reactive center of proteolytic enzymes and their inhibitors. *J. molec. Evol.* **39**: 614–619.

Paine, M. J. I., Desmond, H. P., Theakston, R. D. G. & Crampton, J. M. (1992). Purification, cloning, and molecular characterization of a high molecular weight hemorrhagic metalloprotease, jararhagin, from *Bothrops jararaca* venom. *J. biol. Chem.* **267**: 22869–22876.

Paine, M. J. I., Moura da Silva, A. M., Theakston, R. D. G. & Crampton, J. M. (1994). Cloning of metalloproteinase genes in carpet viper (*Echis pyramidum leakeyi*): further members of the metalloproteinase/disintegrin gene family. *Eur. J. Biochem.* **224**: 483–488.

Perry, A. C. F., Jones, R., Barker, P. J. & Hall, L. (1992). A mammalian epididymal protein with remarkable sequence similarity to snake venom hemorrhagic peptides. *Biochem. J.* **286**: 671–675.

Rosenberg, P. (1986). The relationships between enzymatic activity and pharmacological properties of phospholipases in natural poisons. In *Natural toxins: animal, plant and microbial*: 129–174. (Ed. Harris, J. B.). Clarendon Press, Oxford.

Saitou, N. & Nei, M. (1987). The neighbor-joining method: a new method for reconstructing phylogenetic trees. *Mol. Biol. Evol.* **4**: 406–425.

Warrell, D. A. (1983). Injures, envenoming, poisoning, and allergic reactions caused by animals. In *Oxford textbook of medicine* **1**: 6.35–6.45. (Eds Weatherall, D. J. & Ledingham, J. G. G.). Oxford University Press, Oxford.

Weskamp, G. & Blobel, C. P. (1994). A family of cellular proteins related to snake venom disintegrins. *Proc. natn. Acad. Sci. USA* **91**: 2748–2751.

Woessner, J. F. (1991). Matrix metalloproteinases and their inhibitors in connective-tissue remodelling. *FASEB J.* **5**: 2145–2154.

Yoshida, S., Setoguchi, M., Higuchi, Y., Akizuki, S. & Yamamoto, S. (1990). Molecular cloning of a cDNA encoding MS2 antigen, a novel cell surface antigen strongly expressed in murine monocytic lineage. *Int. Immunol.* **2**: 586–591.

14 Geographical and intraspecies variation in the clinical manifestations of envenoming by snakes

DAVID A. WARRELL

Synopsis

The range and severity of clinical manifestations after a bite by a particular species of snake and hence the prognosis depend on the quantity of venom injected, its composition, the site and depth of the bite, the age and previous health of the victim and the quality of first aid and medical treatment. The quantity and composition of venom will vary with the age/size of the snake, the mechanical efficiency of the bite and the amount of venom present in the venom apparatus immediately before the strike. Variation in venom composition and antigenicity throughout the geographical range of a single 'species' has been documented and, even at a particular geographical locus, composition of venom from individual snakes of the same age/size may not be homogeneous. All these sources of variation make it difficult to predict the outcome of a bite and to provide guidelines for clinical management, even if the species is known. Optimal medical treatment depends on the precise identification of the biting species. Ideally, this will be based on examination of the snake responsible for the bite or on rapid identification of specific venom antigen in the patient's serum or other body fluids, for example by enzyme immunoassay. Failing that, the clinician must decide on the most likely species, based on the circumstances of the bite, the patient's description of the appearance and behaviour of the snake and the symptoms and signs of envenoming. For antivenom production, use of a pool of venom from correctly identified snakes of different ages/sizes and geographical origins is essential in view of the intraspecies variation in venom composition and antigenicity.

Introduction

Importance of snake bite

Snake bites occur in most parts of the world, especially in the tropical developing countries. In parts of the Indian subcontinent, South East Asia, West Africa, Latin America and New Guinea, snake bites are some of the commonest medical emergencies. Only one study, by Swaroop & Grab (1954) of the Statistical Studies

Symp. zool. Soc. Lond. (1997) No. 70: 189–203
ISBN 0 19 854986 5

Section, World Health Organization, has attempted to assess the global incidence of snake bites and snake bite mortality. They estimated that between 30 000 and 40 000 people died of snake bite each year in the whole world, as a result of about a quarter of a million snake bites (Table 1). These authors relied on data provided by hospitals to national ministries of health. A number of more recent studies of snake bite in localized areas have suggested that these figures greatly underestimated the problem. It is clear that, in many parts of the rural tropics, especially in Africa, snake-bite victims do not go to hospital but die at home under the care of traditional healers (Warrell & Arnett 1976; Pugh *et al.* 1980; Sawai *et al.* 1984; Snow *et al.* 1994). Swaroop & Grab (1954) were unable to include data from China, the then USSR and Central European countries and they clearly underestimated the snake-bite mortality in Oceania (Lalloo *et al.* 1995) and Africa (Warrell & Arnett 1976). It now seems likely that more than 50 000 people die each year from snake bite. To reduce this mortality, and the unmeasured but considerable morbidity resulting from snake bite, efforts must be made to improve the first aid and hospital treatment of this common accident. Optimal treatment depends on identification of the snake responsible for the bite, which will allow the doctor to anticipate the evolution of symptoms and signs of envenoming, give appropriate treatment with specific antivenom, arrange necessary ancillary treatments such as mechanical ventilation for respiratory paralysis and be aware of the likely prognosis.

Table 1. Swaroop & Grab's (1954) estimates of annual snake-bite mortality

Area		Mortality
Asia		25 000–35 000
South America		3000–4000
North America and Mexico		300–500
Europe		50
Oceania		10
Africa		400–1000
	Total	30 000–40 000

Species diagnosis of snake bites

The most reliable way of attributing a bite is by examination of the snake responsible, but pursuit and capture of the animal should not be encouraged (for example in community education campaigns) as this creates the risk of further bites. Nor must the killing of snakes be encouraged, although this is bound to happen as a reaction to being bitten. Even if patients are able to bring the snake with them to hospital, diagnosis may not be possible if medical personnel are unable to identify it. In Thailand, young doctors were unable to distinguish two medically important vipers, Malayan pit viper (*Calloselasma rhodostoma*) and Russell's viper (*Daboia russelli siamensis*). Alternatively, immunodiagnosis could be used (Coulter, Sutherland & Broad 1974; Theakston, Lloyd-Jones & Reid 1977; Coulter, Harris &

Sutherland 1980; Ho *et al.* 1986). However, only the Commonwealth Serum Laboratories' kit has been produced commercially and shown to produce a result quickly enough to be useful clinically (Chandler & Hurrell 1982; Hurrell & Chandler 1982). Usually, the diagnosis must be inferred from patients' descriptions of the snake and the circumstances of the bite, the clinical picture of envenoming and results of simple tests such as the 20-min whole blood clotting test (Warrell *et al.* 1977; Sano-Martins *et al.* 1994).

To provide this pattern recognition of clinical features for the diagnosis of a snake-bite case, it is necessary to carry out detailed clinical studies of large groups of patients with bites proven to be by a particular species. However, there are differences in the clinical pattern and severity of envenoming among patients bitten by a single species within a small geographical focus and predictably greater differences between patients bitten by the same species at widely separate locations within the geographical range of that species.

Determinants of intraspecies variation in the pattern and severity of envenoming

Theoretical sources of variation in the pattern and severity of envenoming among a group of patients bitten by the same species of snake are summarized in Table 2.

Table 2. Sources of intraspecies variation in the pattern and severity of envenoming

	Variable	Determinant of variation
Snake-related factors	Venom: amount injected	Mechanical efficiency of bite Amount of venom in the venom apparatus Size/age of the snake Season Geographical region
	Venom: composition	Size/age of the snake Season Geographical region
	Site/depth of bite	Rate of systemic venom absorption Proximity of vital structures
Patient-related factors	Immunity/hypersensitivity to venom	Past exposure/immunization
	Age/size	Dose of venom/body weight or surface area
	Pre-existing state of health	Susceptibility to specific venom effects
	First aid and medical treatment	Efficacy in delaying, neutralizing envenoming

Intraspecies variation in the amount of venom injected

The amount of venom injected at a bite seems to be very variable and a proportion of human victims of snake bite, more than half in some series, showed negligible or no signs of envenoming even though the snake was identified as being venomous and there were obvious puncture marks indicating that the fangs had penetrated the skin (Reid 1968; Myint-Lwin *et al.* 1985; Phillips *et al.* 1988). These so-called 'dry' bites have led to the belief that snakes can somehow control the amount of venom injected and inflict a 'defensive bite' without expending venom when they strike an aggressor rather than a prey species. It seems unlikely that, in the very short period of time between the snake's being trodden upon and its striking the human patient, it could make this calculation. There is disagreement about whether the snake adjusts the dose of venom injected even when it makes elective strikes on its prey (Gennaro, Leopold & Merriam 1961; Allon & Kochva 1974; Russell 1980: 141–142). When humans are bitten, the circumstances of the bite and hence the mechanical efficiency with which the snake is able to bring its venom apparatus into action against an unexpected and abnormal target may be important. The idea that a snake might exhaust its supply of venom by a recent strike, and therefore be less venomous after eating or after striking another patient, has also been refuted. In several reports of consecutive strikes by a single snake on several human victims, the first bitten was not always the most severely envenomed (Fayrer 1869; Russell 1980: 156). In the case of Russell's viper bites in Burma, snakes which had recently fed (those which had a rodent in their stomach), inflicted bites no less venomous than those with empty stomachs (Tun-Pe, Ba-Aye *et al.* 1991). The largest specimens do not always inflict the most dangerous bites, but there is a tendency, in most species which have been studied, for the severity of local and systemic envenoming to increase with the size of the individual snake responsible (Tun-Pe, Ba-Aye *et al.* 1991). It is theoretically possible that, in some species, there may be seasonal and geographical variations in the amount of venom injected, but this has not been documented. Venom yield has been shown to increase with ambient temperature (Kochva & Gans 1967).

Intraspecies variation in venom composition

Age variation in venom composition

In several species of snake, venom composition changes as the snake grows older (Table 3). Minton (1957), in a *Crotalus atrox*, was the first to demonstrate this phenomenon. In several other crotaline species, anti-haemostatic activity is relatively more marked in the venoms of young snakes.

Seasonal variation in venom composition

Seasonal variation in the venom of *Vipera aspis* was suggested by Ambrose Parey (1678) but little evidence has been adduced in support of this phenomenon (Jiménez-Porras 1964; Boche, Chippaux & Courtois 1981).

Geographical variation in venom composition

There is now abundant literature describing variations in venom lethality, composition and activities between individuals of a particular species at the same geographical focus or from different places within the geographical range (Jiménez-Porras 1964; Táborská 1971; Kornalik & Táborská 1973; Gutiérrez, Chaves & Bolaños 1980; Boche *et al.* 1981; Minton & Weinstein 1986; Warrell 1989).

Table 3. Variation in venom composition with age: laboratory studies

Species	Change with increasing age	Reference
Crotalus atrox	Decreasing lethality	Minton (1957)
Bothrops/Atropoides/ Porthidium nummifer	No consistent change	Jiménez-Porras (1964)
Cr. horridus atricaudatus	Decreasing lethality, increasing haemorrhagic activity	Minton (1967)
Agkistrodon contortrix mokasen	None	Minton (1967)
Naja naja	Decreasing lethality	Minton (1967)
Cr. v. viridis	Decreasing lethality, increasing protein concentration	Fiero *et al.* (1972)
Cr. h. horridus	Decreasing procoagulant activity	Bonilla, Faith & Minton (1973)
Cr. atrox	Decreasing lethality and procoagulant activity	Reid & Theakston (1978)
B. asper	Decreasing lethality, proteolytic, haemorrhagic, oedema-forming and lethal effects	Gutiérrez, Chaves & Bolaños (1980)
	Increasing haemolytic and myonecrotic effects	Chaves, Gutiérrez & Brenes (1992)
B. atrox *Echis carinatus* *Naja nigricollis*	Decreasing lethality, increasing protein concentration	Meier & Freyvogel (1980)
B. atrox	Decreasing number of protein constituents	Meier (1986)
Cr. atrox	Decreasing lethality, increasing proteinase, acquisition of Mojave toxin	Minton & Weinstein (1986)
Bothrops sp.	Decreasing lethality and procoagulant activity	Furtado *et al.* (1991)

Intraspecies variation in the site and depth of the bite

In experimental animals, the route of venom injection (intradermal, subcutaneous, intramuscular, intraperitoneal, intravascular) affects the speed, pattern and lethality of envenoming, but this has not been studied in human patients. In anecdotal cases in which unusually rapid systemic effects have developed after a bite with minimal local signs, it has been assumed that introduction of venom by the bite may have been, fortuitously, by the intravenous route.

Clinical significance of variations in venom dose and composition

It was a case of envenoming by a two-month-old *Crotalus atrox* in which the patient became defibrinogenated (Reid 1978) that prompted Reid & Theakston (1978) to study the effect of the snake's age on its venom composition. Similarly, Minton (1967) had read of unusually severe effects of envenoming by neonatal copperhead and canebrake rattlesnake. In Brazil, bites by small/young *Bothrops jararaca* and *B. moojeni* are more likely to produce incoagulable blood than bites by large/adult specimens (Furtado *et al.* 1991; Ribeiro & Jorge 1989; Kouyoumdjian & Polizelli 1989). Clinicians must not underestimate the potential toxicity of small, young or even neonatal specimens of venomous snakes; among *Bothrops* and

Crotalus species, their venoms may have greater lethal toxicity, haemorrhagic and procoagulant activity than those of more mature specimens. In Myanmar, bites were inflicted by two discrete size groups of Russell's vipers; snakes < 375 mm in total length, presumably born that year, and those > 550 mm in length (Tun-Pe, Ba-Aye *et al.* 1991) (Fig. 1). Small snakes were capable of causing local and systemic envenoming, but less frequently than the larger snakes and with a lower risk of a fatal outcome.

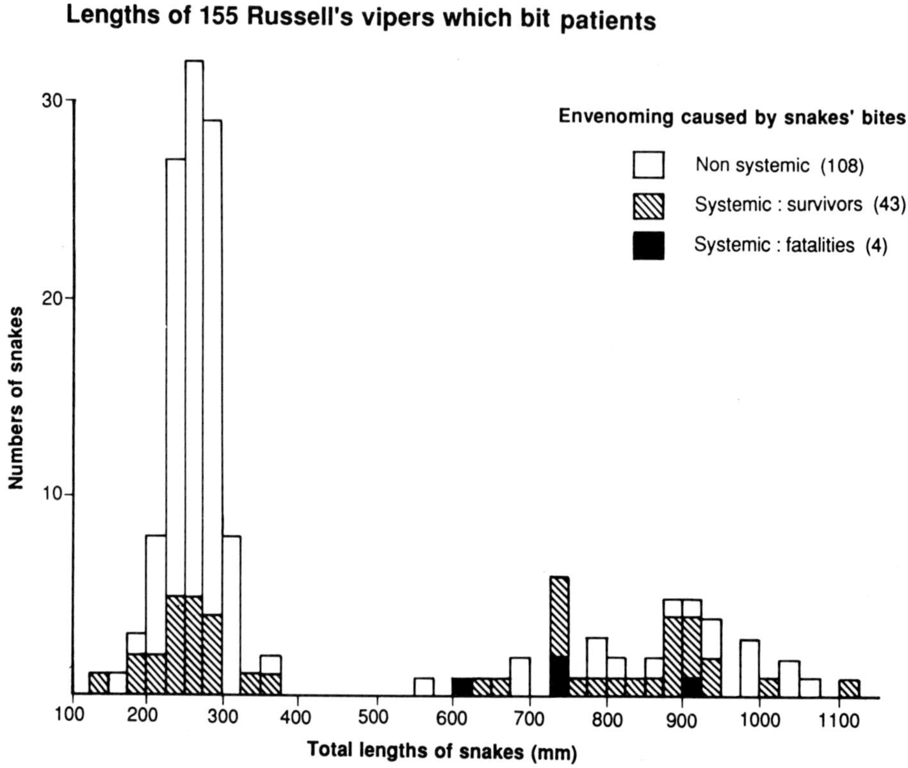

Fig. 1. Distribution of total lengths of Russell's vipers (*Daboia russelli siamensis*) responsible for biting patients in Tharrawaddy, Burma. Two separate size groups of snakes were involved. Small snakes (presumably born that year) were less likely than large snakes to cause systemic envenoming and were not responsible for any fatal cases in this series. (From Tun-Pe, Ba-Aye *et al.* 1991).

Seasonal variations in the case fatality of bites by a single species have been suspected but not convincingly demonstrated.

Geographical intraspecies variations in the pattern and severity of envenoming

Some striking differences have been observed in the clinical effects of envenoming in different parts of the geographical range of several medically important, abundant and widely distributed 'species'. In some cases, these differences have prompted reconsideration of the validity of the 'species' and it seems likely that 'intraspecies' variation will become 'interspecies' variation (Wüster & Thorpe 1991; Wüster *et al.* 1992). Unfortunately, thorough documentation of the clinical features and laboratory abnormalities in adequate-sized groups of patients with proven bites by a particular species of snake has rarely been attempted. Russell's viper (*Daboia* or *Vipera russelli*) has been regarded as comprising five subspecies, *russelli, pulchella, siamensis, formosensis* and *limitis*, distributed discontinuously throughout 10 south Asian countries (Warrell 1989), but the recent studies of Wüster *et al.* (1992) have rejected the validity of several of these subspecies and supported the concept of western (*russelli*) and eastern (*siamensis*) forms. Bites by Russell's viper are a familiar problem to clinicians in most parts of its range. The clinical effects of what has been regarded as the same subspecies (*D. r. siamensis*) are very different in Burma and Thailand. Envenoming is generally more severe in Burma with two distinctive manifestations, haemorrhagic infarction of the anterior pituitary (Fig. 2) and generalized increase in capillary permeability resulting in oedema of the conjunctivae (chemosis) (Fig. 3) and in some patients pulmonary oedema (Warrell 1986, 1989; Tun-Pe, Phillips *et al.* 1987; Myint-Lwin *et al.* 1985) (Fig. 4). In Sri Lanka and southern India, victims of Russell's viper bites (formerly regarded as *D. r. pulchella* in Sri Lanka — Fig. 5 — and *D. r. russelli* in India but now grouped together as *D. r. russelli*) show signs of neurotoxicity (Fig. 6), generalized damage to skeletal muscle and intravascular haemolysis (Phillips *et al.* 1988). The composition of Russell's viper venom was found to differ in different parts of India (Jayanthi & Gowda 1988). Venom from south Indian snakes had a higher phospholipase activity and higher lethal potency. These observations could explain the occurrence of neurotoxicity in Kerala and other parts of south India (C. K. Eapen, N. Chandy & J. K. Joseph pers. comm.).

The content of the neurotoxin crotamine varies between different populations of tropical (South American) rattlesnake, *Crotalus durissus*, in Brazil and Argentina. Crotamine was absent from populations in north and eastern Brazil, present in north-western São Paulo State and adjacent areas of Paraná and in Argentina, while in snakes from south-eastern São Paulo State and Minas Gerais and in Ceará, there were mixed populations, some secreting crotamine and some not (Schenberg 1955). A possible clinical correlation with the presence of crotamine is a particular symptom/sign, 'broken neck' resulting from paralysis of the cervical muscles possibly through the direct action of crotamine. This clinical feature has been reported from various parts of Latin America (Minton & Minton 1971).

Another species complex of great medical importance is the Asian cobras, formerly regarded as subspecies of *Naja naja*, but more recently reclassified as a number of single species (Wüster & Thorpe 1991). Reliable clinical reports suggest considerable variation in the relative frequency of neurotoxicity and local necrosis in victims of cobra bites in various parts of Asia. In Luzon, the Philippines, all but one of a group of 39 patients bitten by *N. philippinensis* developed neurotoxicity, often with startling rapidity, whereas only two showed necrosis at the site of the bite (Watt *et al.* 1988), whereas in north-west Malaysia, among 47 patients bitten by cobras (two by *N. sumatrana*, the rest probably by *N. kaouthia*), only 12% developed neurotoxic

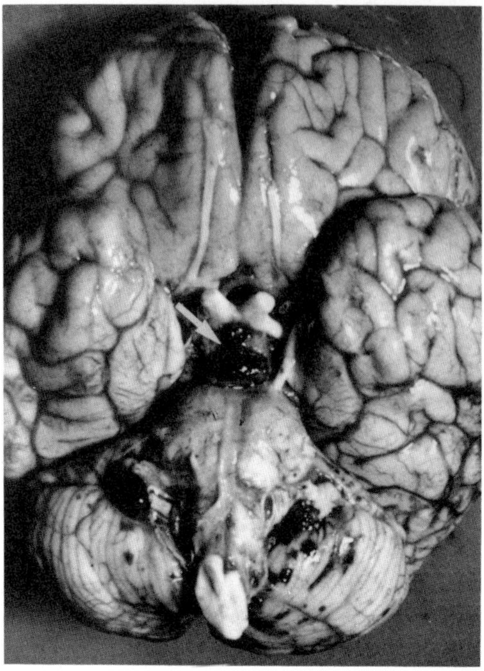

Fig. 2. Acute haemorrhagic infarction of the anterior pituitary (arrow) in a Burmese farmer who died three days after being bitten by a Russell's viper (*Daboia russelli siamensis*). (By courtesy of Dr U Hla Mon.)

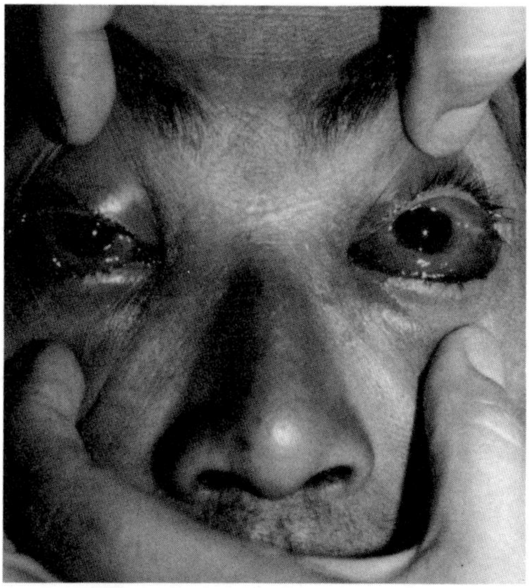

Fig. 3. Burmese farmer showing evidence of generalized increase in capillary permeability (conjunctival oedema — chemosis) after being bitten by a Russell's viper (*Daboia russelli siamensis*). (Copyright D.A. Warrell.)

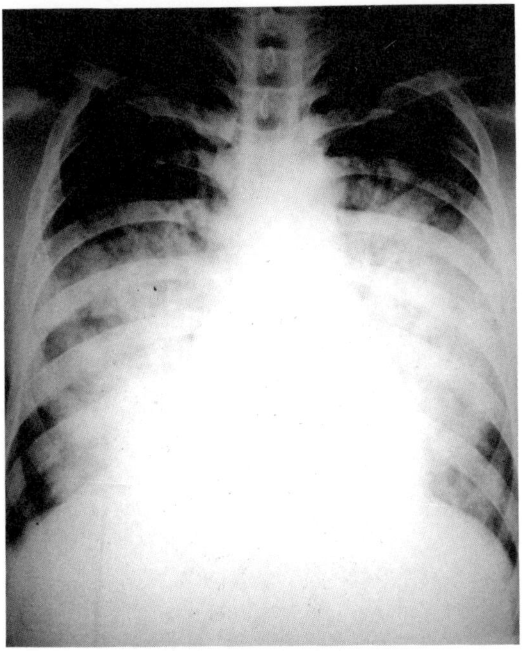

Fig. 4. Burmese farmer who developed acute pulmonary oedema, evidence of a generalized increase in capillary permeability, after being bitten by a Russell's viper (*Daboia russelli siamensis*). (Copyright D.A. Warrell.)

Fig. 5. Sri Lankan Russell's viper (*Daboia russelli russelli*). (Copyright D.A. Warrell.)

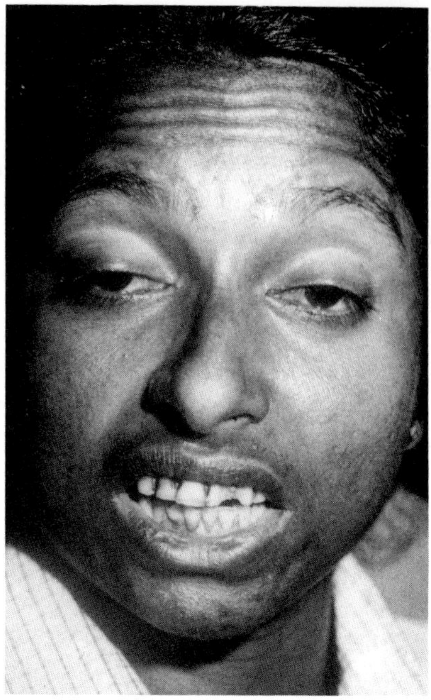

Fig. 6. Sri Lankan woman showing neurotoxic signs after being bitten by a Russell's viper (*Daboia russelli*). Note drooping of the eyelids (ptosis). The patient is attempting to open her eyes by contracting the frontalis muscle (note puckered brow). She is unable to open her mouth and there is evidence of paralysis of the facial muscles. (Copyright D.A. Warrell.)

signs, but 44% showed local necrosis (Reid 1964). The *Echis* complex, which is also under taxonomic revision (Cherlin & Borkin 1990; Cherlin 1990), is distributed throughout the northern part of Africa, the Middle East, north into the former Soviet Union and throughout the Indian subcontinent. Considerable differences in the severity and preponderance of different features of envenoming have been described (Warrell & Arnett 1976; Warrell *et al.* 1977; Gillissen, Theakston, Barth, May, Hatz & Krieg 1994; Gillissen, Theakston, Barth, May, Krieg & Warrell 1994). Venom lethality and activity, judged by the standard WHO assays, show considerable variation among samples collected from different parts of the world (Theakston & Reid 1983; Theakston 1986).

Geographical variations in venom composition: importance in the production of antivenoms

In their selection of venoms for immunizing animals in the production of therapeutic antivenoms, manufacturers must recognize the problem of intraspecific variation in the composition and hence the immunogenicity of snake venoms. Serious problems have arisen when these variations have been ignored. In north-eastern Nigeria, antivenom raised against Iranian *Echis* venom was imported by a mission hospital to treat the large numbers of victims of *E. ocellatus* bites. The antivenom proved ineffective and the resulting increase in hospital case fatality rate discouraged patients from coming to hospital for treatment for several years (Warrell & Arnett 1976). Subsequent studies revealed important differences between the venoms of snakes which were then regarded as being the same species in Iran and Nigeria (M. Latifi pers. comm.). A patient bitten by an *Echis* from Pakistan failed to respond to 16 ampoules of Indian (Haffkine) antivenom (Weis *et al.* 1991) and a patient bitten by a Tunisian '*Echis pyramidum* complex' showed little response to six ampoules of Behring, five of Pasteur and 20 of South African Institute of Medical Research antivenom, all with specific activity against '*Echis carinatus*' (Gillissen, Theakston, Barth, May, Krieg & Warrell 1994). In Sri Lanka, the Indian (Haffkine) antivenom raised against Indian Russell's viper venom was not very effective in treating envenoming by the local 'subspecies' *D. r. pulchella* (Phillips *et al.* 1988). In Ecuador, a different problem was encountered. An antivenom manufactured in Mexico was imported by the Ministry of Public Health because it was purported to have activity against '*Bothrops atrox*' venom from specimens of Mexican origin. Although the taxonomy of Middle and South American *Bothrops* species is under debate, most would probably regard the species in Mexico as being *B. asper* rather than *B. atrox*. In the event, the Mexican antivenom proved to have negligible activity against Ecuadorean *B. asper* or *B. atrox* (Theakston, Laing *et al.* 1995).

Immunological differences between cobra venoms from different parts of Asia have been described, and it is clear that the efficacy of antivenoms must be assessed carefully. Thus, in mice, an antivenom raised against Malaysian spitting cobra venom (*N. sumatrana*) was effective against this venom (ED_{50} 47 µg/mouse) but was ineffective against Thai *N. kaouthia* venom and Indian and Sri Lankan *N. naja* venoms (R. D. G. Theakston unpubl.). Indian anti-cobra antivenom proved ineffective against Philippine cobra venom (Voghtman 1950).

To broaden the geographical range of efficacy of antivenoms, the venom used for immunization should be taken from a large pool including samples from specimens collected from different locations. Venom from snakes of different ages should also be included. To avoid inappropriate purchase and use of antivenoms, the geographical origin of the venoms should be stated in the 'package insert' and the accepted nomenclature should be used. The efficacy of antivenoms against venoms from other geographical areas can be tested by using the WHO standard assays (Theakston & Reid 1983).

Conclusion

Clinicians' diagnosis of the species responsible for a case of snake bite usually depends on their recognition of a characteristic pattern of symptoms, signs and results of laboratory tests. However, variations in the amount of venom injected, its lethality and composition may occur within a single species and can result in clinical confusion. The range of clinical manifestations of bites by a single species throughout its range must be documented by more clinical studies. In treating a case of snake bite, the clinician will choose an antivenom with specific activity against the venom of the snake thought to be responsible. However, antivenoms which are raised against venoms collected from snakes of narrow age and geographical ranges may lack activity against variant venoms possessed by some members of the species. The therapeutic consequences may be disastrous. Antivenom manufacturers should be aware of these variations in venom composition and, in the production process, should choose venom samples from snakes of different ages collected throughout the geographical range of the 'species'.

References

Allon, N. & Kochva, E. (1974). The quantities of venom injected into prey of different size by *Vipera palaestinae* in a single bite. *J. exp. Zool.* **188**: 71–76.

Boche, J., Chippaux, J.-P. & Courtois, B. (1981). Contribution à l'étude des variations biochimiques des venins de serpents d'Afrique de l'Ouest. *Bull. Soc. Path. exot.* **74**: 356–366.

Bonilla, C. A., Faith, M. R. & Minton, S. A. (1973). L-amino acid oxidase, phosphodiesterase, total protein and other properties of juvenile timber rattlesnake (*C. h. horridus*) venom at different stages of growth. *Toxicon* **11**: 301–303.

Chandler, H. M. & Hurrell, J. G. R. (1982). A new enzyme immunoassay system suitable for field use and its application in a snake venom detection kit. *Clinica chim. Acta* **121**: 225–230.

Chaves, F., Gutiérrez, J. M. & Brenes, F. (1992). Pathological and biochemical changes induced in mice after intramuscular injection of venom from newborn specimens of the snake *Bothrops asper* (terciopelo). *Toxicon* **30**: 1099–1109.

Cherlin, V. A. (1990). Taxonomic revision of the snake genus *Echis* (Viperidae). II. An analysis of taxonomy and description of new forms. *Trudy zool. Inst., Leningr.* **207**: 193–223.

Cherlin, V. A. & Borkin, L. J. (1990). Taxonomic revision of the snake genus *Echis* (Viperidae). I. An analysis of the history of study and synonymy. *Trudy zool. Inst., Leningr.* **207**: 175–192.

Coulter, A. R., Harris, R. D. & Sutherland, S. K. (1980). Enzyme immunoassay for the rapid clinical identification of snake venom. *Med. J. Aust.* **1**: 433–435.

Coulter, A. E., Sutherland, S. K. & Broad, A. J. (1974). Assay of snake venoms in tissue fluids. *J. Immun. Meth.* **4**: 297–300.

Fayrer, J. (1869). Deaths from snakebites; a trial, condensed from the sessions' report. *Indian med. Gaz.* **4**: 156–157.

Fiero, M. K., Seifert, M. W., Weaver, T. J. & Bonilla, C. A. (1972). Comparative study of juvenile and adult prairie rattlesnake (*Crotalus viridis viridis*) venoms. *Toxicon* **10**: 81–82.

Furtado, M. F. D., Maruyama, M., Kamiguti, A. S. & Antonio, L. C. (1991). Comparative study of nine *Bothrops* snake venoms from adult female snakes and their offspring. *Toxicon* **29**: 219–226.

Gennaro, J. F., Leopold, R. S. & Merriam, T. W. (1961). Observations on the actual quantity of venom introduced by several species of crotalid snakes in their bite. *Anat. Rec.* **139**: 303.

Gillisen, A., Theakston, R. D. G., Barth, J., May, B., Hatz, C. & Krieg, M. (1994). Defibrinierungssyndrom nach Biss einer nordafrikanischen Sandrasselotter (*Echis carinatus pyramidum*). *Medsche Klinik* **89**: 392–396.

Gillissen, A., Theakston, R. D. G., Barth, J., May, B., Krieg, M. & Warrell, D. A. (1994). Neurotoxicity, haemostatic disturbances and haemolytic anaemia after a bite by a Tunisian saw-scaled or carpet viper (*Echis 'pyramidum'* complex): failure of antivenom treatment. *Toxicon* **32**: 937–944.

Gutiérrez, J. M., Chaves, F. & Bolaños, R. (1980). Estudio comparativo de venenos de ejemplares recien nacidos y adultos de *Bothrops asper*. *Revta Biol. trop.* **28**: 341–353.

Ho, M., Warrell, M. J., Warrell, D. A., Bidwell, D. & Voller, A. (1986). A critical reappraisal of the use of enzyme-linked immunosorbent assays in the study of snake bite. *Toxicon* **24**: 211–221.

Hurrell, J. G. R. & Chandler, H. W. (1982). Capillary enzyme immunoassay field kits for the detection of snake venom in clinical specimens. A review of two years' use. *Med. J. Aust.* **2**: 236–237.

Jayanthi, G. P. & Gowda, T. V. (1988). Geographical variation in India in the composition and lethal potency of Russell's viper (*Vipera russelli*) venom. *Toxicon* **26**: 257–264.

Jiménez-Porras, J. M. (1964). Intraspecific variations in composition of venom of the jumping viper, *Bothrops nummifera*. *Toxicon* **2**: 187–195.

Kochva, E. & Gans, C. (1967). The structure of the venom gland and secretion of venom in viperid snakes. In *Animal toxins*: 195–203. (Eds Russell, F. E. & Saunders, P. R.). Pergamon Press, Oxford.

Kornalik, F. & Táborská, E. (1973). Individual interspecies variability in the composition of some Viperidae venoms. In *Animal and plant toxins*: 99–103. (Ed. Kaiser, E.). Wilhelm Goldmann, München.

Kouyoumdjian, J. A. & Polizelli, C. (1989). Acidentes ofídicos causados por *Bothrops moojeni*: correlação do quadro clínico com o tamanho da serpente. *Revta Inst. Med. trop. S. Paulo* **31**: 84–90.

Lalloo, D. G., Trevett, A. J., Korinhona, A., Nwokolo, N., Laurenson, I., Paul, M., Black, J., Naraqi, S., Mavo, B., Saweri, A., Hutton, R. A., Theakston, R.D.G. & Warrell, D. A. (1995). Snakebites by the Papuan taipan (*Oxyuranus scutellatus canni*): paralysis, haemostatic and electrocardiographic abnormalities and effects of venom. *Am. J. trop. Med. Hyg.* **52**: 525–531.

Meier, J. (1986). Individual and age-dependent variations in the venom of the fer-de-lance (*Bothrops atrox*). *Toxicon* **24**: 41–46.

Meier, J. & Freyvogel, T. A. (1980). Comparative studies on venoms of the fer-de-lance (*Bothrops atrox*), carpet viper (*Echis carinatus*) and spitting cobra (*Naja nigricollis*) snakes at different ages. *Toxicon* **18**: 661–662.

Minton, S. A. (1957). Variation in yield and toxicity of venom from a rattlesnake (*Crotalus atrox*). *Copeia* **1957**: 265–268.

Minton, S. A. (1967). Observations on toxicity and antigenic makeup of venoms from juvenile snakes. In *Animal toxins*: 211–222. (Eds Russell, F. E. & Saunders, P. R.). Pergamon Press, Oxford.

Minton, S. A. & Minton, M. R. (1971). *Venomous reptiles*. George Allen & Unwin Ltd, London.

Minton, S. A. & Weinstein, S. A. (1986). Geographic and ontogenic variation in venom of the western diamondback rattlesnake (*Crotalus atrox*). *Toxicon* **24**: 71–80.

Myint-Lwin, Warrell, D. A., Phillips, R. E., Tin-Nu-Swe, Tun-Pe & Maung-Maung-Lay (1985). Bites by Russell's viper (*Vipera russelli siamensis*) in Burma: haemostatic, vascular and renal disturbances and response to treatment. *Lancet* ii: 1259–1264.

Parey, A. (1678). *The works of that famous chirurgeon Ambrose Parey, translated out of Latin, and compared with the French by Th. Johnson*. Clark, London.

Phillips, R. E., Theakston, R. D. G., Warrell, D. A., Galigedara, Y., Abeysekera, D. T. D. J., Dissanayaka, P., Hutton, R. A. & Aloysius, D. J. (1988). Paralysis, rhabdomyolysis and haemolysis caused by bites of Russell's viper (*Vipera russelli pulchella*) in Sri Lanka: failure of Indian (Haffkine) antivenom. *Q. Jl Med.* **68**: 691–716.

Pugh, R. N. H., Theakston, R. D. G., Reid, H. A. & Bhar, I. S. (1980). Epidemiology of human encounters with the spitting cobra, *Naja nigricollis*, in the Malumfashi area of northern Nigeria. *Ann. trop. Med. Parasit.* 74: 523–530.

Reid, H. A. (1964). Cobra-bites. *Br. med. J.* 2: 540–545.

Reid, H. A. (1968). Symptomatology, pathology, and treatment of land snake bite in India and southeast Asia. In *Venomous animals and their venoms*: 1. *Venomous vertebrates*: 611–642. (Eds Bücherl, W., Buckley, E. E. & Deulofeu, V.). Academic Press, New York & London.

Reid, H. A. (1978). Bites by foreign venomous snakes in Britain. *Br. med. J.* 1: 1598–1600.

Reid, H. A. & Theakston, R. D. G. (1978). Changes in coagulation effects by venoms of *Crotalus atrox* as snakes age. *Am. J. trop. Med. Hyg.* 27: 1053–1057.

Ribeiro, L. A. & Jorge, M. T. (1989). Alteração do tempo de coagulação sangüinea em pacientes picados por serpentes *Bothrops jararaca* adulta e filhote. *Revta Hosp. Clín. Fac. Med. Univ. S. Paulo* 44: 143–145.

Russell, F. E. (1980). *Snake venom poisoning*. Lippincott, Philadelphia.

Sano-Martins, I. S., Fan, H. W., Castro, S. C. B., Tomy, S. C., França, F. O. S., Jorge, M. T., Kamiguti, A. S., Theakston, R. D. G., Warrell, D. A. & Bias, G. (1994). Reliability of the simple 20 minute whole blood clotting tests (WBCT20) as an indicator of low plasma fibrinogen concentration in patients envenomed by *Bothrops* snakes. *Toxicon* 32: 1045–1050.

Sawai, Y., Toriba, M., Itokawa, H., de Silva, A., Perera, G. L. S. & Kottegoda, M. B. (1984). Study on deaths due to snakebite in Anuradhapura district, Sri Lanka. *Snake* 16: 7–15.

Schenberg, S. (1955). Geographical pattern of crotamine distribution in the same rattlesnake subspecies. *Science* 129: 1361–1363.

Snow, R. W., Bronzan, R., Roques, T., Nyamawi, C., Murphy, S. & Marsh, K. (1994). The prevalence and morbidity of snake bite and treatment — seeking behaviour among a rural Kenyan population. *Ann. trop. Med. Parasit.* 88: 665–671.

Swaroop, S. & Grab, B. (1954). Snakebite mortality in the world. *Bull. Wld Hlth Org.* 10: 35–76.

Táborská, E. (1971). Intraspecies variability of the venom of *Echis carinatus*. *Physiologia bohemoslov.* 20: 307–318.

Theakston, R. D. G. (1986). Characterization of venoms and standardization of antivenoms. In *Natural toxins. Animal, plant and microbial*: 287–303. (Ed. Harris, J. B.). Clarendon Press, Oxford.

Theakston, R. D. G., Laing, G. D., Freire Lascano, A., Wüster, W., Touzet, J.-M., Vallejo, F., Guderian, R. H., Nelson, S. J., Rumbea Guzman, J. & Warrell, D. A. (1995). Treatment of snake bites by *Bothrops* species and *Lachesis muta* in Ecuador: laboratory screening of candidate antivenoms. *Trans. R. Soc. trop. Med. Hyg.* 89: 550–554.

Theakston, R. D. G., Lloyd-Jones, M. J. & Reid, H. A. (1977). Micro-Elisa for detecting and assaying snake venom and venom-antibody. *Lancet* ii: 639–641.

Theakston, R. D. G. & Reid, H. A. (1983). The development of simple standard assay procedures for the characterization of snake venoms. *Bull. Wld Hlth Org.* 61: 949–956.

Tun-Pe, Ba-Aye, Aye-Aye-Myint, Tin-Nu-Swe & Warrell, D. A. (1991). Bites by Russell's vipers (*Daboia russelii siamensis*) in Myanmar: effect of the snake's length and recent feeding on venom antigenaemia and severity of envenoming. *Trans. R. Soc. trop. Med. Hyg.* 85: 804–808.

Tun-Pe, Phillips, R. E., Warrell, D. A., Moore, R. A., Tin-Nu-Swe, Myint-Lwin & Burke, C. W. (1987). Acute and chronic pituitary failure resembling Sheehan's syndrome following bites by Russell's viper in Burma. *Lancet* ii: 763–767.

Voghtman, D. B. (1950). The relative efficiency of two types of anti-venom sera in neutralizing cobra venom. *Copeia* 1950: 225–228.

Warrell, D. A. (1986). Tropical snake bite: clinical studies in South East Asia. In *Natural toxins. Animal, plant and microbial*: 25–45. (Ed. Harris, J. R.). Clarendon Press, Oxford.

Warrell, D. A. (1989). Snake venoms in science and clinical medicine. 1. Russell's viper: biology, venom and treatment of bites. *Trans. R. Soc. trop. Med. Hyg.* **83**: 732–740.

Warrell, D. A. & Arnett, C. (1976). The importance of bites by the saw-scaled or carpet viper (*Echis carinatus*). Epidemiological studies in Nigeria and a review of the world literature. *Acta trop.* **33**: 307–341.

Warrell, D. A., Davidson, N. McD., Greenwood, B. M., Ormerod, L. D., Pope, H. M., Watkins, B. J. & Prentice, C. R. M. (1977). Poisoning by bites of the saw-scaled or carpet viper (*Echis carinatus*) in Nigeria. *Q. Jl Med.* **46**: 33–62.

Watt, G., Padre, L., Tuazon, M. L., Theakston, R. D. G. & Laughlin, L. (1988). Bites by the Philippine cobra (*Naja naja philippinensis*): prominent neurotoxicity with minimal local signs. *Am. J. trop. Med. Hyg.* **39**: 306–311.

Weis, J. R., Whatley, R. E., Glenn, J. L. & Rodgers, G. M. (1991). Prolonged hypofibrinogenemia and protein C activation after envenoming by *Echis carinatus sochureki*. *Am. J. trop. Med. Hyg.* **44**: 452–460.

Wüster, W., Otsuka, S., Malhotra, A. & Thorpe, R. S. (1992). Population systematics of Russell's viper: a multivariate study. *Biol. J. Linn. Soc.* **47**: 97–113.

Wüster, W. & Thorpe, R. S. (1991). Asiatic cobras: systematics and snakebite. *Experienta* **47**: 205–209.

15 Snake venom and snakebite in Australia

V. WILLIAMS and J. WHITE

Synopsis

Australian snakes are amongst the most toxic in the world, but because of the distribution of the human and snake populations and the availability of excellent medical care, of the 500–3000 cases of suspected snakebite per year, only two on average have a fatal outcome. Brown snakes (*Pseudonaja* spp.) are responsible for the majority of bites and have been associated with sudden rapid death.

Victims of snake bite should have details of the bite recorded and a relevant medical history gathered, with particular reference to allergic responses and previous exposure to antivenom. In situations where the snake responsible for a bite is unknown, the CSL Snake Venom Detection Kit may identify the culprit and assist in determining the correct antivenom. Laboratory investigations are important, particularly coagulation tests, which may be crucial in optimizing treatment.

The venoms contain a number of biologically active components including neurotoxins, myolysins, haemotoxins and possibly a nephrotoxin and these are responsible for the major clinical features: defibrination coagulopathy, paralysis, myolysis and renal damage.

The diversity of snakes that may be responsible for envenomation makes the task of identification difficult; however, knowledge of local venom composition and symptomatology may assist, as does venom detection. We examined the venom of brown snakes to determine the influence geographical origin may have on venom make-up, but failed to find any evidence of a geographically predictable venom composition. Individual variability appeared to be the biggest influence on venom composition.

In the isolated island populations of black tiger snakes, however, distinct venom variation was noted between groups and this variability was not related to prey or to distance of separation between populations, but directly correlated with the length of time of isolation from the mainland breeding population. Individual variability was minimized in these small populations, possibly owing to the reduced gene pool, a homogeneous venom composition being found in the individuals from one island.

We have also examined the variability in a single component, the procoagulant from brown snakes, *Pseudonaja* spp. Although some minor differences were found between the four species examined, *P. textilis, P. nuchalis, P. affinis, P. inframacula*, the probable selective advantage of the procoagulant in the venom is reflected in its conservation across the *Pseudonaja* genus.

Symp. zool. Soc. Lond. (1997) No. 70: 205–217
ISBN 0 19 854986 5

Introduction

Australia reputedly has some of the most toxic snakes in the world. Because of the distribution of both the human and snake populations the incidence of snakebite is lower than among Australia's near neighbours, while the implementation of an effective first-aid response, the availability of antivenom, retrieval services and well-equipped intensive care units have also reduced morbidity and mortality rates.

There are between 500 and 3000 cases of suspected snakebite reported in Australia per year and on average two deaths. The cases originate from predominantly rural areas, although snakes are commonly found in the outer suburbs. There are also a number of bites reported annually that have been inflicted on snake handlers/keepers.

Brown snakes are the most likely to have been responsible for a bite, but bites are reported each year from all the dangerous snake genera (Table 1). The variety of snakes that may have been responsible for a bite can complicate the issue although the use of appropriate first aid (Sutherland, Coulter & Harris 1979) and antivenom as required should see the victim make a full recovery in the majority of cases. Knowledge of venom variability and perhaps regional peculiarities in venom composition may assist in both the diagnosis and the treatment of snakebite.

Table 1. Decreasing order of frequency of reported bites by Australian snakes

Brown snakes, *Pseudonaja* spp.
Black snakes, *Pseudechis* spp.
Tiger snakes, *Notechis* spp.
Taipans, *Oxyuranus* spp.
Death adders, *Acanthophis* spp.
Rough-scaled snakes, *Tropidechis carinatus*
Copperheads, *Austrelaps* spp.

Snakebite

Envenomation

The occurrence of a snakebite is perhaps more difficult to diagnose than would be expected, and anecdotally this would appear to be the case. There are a number of cases where snakebite was least suspected (Hilton 1989). Confusion may arise with a possible spiderbite, or 'bites' occurring in thickly vegetated areas may be nothing more than a wound inflicted by the undergrowth, and the influence alcohol can play in the perception of the victim should not be underestimated (Sutherland 1992; White 1992). When young children are the victims, a history of a bite may be unavailable; however, the possibility of snakebite should not be ignored (White *et al.* 1983–84; Underhill 1987).

A patient admitted for suspected snakebite should be questioned about the circumstances surrounding the bite or a witness's account should be recorded. The patient's clinical history should be obtained if possible and any previous convulsions, renal or cardiac problems and allergies, including asthma, noted (Table 2). It is advantageous to determine if the patient has received antivenom therapy on a previous occasion.

Table 2. Key points of clinical history that should be recorded

Bite details	When, how, why and by what?
	First aid:
	  Type of first aid
	  How long after bite was it instituted?
	Activity of the patient
	Treatment of bite site

Clinical history	Symptoms:
	  headache, nausea, vomiting, collapse, convulsion etc.
	Specific symptoms of:
	  paralysis (drooping eyelids, weakness, respiratory distress)
	  myolysis (dark urine, muscle pain)
	  coagulopathy (bite site ooze)
	Renal/cardiac problems
	Allergies
	Current medication

Clinical investigations

A patient hospitalized with suspected or confirmed snakebite may require a venom identification test to confirm the presence of venom and/or to identify the species involved to allow the correct choice of antivenom to be made.

Coagulation studies should be performed to determine the presence of a coagulopathy. A prothrombin time and international normalized ratio (INR), activated partial thromboplastin time (APTT), fibrinogen level and cross-linked fibrin degradation products are routinely tested for in most laboratories. However, if access to a coagulation laboratory is not possible, a test for whole-blood clotting time performed in a simple glass tube can be employed to give at least a gross idea of the patient's coagulation status. If the situation calls for it, this technique can also be applied to determine gross neutralization by the antivenom of the coagulopathy, although the more sensitive laboratory tests are recommended.

If myolysis is a possibility, muscle weakness and soreness should be noted while plasma creatine kinase and urinary myoglobin should be determined. Renal function can be assessed from urea, creatinine and electrolyte levels. At this time, respiratory function should be monitored, while chest X-rays and an ECG should also be considered.

Snake venom: composition

The venoms are a complex mixture of proteins, nucleotides and inorganic ions. These combinations have produced a formidable array of toxic properties in the venom.

Neurotoxins are of great importance and both pre- and postsynaptic activities can be found. Neurotoxicity in a severely envenomed patient can be life-threatening, although respiratory paralysis is unlikely to develop in under 12 h and modern intensive care facilities can provide life-supporting ventilation if it does. The paralytic effects of the neurotoxins are first detected in the muscles of the face, with the patient showing signs of ptosis and ophthalmoplegia.

Myolysins attack skeletal muscle and result in selective damage to muscle fibres. The attachment occurs early but the evidence of damage may not be apparent for some hours. The patient will experience muscle weakness and pain on movement. The patient may also have myoglobinuria, which may be mistaken for haematuria. If there is any doubt, urinary myoglobin determinations should be performed immediately. A recent fatal case due to secondary hyperkalaemia emphasizes the importance of myolysins.

Haemotoxins, or components that affect blood, have their activity most dramatically displayed by the effect the prothrombin converters have on blood both *in vivo* and *in vitro*. *In vitro* plasma is rapidly clotted by the venom, while *in vivo* there is massive fibrinogen consumption resulting in unclottable blood and the potential for catastrophic haemorrhage.

Taipans (*Oxyuranus* spp.)

The common taipans (*Oxyuranus scutellatus*) are found across northern and north-eastern Australia, while the inland taipan (*Oxyuranus microlepidotus*) is found in an ill-defined region of eastern central Australia. They have a potent venom with strong neurotoxic effect (Fohlman *et al.* 1976) and a powerful procoagulant component (Walker, Owen & Esmon 1980; Govers-Riemslag *et al.* 1985). The snakes have long fangs and the capacity to deliver large quantities of venom.

The bite sites may appear minor although some local pain or initial bruising may be evident. The effective delivery of venom results in rapid development of systemic envenoming with the associated clinical signs of neurotoxicity and coagulation disturbance.

Tiger snakes (*Notechis* spp.)

The tiger snakes are found across the temperate southern regions of Australia and have a preference for wetlands. There are also numerous isolated island populations across the southern coastline. The venom again is potent, showing neurotoxic effects (Karlsson, Eaker & Rydén 1972), an active procoagulant component (Tans *et al.* 1985) and a distinct myolytic action with resultant severe muscle wasting in severe cases (Hood & Johnson 1975). The delivery of the venom is effective although the fangs are small to medium in size.

The bite sites are more distinctive with the likelihood of pain and swelling, redness and bruising. Generally, the envenomed patient will have headaches, nausea, vomiting and possibly collapse. Again, paralysis, coagulopathy and evidence of rhabdomylosis will be noted.

Brown snakes (*Pseudonaja* spp.)

The brown snakes have adapted to all areas of Australia and are not uncommon in urban regions.

The venom contains an extremely powerful coagulant activity (Masci, Whitaker & de Jersey 1988; Stocker *et al.* 1994) and there is the possibility of renal damage through an uncertain mechanism, perhaps a direct nephrotoxin (White & Fassett 1983; Acott 1988). Neurotoxins are also present, although less dramatic in effect than those of the taipans and tiger snakes (Coulter, Broad & Sutherland 1979). The venom delivery is poor, with the animals having fangs only 2–3 mm long, while the amount of venom available for delivery has been considered to be small, although some evidence is emerging to dispute this (Williams, White & Mirtschin 1994).

Many bite victims do not develop systemic envenoming, and this is most probably due to the ineffective venom delivery. The bite sites are generally unremarkable although 'scratching' from the fangs may be noted. The patient displays signs similar to those of tiger snake victims, with headache, nausea, vomiting and a likelihood of early collapse. The brown snakes are now regarded as the most dangerous snakes in Australia because their venom is associated with early unexpected collapse and death (Sutherland 1992). Coagulopathy may be obvious with wound ooze and haematuria, while renal function should be carefully monitored for any early signs of damage.

Snake venom: variability

Identification of the snake responsible for a bite has to some extent been made a lot easier by the introduction of the venom detection kit. The kit indicates the presence of venom, the type of snake responsible, and hence the most appropriate monovalent antivenom.

Should the kit be unavailable, determining the most appropriate antivenom is a major problem facing the clinician, owing to the diversity of snake genera, the large number of species possibly responsible and the conflicting clinical presentations which are most probably due to individual venom variability.

Diversity of snakes

Although our interest in snakebite predominantly relates to the taipans, tiger snakes and brown snakes, black snakes, copperheads, rough-scaled snakes and death adders are also responsible for bites. Apart from the distinctive morphological characteristics of the death adder (Fig. 1) the remaining genera and their species may be difficult to distinguish even for an expert.

The distribution of some genera can lead to their elimination as a likely culprit in snakebite cases; however, the wide distribution of the *Pseudonaja* genus presents at least two possibilities in most cases. The colour morphs within a species alone illustrate the difficulties encountered in snake identification.

The clinical presentation of the patient may be of little value and, as we have seen, the (non-specific) presence of headache, nausea, vomiting and neurotoxic symptoms of paralysis coupled with an obvious coagulopathy could quite easily have arisen from a bite by any number of snakes.

Fig. 1. Left to right, top row: common death adder, common tiger snake, inland taipan; bottom row: mulga snake, black snake, brown snake (all photos © Dr J. White).

Venom variability

Geographical variability

The influence that geographical origin of a snake can have on both clinical presentation and venom composition is perhaps best illustrated with the presence or absence of Mojave toxin in the venom of *Crotalus scutulatus scutulatus* specimens in Arizona (Glenn & Straight 1978). Neurotoxic symptoms dominate envenomations by the population containing Mojave toxin, whereas tissue damage is the predominant feature of the other population (Hardy 1983).

Clinical assessment of brown snake bite victims suggested the possibility that whether neurotoxic or coagulant features predominate depends on the geographical origin of the snake. We examined the venoms of five specimens of *Pseudonaja textilis* and nine specimens of *Pseudonaja nuchalis* collected from areas to give a north–south coverage and a comparative cluster within one area. We tested clotting strength, phosphodiesterase and 5′nucleotidase activity and the conversion of the chromogenic substrate S2222. (Fig. 2). The results indicated that the variability in activity between snakes was no greater across the full geographical span of the study than it was between the clustered specimens.

In the case of *Pseudonaja*, which is wide-ranging with possibly very little in the way of barriers to disrupt a continuous interbreeding population, the resultant venoms may be very similar from all geographical points. Individual variability would appear to be a greater influence on venom composition and presumably the predominance of a particular symptom in a bite victim is most likely to be a reflection of this.

Variability in geographically isolated populations

In isolated populations, such as the black tiger snakes (*Notechis ater niger*) on the islands off the coast of South Australia, the composition of the venom may be influenced by a number of factors, such as prey availability and distance of separation between one population and another. The venom compositions were compared by gel filtration chromatography and by sodium dodecyl-sulphate, polyacrylamide gel electrophoresis (SDS PAGE) of the whole venoms. The islands themselves are homogeneous habitats with low scrub, and examination of the electrophoretic banding patterns from individuals from one of these islands showed a homogeneous venom composition which most probably is the result of the restricted gene pool (Williams & White 1987).

Manipulation of the banding pattern results allowed us to visualize the differences/similarities of the populations through principal component analysis and a minimum spanning tree was also calculated (Fig. 3). Islands that were close together tended to produce similar banding patterns while considerable variation was noted with increasing distance between populations. The diversity of prey among like-banding islands eliminated the prey as being the influence on variability; the distance by which the populations were geographically separated appeared likely to be the cause.

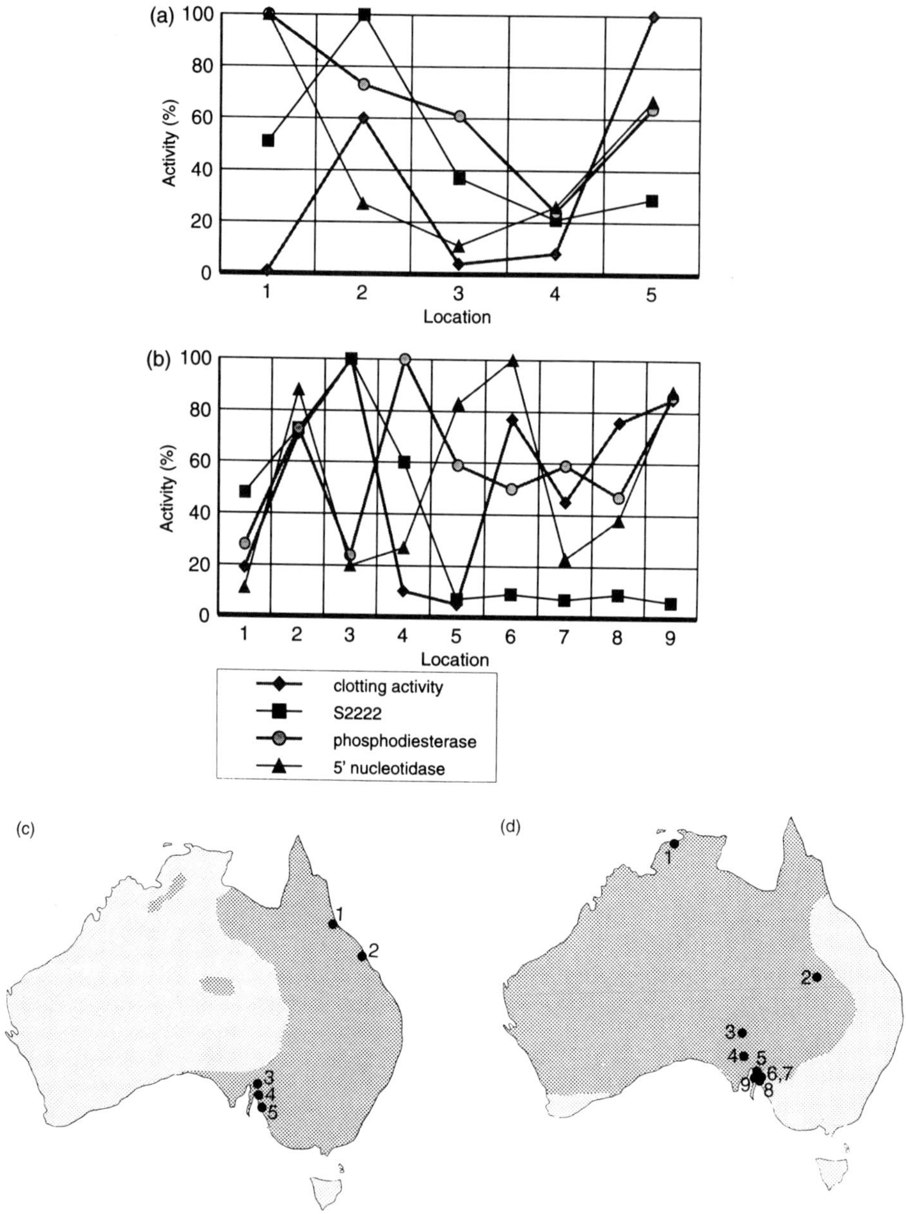

Fig. 2. Geographical variability in clotting activity, S2222 conversion, phosphodiesterase and 5′ nucleotidase activity of *Pseudonaja textilis* venom (upper graph: 1 = Townsville, 2 = Bundaberg, 3 = Crystal Brook, 4 = Wanderah, 5 = Adelaide) and *Pseudonaja nuchalis* venom (lower graph: 1 = Darwin, 2 = Charleville, 3 = Roxby, 4 = Depot Creek, 5 = Napperby, 6,7 = Pt Pirie, 8 = Wanderah, 9 = Weeroona). Activity for each population determined against highest responder (100% activity). Maps illustrate approximate positions of each locality and species distribution (left = *P. textilis*, right = *P. nuchalis*).

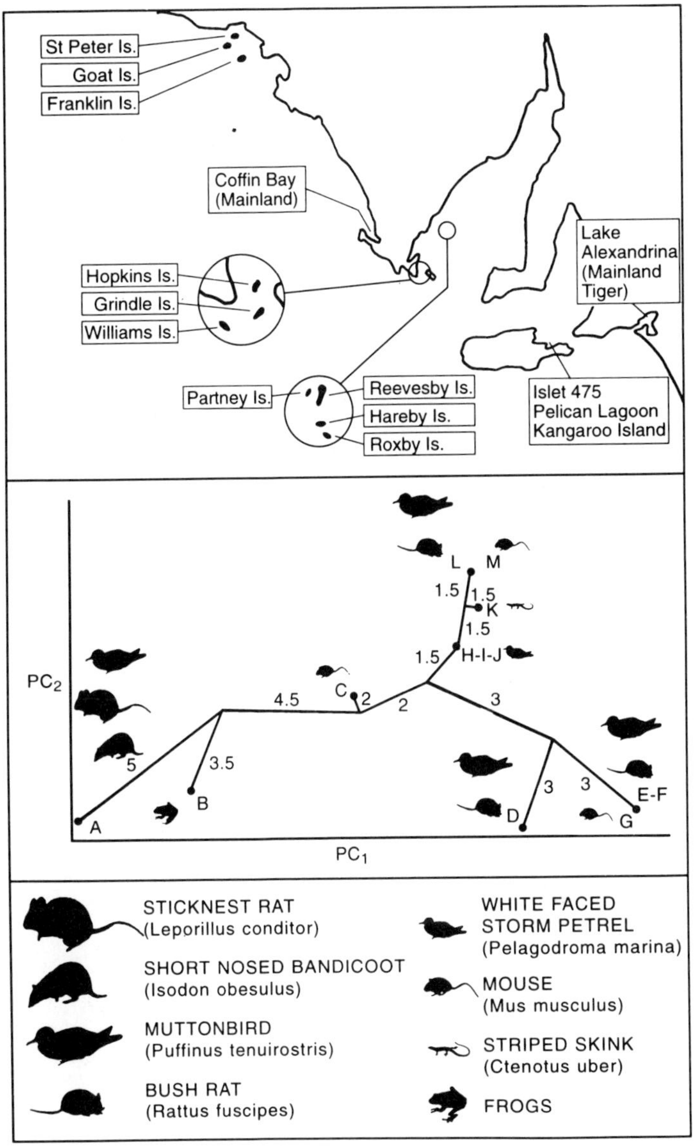

Fig. 3. Diagram of results of principal components and minimum spanning tree analyses, and comparisons of prey types for points representing populations studied. PC1 and PC2 are the first two principal components axes upon which the populations (A–M) are projected according to the analysis of the polyacrylamide banding patterns of the whole venoms from each location.

A = Franklin Is.; B = Lake Alexandrina; C = Islet 475 (Pelican Lagoon, Kangaroo Island). D = Williams Is.; E,F,G = Goat Is., St. Peter Is., Coffin Bay respectively; H,I,J = Reevesby Is., Partney Is., Hareby Is., respectively; K = Roxby Is.; L = Hopkins Is.; M = Grindle Is.

Further examination of the data, however, showed that two like-banding populations were separated by some 300 km while a population only 21 km away showed a quite different banding pattern. Some alternative explanation was required and this was found in sea-depth measurements which correlated with the length of time each population had been isolated from others since the last ice-age (Williams, White, Schwaner & Sparrow 1988). Thus the prey may have very little influence on the presence or absence of venom components, while the composition appears to be genetically determined, as illustrated by the identical venoms from a single isolated population.

Interspecies variability in a single venom component

Venom composition would appear to be genetically determined and this may also influence the amount of any individual component in a given sample, although other factors (e.g. season, age) may contribute. It is this individual variability that may be responsible for the clinical manifestations noted. Our interest lies particularly with the procoagulants found in the venoms and our studies with tiger and brown snakes indicate that the procoagulants are highly conserved across these genera and suggest that the procoagulants are of some selective advantage.

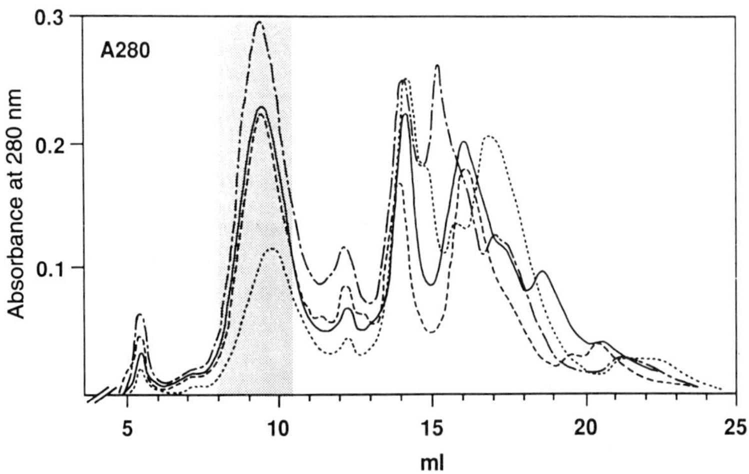

Fig. 4. Gel filtration chromatogram of the whole venom from *Pseudonaja textilis, P. nuchalis, P. affinis* and *P. inframacula* showing the co-elution of the procoagulant activity (shaded area). - - - *P. affinis* —— *P. inframacula* · · · *P. nuchalis* – · – · *P. textilis.*

Clinical reports suggest that bites in western Australia by snakes of the *Pseudonaja* genus require far more antivenom to neutralize the coagulopathy than in eastern Australia. The possibility that the procoagulant from the local species (*P. affinis*) was more potent than that from *P. textilis, P. nuchalis* or *P. inframacula*, the other clinically significant species, was considered. We performed a comparative study on the procoagulants from each of these species (Williams, White & Mirtschin 1994). The procoagulants co-eluted with gel filtration chromatography (Fig. 4); immunoprecipitin lines, developed between the procoagulants and brown snake antivenom, showed identity (Fig. 5). Some difference was noted in the specific activity and the PAGE banding for the procoagulant but these did not seem to be sufficient to explain the suggested clinical differences.

Preliminary investigation of venom volume milked from these species indicates that this may be the key to explaining the variability in clinical presentation.

Fig. 5. Immunoprecipitin lines developed between brown snake antivenom (central well) and (from left to right) *P. affinis, P. inframacula, P. nuchalis, P. textilis.*

Conclusion

Snakebite in Australia has the recipe for catastrophe, with a lot of the genera responsible for bites having venoms containing neurotoxins, haematologically active components, myolysins and possibly direct nephrotoxins.

Established first-aid measures when applied correctly reduce the potential severity of symptoms and in most cases with modern intensive care unit facilities and monospecific antivenom the likelihood of a fatal outcome is reduced dramatically. However, fatalities still do occur.

The venoms themselves appear to conserve the neurotoxins and coagulants within species even in isolated populations, and imply their selective advantage to the animals. The variability seen in clinical presentations may be directly related to the bite — whether it was a 'dry' bite, glancing or substantial — and the quantity of venom injected. The other significant influence would appear to be individual variability in venom composition and we would suggest that this should not be underestimated as a source of clinical inconsistencies and frustration in research programmes.

References

Acott, C. J. (1988). Acute renal failure after envenomation by the common brown snake. *Med. J. Aust.* **149**: 709–710.

Coulter, A. R., Broad, A. J. & Sutherland, S. K. (1979). Isolation and properties of a high molecular weight neurotoxin from the eastern brown snake (*Pseudonaja textilis*). In *Neurotoxins: fundamental and clinical advances*: 260. (Eds Chubb, I. W. & Geffen, L. B.). Adelaide University Union Press, Adelaide.

Fohlman, J., Eaker, D., Karlsson, E. & Thesleff, S. (1976). Taipoxin, an extremely potent presynaptic neurotoxin from the venom of the Australian snake taipan (*Oxyuranus scutellatus scutellatus*). Isolation, characterization, quaternary structure and pharmacological properties. *Eur. J. Biochem.* **68**: 457–469.

Glenn, J. L. & Straight, R. C. (1978). Mojave rattlesnake *Crotalus scutulatus scutulatus* venom: variation in toxicity with geographical origin. *Toxicon* **16**: 81–84.

Govers-Riemslag, J. W. P., Speijer, H., Zwaal, R. F. A. & Rosing, J. (1985). Purification and characterization of the prothrombin activator from *Oxyuranus scutellatus* (Taipan snake). In *Hemostasis and animal venoms:* 41–53. (Eds Pirkle, H. & Markland, F. S.). Marcel Dekker, New York.

Hardy, D. L. (1983). Envenomation by the Mojave rattlesnake (*Crotalus scutulatus scutulatus*) in southern Arizona, U.S.A. *Toxicon* **21**: 111–118.

Hilton, J. M. N. (1989). A tragic case of snakebite. *Med. J. Aust.* **151**: 600–601.

Hood, V. L. & Johnson, J. R. (1975). Acute renal failure with myoglobinuria following tiger snake bite. *Med. J. Aust.* **2**: 638–641.

Karlsson, E., Eaker, D. & Rydén, L. (1972). Purification of a presynaptic neurotoxin from the venom of the Australian tiger snake *Notechis scutatus scutatus*. *Toxicon* **10**: 405–413.

Masci, P. P., Whitaker, A. N. & de Jersey, J. (1988). Purification and characterization of a prothrombin activator from the venom of the Australian brown snake, *Pseudonaja textilis textilis*. *Biochem. Int.* **17**: 825–835.

Stocker, K., Hauer, H., Muller, C. & Triplett, D. A. (1994). Isolation and characterization of Textarin, a prothrombin activator from Eastern brown snake (*Pseudonaja textilis*) venom. *Toxicon* **32**: 1227–1236.

Sutherland, S. K. (1992). Deaths from snake bite in Australia, 1981–1991. *Med. J. Aust.* **157**: 740–746.

Sutherland, S. K., Coulter, A. R. & Harris, R. D. (1979). Rationalisation of first-aid measures for elapid snake bite. *Lancet* **i**: 183–186.

Tans, G., Govers-Riemslag, J. W. P., van Rijn, J. L. M. L. & Rosing, J. (1985). Purification and properties of a prothrombin activator from the venom of *Notechis scutatus scutatus*. *J. biol. Chem.* **260**: 9366–9372.

Underhill, D. (1987). *Australia's dangerous creatures*. Readers Digest, Surry Hills.

Walker, F. J., Owen, W. G. & Esmon, C. T. (1980). Characterization of the prothrombin activator from the venom of *Oxyuranus scutellatus scutellatus* (taipan venom). *Biochemistry* **19**: 1020–1023.

White, J. (1992). A review of snakebites and suspected snakebites treated in South Australia with particular reference to reptile handlers. In *Recent advances in toxinology research* **1**: 367–377. (Eds Gopalakrishnakone, P. & Tan, C. K.). National University of Singapore, Singapore.

White, J. & Fassett, R. (1983). Acute renal failure and coagulopathy after snakebite. *Med. J. Aust.* **2**: 142–143.

White, J., Tomkins, D., Steven, I. & Williams, V. (1983–84). Tiger snake bite. *Rec. Adelaide Child. Hosp.* **3**: 169–173.

Williams, V. & White, J. (1987). Variation in venom constituents within a single isolated population of peninsula tiger snake (*Notechis ater niger*). *Toxicon* **25**: 1240–1243.

Williams, V., White, J. & Mirtschin, P. J. (1994). Comparative study on the procoagulant from the venom of Australian brown snakes (Elapidae; *Pseudonaja* spp.). *Toxicon* **32**: 453–459.

Williams, V., White, J., Schwaner, T. D. & Sparrow, A. (1988). Variation in venom proteins from isolated populations of tiger snakes (*Notechis ater niger, N. scutatus*) in South Australia. *Toxicon* **26**: 1067–1075.

16 Russell's viper in Indonesia: snakebite and systematics

PAUL J. BELT, ANITA MALHOTRA*, ROGER S. THORPE, DAVID A. WARRELL and WOLFGANG WÜSTER

Synopsis

Russell's viper has a widespread but discontinuous distribution across 10 Asian countries. The presence of the species in southern Indonesia is of particular interest, as these populations are isolated by more than 2000 km from the nearest conspecific populations in Thailand. Formerly divided into five subspecies, the population systematics of this species has recently been revised as a result of a multivariate analysis of scalation and colour pattern characteristics. The analysis reveals two morphologically distinct forms occupying the eastern and western part of the range of the species respectively. While the populations from the Lesser Sunda Islands (Nusa Tenggara Timur) group with the eastern form, they appear to be somewhat divergent from the other populations in Java and continental south-east Asia. In the species as a whole there is clear lack of concordance between morphological and venom variation, but data on venom effects, or indeed on the incidence of bites by this species, is lacking for the Indonesian populations.

In this paper, fresh distributional data from recent expeditions and the results of a retrospective case study of Russell's viper bites ($n = 26$) from the Lesser Sunda Islands are presented. The magnitude of the public health problem and effect of medical and traditional treatments on outcome are assessed. The case fatality of Russell's viper bites in Indonesia is 38.5%. The mean interval between snake bite and death is 14.5 h, which is considerably shorter than that reported from other Asian populations. Incoagulable blood caused by procoagulant effects of the venom and the ensuing consumption coagulopathy is the most common feature across the entire geographical range. Spontaneous systemic haemorrhage occurred in 34.6% of the victims in Indonesia and represents a significant indicator of poor prognosis. There is a higher incidence of significant local tissue destruction and necrosis (23%) than in other populations, and some evidence of neurotoxic symptoms is present. These data strongly suggest that the effects of envenoming by Russell's viper in Indonesia are different from those of other populations, and the geographical variation in the clinical manifestations of Russell's viper bite is reviewed in the light of this new information.

Symp. zool. Soc. Lond. (1997) No. 70: 219–234
ISBN 0 19 854986 5

*Author to whom correspondence should be addressed.

Introduction

Russell's viper (*Daboia russelli*) is one of the world's major causes of snakebite morbidity and mortality (Aung-Khin 1980; Warrell 1989; Looareesuwan, Viravan & Warrell 1988; Phillips *et al.* 1988; Aye 1990; Viravan *et al.* 1992). Formerly known as *Vipera russelli*, recent research on the phylogeny of viperine snakes (Herrmann, Joger & Nilson 1992) has placed it in the monotypic genus *Daboia*. It has a widespread distribution extending across 10 Asian countries, including parts of mainland China, western Cambodia, Thailand, Burma, Bangladesh, southern Indonesia, Taiwan, Pakistan, India and Sri Lanka. There is considerable discontinuity in this distribution, with near-ubiquitous occurrence in India, a scattered distribution in China and Indo-China and a complete absence in most of Java, Sumatra, Borneo and the Malayan Peninsula. Isolated populations occur in Indonesia, in eastern Java, Komodo, Flores and Lembata.

Russell's viper in Indonesia

The presence of Russell's viper in Indonesia was only reliably established in the 20th century, and its distribution in the archipelago is still being elucidated. In the following paragraphs, we briefly review the literature on the occurrence of Russell's viper in Indonesia.

There are two 19th century records in the literature reporting the presence of Russell's viper on Java (Duméril, Bibron & Duméril 1854; Boulenger 1896). However, there is evidence that the samples on which these reports are based did not in fact originate from Indonesia (Brongersma 1958). In view of their high ventral scale counts, it is more probable that the snakes originated from India. The first authentic record of Russell's viper from Java was published by Neuhaus (1935), from the region of Surabaya in east Java. Since then, the viper has repeatedly been found in east Java, especially in the neighbourhoods of Sepandjang and Kembang Kuning on the outskirts of Surabaya (Kopstein 1936: van Hoesel 1954). Kopstein (1936) described the Javan population as a new subspecies, *V. r. sublimitis*. Van Hoesel (1959) reported the finding of seven full-grown specimens in a small field at the cemetery at Kembang Kuning, Surabaya, and also noted the occurrence of the species in the limestone hills west of the city. Hodges (1993) recorded six specimens of Russell's viper during a three-year period which had been killed by cars on the toll road close to the Kembang Kuning military cemetery. He also mentioned reported sightings in the forested areas around Gresik and Bojonegoro, although he suggested that these might have been a result of confusion with the Malayan pitviper (*Calloselasma rhodostoma*), which is known to occur in the area.

The first evidence of Russell's viper in the Lesser Sunda Islands (Nusa Tenggara Timur), and indeed in Indonesia, was provided by the Douglas Burden expedition to the island of Komodo in the summer of 1926 (Dunn 1927), which collected two specimens from the lower hills of the island. In 1927, Mertens collected a specimen from Ende, a small island off the southern coast of Flores, which he described as the subspecies *Vipera russelli limitis* (Mertens 1927). Van Hoesel (1958) was the first to report Russell's viper from the island of Flores itself, as well as from Lembata. He commented on how common the species was on Flores and listed many regional dialect names for it (Table 1). Over 40 specimens were collected from altitudes of up to 1200 m on Flores. According to van Hoesel, the inhabitants of Adonara, Solor, Alor and Sumba claimed that the species was well known on these islands, but no specimens were ever obtained. The only evidence of the medical significance of Russell's viper in these islands comes from Auffenberg (1980), who stated that the snake caused more deaths on Komodo than any other species. It certainly appeared to be common when his collection was made in the 1960s; 59 specimens were collected from Komodo below an altitude of 120 m, in *Zizyphus* savanna and the beachside *Pandanus* community.

Table 1. Local names for Russell's viper and other snakes in specific regions of Indonesia

Island	Region	Russell's viper *Daboia russelli*	Green tree viper *Trimeresurus albolabris*	Other
Komodo		Pupu		
		Misa[3]		
Rinca		Kaka botek		
		Pupu		
Flores	Ruteng		Kaka taa	
	Ngada District	Barat[1]		
	Bajawa	Mbala		
		Bara[1]		
		Baka		
	Riung	Mbarat	Likur	
	Mbay	Spret[2]		
		Sapret[2]		
	Ende Town	Nipakaru[1]		
		Ular heutan		Cobra = Pupurupi[2]
	Maumere	Blarat[1]	Mea	Cobra = Puput
	Larantuka	Kramat	Kamea	
		Kamek[1]		
		Temuk lado		
Lembata		Kramek[1]		
		Krama	Toha	Python = Punay
Adonara	Podor	Kramek[1]		
		Krame		
Java	East Java	Bandotan puspo[2]	Ular daun[2]	Krait = Weling[2]
			Luwuk	Cobra = Ular sendok[2]
				Python = Ular sawa[2]

Source: [1]van Hoesel (1958); [2]AM/RST; [3]Auffenberg (1980); Remainder from PJB.

Geographic variation in Russell's viper

Morphological variation

Conventionally, Russell's viper has been divided into five subspecies; *D. r. russelli* from India, Pakistan and Bangladesh, *D. r. pulchella* from Sri Lanka, *D. r. siamensis* from Burma, Thailand, and southern China, *D. r. formosensis* from Taiwan and *D. r. limitis* from southern Indonesia. In addition, some workers have recognized the subspecies *D. r. sublimitis* (Kopstein 1936) from eastern Java and *D. r. nordicus* (Deraniyagala 1945) from northern India. These subspecies were based primarily on variation in the number of rows of dorsal spots and a few other colour pattern characters and their validity had been questioned (Brongersma 1958). Wüster, Otsuka, Malhotra & Thorpe (1992) carried out a multivariate analysis of scalation and colour pattern characters, as a result of which they recognized only two subspecies, a western form (*D. r. russelli*) from Pakistan, India, Sri Lanka and Bangladesh, and an eastern form (*D. r. siamensis*) from Burma, Thailand, China, Taiwan and Indonesia.

Venom variation

Clinically, Russell's viper is of particular interest in that the composition and effects of its venom vary considerably between different areas (e.g., Jayanthi & Gowda 1988; Woodhams *et al.* 1990; Warrell 1989). Envenoming by this species has been studied in Burma (Myint-Lwin *et al.* 1985), Thailand (Mahasandana, Rungruxsirivorn *et al.* 1980; Kanjanajatanee & Visutipant 1984; Mahasandana, Pochanugool *et al.* 1990) and Sri Lanka (Jeyarajah 1984; Phillips *et al.* 1988).

Clinical effects of Russell's viper bites are local, in the bitten limb, and systemic. Local symptoms include pain, swelling, bruising, blistering, necrosis and local lymphadenopathy. Systemic effects include disseminated intravascular coagulation resulting from a number of venom procoagulant enzymes, notably those activating factor V and factor X, with fibrin deposition in small blood vessels in the kidney, lungs and anterior pituitary. There is spontaneous systemic bleeding from the gingival sulci and nose and into the skin, retroperitoneal tissues and, most dangerous, into the brain and gastrointestinal tract. Acute renal failure (acute tubular necrosis and bilateral renal cortical necrosis) is more common than with other species of snake. In Sri Lanka and parts of India there is generalized rhabdomyolysis, intravascular haemolysis and neurotoxicity attributable to a pre-synaptic toxin (phospholipase A_2). In Burma, Russell's viper bites cause a generalized increase in capillary permeability causing chemosis, periorbital oedema, pulmonary oedema and heavy proteinuria. Various pathophysiological processes, including hypovolaemia, vasodilation and acute pituitary failure (in Burma) contribute to shock (Warrell 1995).

There is a marked variation in the clinical manifestations of Russell's viper bite across its discontinuous distribution, which probably reflects differences in venom composition. Approximately 90% of the venom is protein, and up to 70% of the protein is phospholipase A_2, of which seven different isoenzymes have been identified. This enzyme group exhibits a large spectrum of activities, which may explain the wide symptom profile of the venom.

Congruence between morphology and venom

It has been noted that there is little relationship between morphological differentiation and venom variation in this species (Warrell 1989). Unlike the case in other venomous snake taxa (e.g. cobras: Wüster & Thorpe 1989, 1991), a systematic revision has failed to clarify the matter as there appears to be as much variation in venom effects within the eastern and western subspecies as there is between them (Wüster, Otsuka, Malhotra & Thorpe 1992). Although more detailed analysis has revealed some morphological differences between some populations of the eastern subspecies having different venom effects but formerly considered very similar (Wüster, Otsuka, Thorpe & Malhotra 1992), the link between morphological variation and venom variation appears weak in this species.

Antivenom effectiveness in Russell's viper also does not appear to be related to the principal division into eastern and western forms; it is also not necessarily related to differences in the effects of the venom. Despite profound differences in clinical effects, antivenom against Burmese Russell's viper (eastern form) was found to be effective against the venom of Thai vipers (eastern form), and even against the venom of Russell's viper from India (western form); however, it was ineffective in neutralizing the venom of Sri Lankan vipers (western form) (Phillips *et al.* 1988).

This considerable variation in venom effects and composition between different geographical areas makes Russell's viper of particular clinical interest. However, detailed studies of envenoming by this species have only been carried out in Sri Lanka, Burma and Thailand. Few details are known of the symptoms of envenoming in China, Taiwan, and Pakistan and, until now, nothing was known about the effects of envenoming by the isolated Indonesian populations. In this paper, the epidemiology and symptomatology of snakebites by Russell's viper in Indonesia will be discussed, and new information on the distribution of the species in the Indonesian Archipelago will be presented.

Materials and methods

Systematic relationships of the Indonesian populations

The population systematics of Russell's viper were studied by means of multivariate morphometric analysis of 22 morphological characters. For details of the characters and methods see Wüster, Otsuka, Malhotra & Thorpe (1992). Indonesian populations were represented by three males and two females from Java, 35 males and 18 females from Komodo, three females from Flores and two females from Lembata.

Geographical distribution of Russell's viper in Indonesia

The islands where Russell's viper had been previously recorded were visited by RST and AM in 1993 (except Rinca and Lembata). Photographs of Russell's viper as well as of snakes with similar appearance were carried to avoid confusion. These include the reticulated python (*Python reticulatus*) throughout the islands, and also the Malayan pitviper (*Calloselasma rhodostoma*) and the many-spotted cat snake (*Boiga multomaculata*) in east Java. Promising areas identified from the literature or through talking to local experts were visited and local people questioned. Searches were made in likely areas as the expedition had been timed to coincide with the expected period of greatest activity by Russell's viper (the dry season). The following year, PJB revisited the areas identified as likely to have a high incidence of snakebite, as well as islands not visited in the previous year.

Epidemiology of Russell's viper bite in Indonesia

A questionnaire (designed by PJB) was used to compile retrospective data from medical practitioners, paramedics, herbalists, local village elders and villagers regarding all aspects of Russell's viper bite, treatment and outcome in that area. The questionnaire recorded the age, sex and occupation of snakebite victims, the circumstances of bites, the location of bites on the body, the identification of the snake, the symptoms experienced, the treatment given, and the outcome of any bites. All of the subjects ($n = 68$) were interviewed in the presence of a translator. Only those subjects who were able to give precise details regarding the bite(s) and victim(s) were used for subsequent analysis.

Results

Systematic relationships of Indonesian Russell's viper populations

A canonical variate analysis of populations across the entire range of Russell's viper revealed a clear morphological distinction between Russell's viper populations from west and east of the Bay of Bengal, suggesting the existence of a western and an eastern taxon in this species. However, within the eastern taxon, there was some indication of north–south differentiation (Fig. 1a). This was investigated further by running a canonical variate analysis of eastern populations only (Fig. 1b). Although the number of male specimens from the Lesser Sunda Islands is small, and the sample is restricted to Komodo, both males and females show populations from the Lesser Sunda Islands are clearly distinct from the other populations. Interestingly, the population from east Java is not morphologically distinct from the eastern mainland populations despite being separated from them by a large distributional gap.

Geographical distribution of Russell's viper in Indonesia

Although the earliest reported (and substantiated) population from east Java, in the vicinity of the Kembang Kuning cemetery near Surabaya, appears to be under threat from the expansion of the city, guards described having seen a snake that fitted the description in the cemetery itself recently. Russell's viper, known locally as 'bandotan puspo', may be more widespread than formerly thought in east Java (Fig. 2a). Possible locations of the viper include Lamongan, Nganjuk and Bojonegoro in the Kendeng foothills west of Surabaya, Tuban in the Blora-Drajat karstic hills, and Jember and Banyuwangi in the southern dissected plateau of east Java (Fig. 2a). Most of these were anecdotal reports, but eight Russell's viper, some of extremely large size, were seen by PJB at a Surabayan snake dealer's premises. These had reportedly been captured in Tuban. However, no reports of bites were obtained from eastern Java.

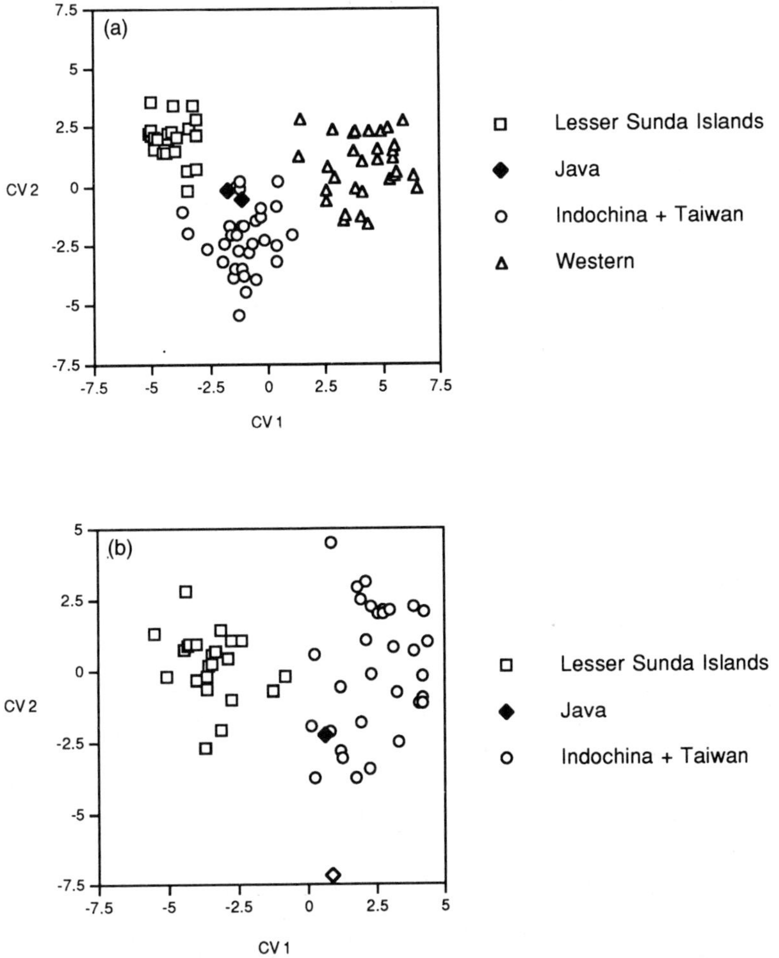

Fig. 1. Ordination of female Russell's viper populations. (a) All populations, showing the primary division into eastern and western groups along CV1 and the indication of a north–south differentiation within the eastern group along CV2; (b) Eastern group only, showing that the Lesser Sunda populations are somewhat distinct, while the east Javan specimens are morphologically indistinguishable from other mainland populations.

In Nusa Tenggara Timur (Fig. 2b), no recent cases of snakebite were reported by the National Park Rangers in Komodo, although the species was stated to be occasionally encountered in the grounds of the Loh Liang camp. They also indicated that it was present on the neighbouring island of Rinca, and when PJB visited this island the following year, local rangers were able to accurately describe the snake and reported seeing one two days prior to his visit. Whatever the current status of the species in Komodo, little evidence was seen of 'the beachside *Pandanus* community' in Loho Lavi bay, where Auffenberg (1980) states he caught the majority of his specimens. On Flores, the presence of the viper in western Flores was suggested by local reports (especially around the town of Lembor in the Cancar region) but these

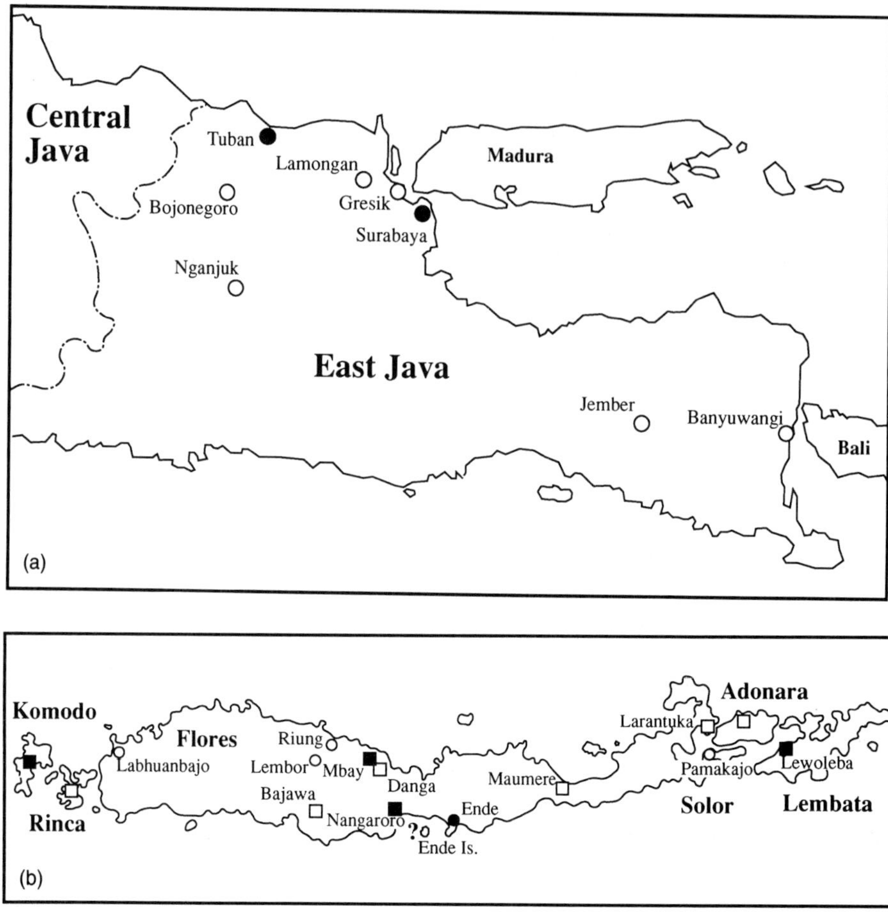

Fig. 2. Map showing the places mentioned in this study in (a) east Java, (b) Nusa Tenggara Timur (Lesser Sunda Islands). Positive records of Russell's viper are indicated by solid symbols. Localities at which the species may possibly occur (as judged by a variety of circumstantial evidence, see text) but at which its presence has not been positively confirmed are indicated by empty symbols. Localities at which histories of snakebite attributable to Russell's viper have been collected are marked with squares, others are marked with circles. Ende Island is marked with a question mark as the species has been previously recorded but is now possibly extinct there (see text).

were not confirmed. It now appears to be comparatively rare around the town of Ende on Flores, although a specimen was sighted in Nangaroro (40 km to the west of Ende town).

However, no evidence at all was found of the continuing presence of Russell's viper on Ende Island itself (the type locality for *V. r. limitis*), where it was reportedly very common 30 years ago. The island is small, and most parts of it were visited by RST and AM, who found that no part of the island contained habitat which might be expected to harbour the species. The majority of the island is given over to growing coconuts with an understorey of cassava and there is virtually no ground cover. Farmers on the island could not distinguish between the photographs of pythons and

Russell's viper and the only snakes they could describe were likely to be pythons because of their large size, arboreal habits, season of activity (wet season) and the lack of concern displayed by the farmers. We conclude that Russell's viper is now extremely rare, or extinct, on Ende island.

While in Ende, reports of the viper in an area of the island in which it has not previously been recorded, the Ngada district on the northern coast (near the towns of Riung and Mbay), were obtained. On visiting the area, it became clear that Russell's viper may be a significant public health problem. Specimens were encountered near Riung by both parties. Many of the people in the area knew someone who had died, and the symptoms described seemed accurate. The victims did not appear to consider admission to hospital for treatment, and the nearest hospital may be quite distant from some of the villages. Because of this the true scale of the problem is largely unknown. One aim of the study was therefore to attempt to gauge the true extent of the problem. Other areas in east Flores, such as Maumere and Larantuka, also appear to have a high incidence of Russell's viper bite.

Russell's viper was also seen by PJB on Lembata, and numerous reported sightings and case histories of bites were obtained on that island. Intriguing evidence that the species has a wider distribution east of Flores than previously recorded was also found. A museum in Larantuka contained a specimen captured on 1 July 1975 by Father M. Clemens from Pamakajo on Solor, which suggests that this species is, or had been, present there. Additionally, a missionary from the town of Podor on Adonara was able to give highly accurate descriptions of the snake and claimed it was responsible for approximately two deaths on the island per year (population size 4500). However, its presence could not be confirmed by the authors. A recent survey of reptiles in Nusa Tenggara by the Western Australian Museum and the Museum Bogoriense found only a single specimen of *D. russelli*, from Lembata (R. How, pers. comm.).

Effects of envenoming by Russell's viper in Flores

Epidemiological data on Russell's viper envenoming

A total of 26 case histories were obtained. The average age was 42.5 years. Eighteen were male, eight female. The majority were farmers (21), the remainder consisting of a fisherman, a student, an army private, a clerk, and a businessman. The geographical distribution of bites is shown in Table 2. Twenty-one of the bites were on the foot, two on the thigh, one on the ankle, one on the calf and one on the hand. Most of the bites occurred in remote villages and plantations, with 14 occurring in the plantations, six on the main village road, three in the village itself, two on a hillside and one in a house. The bites occurred most frequently at dawn and dusk, when poor light conditions cause the snake to be accidentally stepped upon. The colour and general morphological descriptions were accurate and added weight to the evidence for the correct identification of the snake. The reported average length of the snake was 38 cm (range 15–130 cm).

Table 2. Regional breakdown of Russell's viper bite incidence in the Lesser Sunda Islands

Location	Number of victims	Date of incident	Outcome	Population size
Komodo	1	1964	Died	400
Rinca	1	1989	Died	300
Flores				
Labhuanbajo	1	1993	Survived	2500
Bajawa	3	1993	All survived	10 000
Danga (Mbay)	1	1980	Died	1000
	2	1992	One died, one survived	
	1	1993	Survived	
Nangaroro	1	1970	All survived	250
	1	1992		
	1	1994		
Maumere	1	1972	Died	300 in villages,
	1	1978	Survived	40 000 in region
	1	1979	Survived	
	1	1992	Survived	
Larantuka	1	1993	Survived	25 000
Lembata				
Lewoleba	1	1967	Survived	Twelve villages,
	1	1982	Survived	total population
	1	1990	Survived	11 977
	2	1991	Both died	
	2	1994	Both died	
Adonara	1	1985	Died	4500

Table 3. Interval between snakebite and death in different populations of Russell's viper

Population	Cases	Median	Range	Source
India	7	41.3	0.25–216	Reid (1968)
Burma	12	72	15–112	Myint-Lwin *et al.* (1985)
Thailand	7	72	2.5–264	Looareesuwan *et al.* (1988)
Indonesia (Lesser Sunda Islands)	10	14.5	0.5–46	This study

Morbidity and mortality

There were ten deaths amongst the 26 cases, giving a mortality rate of 38.5%. The average time to death (Table 3) was 14.5 h (range 30 min to 46 h). One of the subjects died one month after envenoming owing to secondary infection (this fatal case was not included in the calculation of mean time to death). Several symptoms were recorded only in the fatally envenomed cases, including haematemesis (nine cases), severe bleeding from the bite site (eight cases), generalized swelling (five cases) and systemic bleeding from other sites (two cases). Sixteen patients survived, and only one was left with a significant handicap (inability to bear weight due to necrosis of the calf muscles). The average time to make a full recovery was nine days.

Traditional treatments

Most victims of snakebite do not seek medical attention, but tend to approach village elders or herbalists. This group was most valuable for assessing incidence of bites and the prevalence of different species in that area. The main difference in the symptom profile reported was an additional inability of patients to open their mouths (reported by all herbalists interviewed). However, as it is very unusual for pure neurological sequelae to follow Russell's viper bite, it is possible that this arises out of the unwillingness of patients to swallow the substances offered to them as therapy (which include the local alcoholic beverage, arak, the water in which a dead person's femur has been boiled, banana water, one's own urine, betel nuts and the leaves and roots of certain trees). Treatment also includes the topical application of palm leaves, lime, tree roots and spit. Infections following the application of these substances may be a contributory factor to mortality and morbidity from snake bites, but this is difficult to assess.

Hospital treatment

Of the 26 victims, 19 received hospital treatment. This included nine of the 10 fatal cases, and 10 of the 16 surviving patients. Seven patients received no hospital treatment. The following treatments were given: steroid injection (three cases), antibiotics (four cases), application of a tourniquet (one case), analgesia (one case) and attachment of a black stone, 'to suck out the venom' (one case).

Medical practitioners' treatment and estimates of morbidity and mortality

Seven doctors and one paramedic were questioned regarding their management of snakebites. Each practitioner saw an average of 7.3 Russell's viper victims per year. The following treatments were given to snakebite victims (numbers in brackets refer to the number of practitioners performing the treatment): steroid administration (6), non-specific (not raised against Russell's viper venom) antivenom (5), antibiotics (5), artificial ventilation (3), tourniquets (3), cross-shaped skin incisions (3), local suction/irrigation (1), antihistamines (1), anti tetanus prophylaxis (1), black stone (1), diuretics (1) and fasciotomy (1).

In addition to the symptoms and signs reported by the villagers (Table 4), the medical practitioners also reported evidence of skin anaesthesia, shock, compartment syndrome and bleeding into the skin and mucous membranes. Death was reported as being more common with late presentation and a bleeding tendency. One medical practitioner estimated the average survival rate after Russell's viper envenoming at 80%.

Discussion

Population systematics of Russell's viper

In view of the advantages of preserving nomenclatural stability for such a medically important animal, the two main morphologically distinct forms (eastern and western) have been treated as subspecies of a single species (*Daboia russelli russelli* and *D. r. siamensis*, respectively). An analysis of the phylogenetic relationships of

Table 4. Russell's viper envenoming symptom profile in all bite case histories ($n=26$)

Symptom	Number of cases with symptom
Local symptoms	
Swelling	23
Tissue destruction and necrosis	6
Local cyanosis	5
Prolonged local bleeding	4
Regional lymphadenopathy	4
Blister formation	2
Temporary loss of use of bitten limb	11
Systemic symptoms	
Fever	6
Thirst	4
Rigors	1
Respiratory symptoms	
Dyspnoea	12
Gastrointestinal symptoms	
Vomiting	19
Genitourinary symptoms	
Haematuria	2
Anuria	6
Neurological symptoms	
Facial weakness	4
Eyelid ptosis	4
Dizziness	3
Confusion	3
Headache	1
Haematological symptoms	
Easy bruising	2

Russell's viper populations, using mitochondrial DNA sequence information, is currently in progress at Bangor, and this may provide further evidence regarding the status of these groups, as well as on the status of the Lesser Sunda populations.

It is clear from previous analyses of geographic variation in the effect of Russell's viper venoms that there is as much variation within the eastern and western forms as there is between them. This applied even within the conventional subspecies recognized before this study, e.g. between the Burmese and Thai populations of *D. r. siamensis* (Warrell 1989) and between populations of *D. r. russelli* from different parts of India (Jayanthi & Gowda 1988). Consequently, it is of crucial importance that toxinologists and physicians ascertain the locality of origin of any Russell's viper specimens or venoms they may be dealing with, irrespective of stated taxonomic affinities. Similarly, venom suppliers must make locality information available to their customers, as failure to do so can lead to considerable confusion and misleading results.

Geographical variation in venom effects — Indonesia and mainland Asia

Local envenoming

The local signs of envenoming by Russell's viper are mild compared with those of some other Viperidae such as *Calloselasma rhodostoma* and are usually of less medical importance than systemic effects. However, 88.5% of the Indonesian patients bitten suffered from notable local symptoms, the most common being swelling. As with other populations of this species, the incidence of systemic envenoming increased with the extent of local swelling and enlargement of regional lymph nodes. All the patients with systemic sequelae showed local signs, in contrast to data from Burma (Myint-Lwin *et al.* 1985). The incidence of local blistering was 7.7%, compared to 2% and 6% in Burma and Sri Lanka respectively (Myint-Lwin *et al.* 1985; Phillips *et al.* 1988). Twenty-three per cent of all bite victims suffered from significant local tissue destruction and necrosis. It therefore appears that bites by the Lesser Sunda populations of this species result in more severe local envenoming.

Table 5. Geographic variation in the clinical manifestations of Russell's viper envenomation. Data from this study are added to the data summarized in Warrell (1989)

	Western form		Eastern form			
Symptom	Sri Lanka	India	Burma	Thailand	Taiwan	Indonesia
Bleeding and coagulopathy	+	++	++	++	?	++
Renal failure	++	+	++	+	+	+
Pituitary infarction	-	+	++	-	?	?
Intravascular haemolysis	++	+	-	+	?	?
Neuromyotoxicity	++	+	-	-	?	+
Generalized capillary permeability	-	-	++	-	?	?+
Primary shock	-	+	++	-	?	?

Systemic envenoming

Incoagulable blood caused by defibrination and the ensuing consumption coagulopathy is the most common feature of Russell's viper envenoming across the entire geographical distribution of the species (Warrell 1989). However, haemostatic abnormalities are less commonly encountered in Sri Lanka. Spontaneous systemic haemorrhage occurred in 34.6% of the victims in Indonesia. This compares with 46% and 22% of the patients in Burma and Sri Lanka respectively. The sites of haemorrhage in these latter countries include the gingival sulci, the gastrointestinal tract (haematemesis and melaena), conjunctivae and skin (discoid haemorrhages), the genitourinary tract (haematuria), respiratory tract (haemoptysis) and the reproductive tract (menorrhagia and pre- and post-partum haemorrhage). In Indonesia the commonest sites of haemorrhage were the gastrointestinal tract (34.6%) and the genitourinary tract (7.7%).

Acute renal failure has a higher incidence and subsequent mortality in Russell's viper than in envenoming by any other species of snake. Anuria occurred in 23% of the Indonesian bites, compared to oliguria in 44% of the Burmese group. The incidence of neurotoxicity in the Indonesian group was considerably less than that experienced in Sri Lanka. Facial weakness was reported in 15.3% of bites in Indonesia (23% in Sri Lanka). Ptosis (drooping of the eyelids) was experienced in 11.5%, considerably less than the Sri Lankan group (77%). It is possible that the venom from Indonesia may have neurotoxic properties, which were previously believed to be unique to Sri Lanka and Southern India. Conjunctival and facial oedema would appear to remain unique to Burma. It is interesting to note that respiratory distress and failure are reported to be relatively rare after Russell's viper bites elsewhere, but 46% of Indonesian bite victims reported dyspnoea.

This study consisted of a retrospective analysis, with a relatively small sample size, thus the descriptions of local and systemic envenoming may be liable to distortion resulting from subject exaggeration and interpretation in translation. Furthermore, the study has no reliable laboratory or clinical evidence to support the symptomatology. There is, however, sufficient evidence to suggest that there are differences in the clinical effects of venom of Russell's viper from the Lesser Sunda Islands, compared to the venom from other populations of both the eastern and western groups of the species (Table 5). Further work is required to assess if the antivenoms raised against Russell's viper from other parts of its range are effective for the treatment of envenoming by the Lesser Sunda population.

Treatment conclusions

Many bite victims do not consider seeking medical attention and the local medical services will tend to underestimate the true extent of the problem. There is no proven effective antivenom to treat Russell's viper envenoming in the Lesser Sunda Islands population, and local practitioners are only able to offer supportive treatment. Many of the doctors questioned are using treatments that are contraindicated and highly dangerous, for example incising skin in patients with incoagulable blood, or use of antivenoms which were not raised against Russell's viper venom and which carry a high rate of anaphylaxis.

Acknowledgements

This paper is the result of a collaboration between the Nuffield Department of Medicine, University of Oxford and the Venomous Snake Research Group at the School of Biological Sciences, University of Wales, Bangor. The systematic study and review of the current distribution of Russell's viper in Indonesia was carried out by the latter and the epidemiological investigation and additional distributional information was provided by the former. RST, WW and AM acknowledge the financial assistance of the Leverhulme Trust in funding this work, and would like to thank Satoko Otsuka for recording the morphological data for the systematic study of Russell's viper. They also thank Professor Sangkot Marzuki, Dr Aucky Hinting, Vincen Khartono and Chriss Lay for their invaluable assistance in Indonesia. PJB wishes to thank Chriss Lay for helping to translate and collate the questionnaires and for invaluable guide services; Dr George Follows for translation skills; Dr Lauren Padmore for computing facilities and support; and Scott Jaeger.

References

Auffenberg, W. (1980). The herpetofauna of Komodo, with notes on adjacent areas. *Bull. Fla St. Mus. biol. Sci.* **25**: 39–156.

Aung-Khin, M. (1980). The problem of snakebites in Burma. *Snake* **12**: 125–127.

Aye, M. M. (1990). Venomous snakes of medical importance in Burma. In *Snakes of medical importance (Asia–Pacific region):* 211–241. (Eds Gopalakrishnakone, P. & Chou, L. M.). Venom and Toxin Research Group, National University of Singapore, Singapore.

Boulenger, G. A. (1896). *Catalogue of the snakes in the British Museum (Natural History)* **3**. British Museum (Natural History), London.

Brongersma, L. D. (1958). Note on *Vipera russelii* (Shaw). *Zool. Meded., Leiden* **36**: 55–76.

Deraniyagala, P. E. P. (1945). Some new races of the python, *Chrysopelea*, binocellate cobra and Tith-polonga inhabiting Ceylon and India. *Spolia zeylan.* **24**: 103–113.

Duméril, A. M. C., Bibron, G. & Duméril, A. H. A. (1854). *Erpétologie générale ou histoire naturelle complète des reptiles* 7 (2). Librairie Encyclopédique de Roret, Paris.

Dunn, E. R. (1927). Results of the Douglas Burden expedition to the island of Komodo. II. Snakes from the East Indies. *Am. Mus. Novit.* **287**: 1–7.

Herrmann, H.-W., Joger, U. & Nilson, G. (1992). Phylogeny and systematics of viperine snakes. III: resurrection of the genus *Macrovipera* (Reuss, 1927) as suggested by biochemical evidence. *Amphibia-Reptilia* **13**: 375–392.

Hodges, R. J. (1993). Snakes of Java with special reference to East Java Province. *Bull. Br. herpet. Soc.* **43**: 15–32.

Hoesel, J. K. P. van (1954). *Vipera russelli* — its zoogeographical range and local distribution in Indonesia. *Trop. Natuur* **33**: 133–139.

Hoesel, J. K. P. van (1958). Notities over *Vipera russellii* en enkele andere slangen van Flores. *Lacerta* **16**: 32–36.

Hoesel, J. K. P. van (1959). *Ophidia Javanica.* Museum Zoologicum Bogoriense, Bogor.

Jayanthi, G. P. & Gowda, T. V. (1988). Geographical variation in India in the composition and lethal potency of Russell's viper (*Vipera russelli*) venom. *Toxicon* **26**: 257–264.

Jeyarajah, R. (1984). Russell's viper bite in Sri Lanka. A study of 22 cases. *Am. J. trop. Med. Hyg.* **33:** 506–510.

Kanjanajatanee, J. & Visutipant, S. (1984). Russell's viper bite: clinical manifestations and treatment. *Thai med. Coun. Bull.* **13:** 25–38.

Kopstein, F. (1936). Über *Vipera russellii* von Java. *Treubia* **15**: 259–264.

Looareesuwan, S., Viravan, C. & Warrell, D. A. (1988). Factors contributing to fatal snake bite in the rural tropics: analysis of 46 cases in Thailand. *Trans. R. Soc. trop. Med. Hyg.* **82**: 930–934.

Mahasandana, S., Pochanugool, C., Limthongkul, S., Meemano, K., Jintakune, P. & Sitprija, V. (1990). Venomous snakes and snake bite in Thailand — viper bite. In *Snakes of medical importance (Asia–Pacific region):* 570–579. (Eds Gopalakrishnakone, P. & Chou, L. M.). Venom and Toxin Research Group, National University of Singapore, Singapore.

Mahasandana, S., Rungruxsirivorn, Y. & Chantarangkul, V. (1980). Clinical manifestations of bleeding following Russell's viper and green pit viper bites in adults. *S.E. Asian J. trop. Med. publ. Hlth* **11**: 285–293.

Mertens, R. (1927). Zur Verbreitung von *Vipera russelii* Shaw. *Senckenberg. Biol.* **9:** 182–184.

Myint-Lwin, Warrell, D. A., Phillips, R. E., Tin-Nu-Swe, Tun-Pe & Maung-Maung-Lay (1985). Bites by Russell's viper (*Vipera russelli siamensis*) in Burma: haemostatic, vascular and renal disturbances and response to treatment. *Lancet* **ii**: 1259–1264.

Neuhaus, H. (1935). Neunachweis von *Vipera russellii* auf Java. *Treubia* **15**: 49–50.

Phillips, R. E., Theakston, R. D. G., Warrell, D. A., Galigedara, Y., Abeysekera, D. T. D. J., Dissanayaka, P., Hutton, R. A. & Aloysius, D. J. (1988). Paralysis, rhabdomyolysis and haemolysis caused by bites of Russell's viper (*Vipera russelli pulchella*) in Sri Lanka — failure of Indian (Haffkine) antivenom. *Q. Jl Med.* **68**: 691–717.

Reid, H. A. (1968). Symptomatology, pathology and treatment of land snake bites in India and South-east Asia. In *Venomous animals and their venoms*: 611–642. (Eds Bücherl, W., Buckley, E. E. & Deulofeu, V.). Academic Press, New York.

Viravan, C., Looareesuwan, S., Kosskarn, W., Wuthiekanun, V., McCarthy, C. J., Stimson, A. F., Bunnag, D., Harinasuta, T. & Warrell, D. A. (1992). A national hospital-based survey of snakes responsible for bites in Thailand. *Trans. R. Soc. trop. Med. Hyg.* **86**: 100–106.

Warrell, D. A. (1989). Snake venoms in science and clinical medicine, 1. Russell's viper: biology, venom and treatment of bites. *Trans. R. Soc. trop. Med. Hyg.* **83**: 732–740.

Warrell, D. A. (1995). Clinical toxicology of snakebite in Asia. In *Handbook of clinical toxicology of animal venoms and poisons*: 493–594. (Eds Meier, J. & White, J.). CRC Press, Boca Raton.

Woodhams, B. J., Wilson, S. E., Bao-Chen Xin & Hutton, R. A. (1990). Differences between the venoms of two sub-species of Russell's viper: *Vipera russelli pulchella* and *Vipera russelli siamensis*. *Toxicon* **28**: 427–433.

Wüster, W. & Thorpe, R. S. (1989). Population affinities of the Asiatic cobra (*Naja naja*) species complex in south east Asia: reliability and random resampling. *Biol. J. Linn. Soc.* **36**: 391–409.

Wüster, W. & Thorpe, R. S. (1991). Asiatic cobras: systematics and snakebite. *Experientia* **47**: 205–209.

Wüster, W., Otsuka, S., Malhotra, A. & Thorpe, R. S. (1992). Population systematics of Russell's viper: a multivariate study. *Biol. J. Linn. Soc.* **47**: 97–113.

Wüster, W., Otsuka, S., Thorpe, R. S. & Malhotra, A. (1992). Morphological variation in Russell's viper in Burma and Thailand. *Herpet. J.* **2**: 99–101.

17 Toxic phospholipases in snake venom: an introductory review

JOHN B. HARRIS

Synopsis

Phospholipases of the class A_2 (PLA_2; EC 3.1.1.4) are common constituents of snake venoms. Some venom phospholipases A_2 seem to have only a digestive function but others have evolved into highly toxic compounds, typically expressing both presynaptic neurotoxicity and myotoxicity.

The toxic phospholipases A_2 are major contributors to morbidity and mortality following snake bite by some snakes of all venomous families. Particularly noteworthy are the myotoxicity accompanying bites by the Australian elapid *Notechis scutatus*, the neurotoxicity and muscle damage caused by the South American rattlesnake *Crotalus durissus terrificus* and the long-term neurotoxicity following bites by the kraits of south-east Asia and the taipan of Papua New Guinea.

Many snakes elaborate several phospholipases A_2 and some elaborate phospholipase A_2 homologues with little or no biological activity. Since each phospholipase A_2 homologue is the product of a single gene, the animals clearly expend a great deal of metabolic energy in maintaining the expression of so many similar genes. Perhaps this suggests that there is a selective advantage to the production of such a multiplicity of protein homologues.

Analysis of the structure of the phospholipases A_2 isolated from snake venoms suggests that the enzymes of viperid/crotalid snakes and elapid/hydrophid snakes respectively have evolved from different precursors.

The toxic phospholipase A_2 components of venom are of great clinical significance; the long-term neurotoxicity that results from many bites can be fatal and is a therapeutic challenge; myotoxicity is potentially life-threatening and difficult to control.

The source and structure of venom phospholipases A_2

Phospholipases are found in virtually all snake venoms, regardless of family, genus or species. The vast majority of phospholipases in venom are of the class A_2 (EC 3.1.1.4). Phospholipases A_2 hydrolyse the ester bond at the C_2 position of 3-*sn*-phosphoglycerides to liberate free fatty acids and to produce the appropriate lysophosphatide (Fig. 1). As with most other phospholipases A_2, venom phospholipases A_2 are Ca^{2+}-dependent (Harris & MacDonell 1981).

Symp. zool. Soc. Lond. (1997) No. 70: 235–250
ISBN 0 19 854986 5

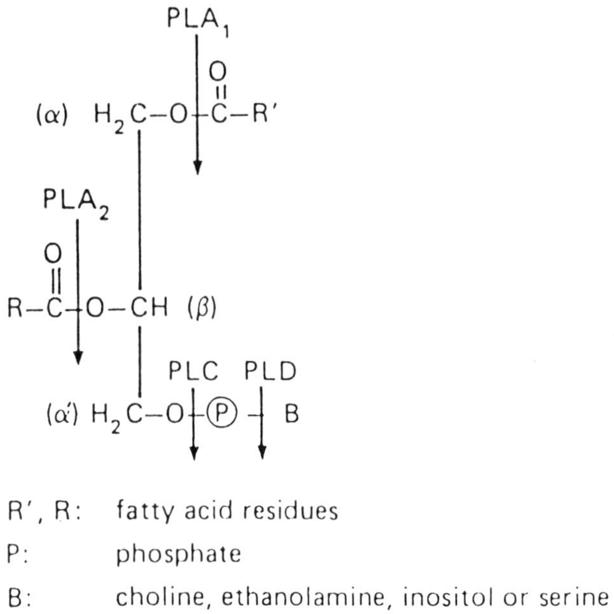

Fig. 1. The classification of phospholipases type A_1, A_2, C, D according to the site of hydrolysis (arrows) of 3-*sn*-phosphoglycerides.

Sequence analysis of venom phospholipases A_2 allows the identification of two classes of enzyme. Class 1 phospholipases A_2 consist of approximately 120 residues cross-linked by seven disulphide bridges. They are characterized by a leucine at homology position 41 and a disulphide bridge between Cys residues at homology positions 11 and 78. Class 1 phospholipases A_2 are found in the venoms of snakes of the families Elapidae and Hydrophiidae, and are homologous with mammalian pancreatic phospholipases A_2.

Class 2 phospholipases are slightly longer (125–130 residues) as a result of additional residues at the C-terminus, lack the leucine at homology position 41 and the disulphide bridge between Cys residues 11 and 78. They contain, instead, a disulphide bridge between Cys residues at homology positions 51 and 134. Class 2 phospholipases A_2 are found in the venoms of snakes of families Viperidae and Crotalidae and are homologous with mammalian tissue phospholipases A_2.

Most snake venoms contain multiple isoforms of phospholipases A_2. In general, these isoforms differ in terms of amino-acid sequence. Although this may lead to significant changes in biological activity, the sequence changes cannot arise as a result of post-translational activity; the only sustainable explanation is that a family of enzymes is produced as a result of gene duplication, each isoform being encoded by a different gene. A typical example is that of the habu, *Trimeresurus flavoviridis*. Six phospholipase A_2 genes have been isolated from the venom gland of *Trimeresurus flavoviridis*. The genes each consist of four exons and three introns (a structure of the gene that is probably typical of all phospholipases A_2). The protein coding regions of the genes are much less homologous than the non-coding regions, suggesting that there must be selective pressure during the evolution of phospholipases A_2 that encourages and maintains gene duplication whilst maintaining hydrolytic activity of the gene product. In the case of *Trimesurus flavoviridis* not all of the protein products have yet been identified and so there remains a possibility that not all genes are fully translated. Whether this is a phenomenon that is common in the venomous snakes is not known.

The secondary structure of phospholipases A_2, regardless of source or class, is remarkably conserved (see Mebs & Claus 1991) and because of the extensive cross-linking with disulphide bridges, the tertiary structure is also conserved. The molecular organization is, however, highly variable. Some of the venom phospholipases A_2 are single chain polypeptides. Typical of this group are the notexin group of phospholipases A_2 of *Notechis scutatus*, pseudexin and its isoforms from the venom of *Pseudechis porphyriacus* and ammodytoxin and its isoforms from the venom of *Vipera ammodytes*. Single chain phospholipases A_2 are found in venoms of elapid, hydrophid and viperid snakes.

A second group of venom phospholipases A_2 consist of two or more unlinked subunits. Typical of such phospholipases A_2 is taipoxin, isolated from the venom of the Australian elapid *Oxyuranus scutellatus*, which is a complex of three subunits. In the natural complex, the subunits form a 1:1:1 ternary complex. α-Taipoxin is a strongly basic subunit, β-taipoxin is neutral and γ-taipoxin is acidic and glycosylated. Textilotoxin is a similar phospholipase A_2 isolated from the venom of the Australian elapid *Pseudonaja textilis*. The toxin consists of five subunits, three non-covalently bound phospholipase A_2 homologues and two other homologues that are covalently linked. No such complex of unlinked subunits has been identified in the venom of any viperid or crotalid snake.

The venoms of a number of crotalid snakes contain a very particular kind of multi-unit phospholipase A_2. Crotoxin, from the venom of the South American rattlesnake *Crotalus durissus terrificus*, is the typical member of this class of phospholipases A_2. Crotoxin consists of a basic single chain polypeptide of 122 residues, CB, linked non-covalently to a second polypeptide, the acidic CA. The subunit CA consists, in turn, of three chains A, B and C, cross-linked by seven disulphide bridges (Fig. 2). Sequence analysis of chains A, B and C suggests strongly that CA arises as a result of the post-translational processing of a CB homologue (Aird *et al.* 1985; Bouchier *et al.* 1988). Several isoforms of both CB and CA exist. As a result a single sample of venom may contain as many as 15 isoforms of crotoxin (Faure & Bon 1987). Similar phospholipases A_2 have been identified in other rattlesnake venoms — including *Crotalus scutulatus scutulatus*, *Crotalus vegrandis*, *Crotalus horridus atricaudatus* and *Crotalus viridis concolor* — and may be present in the venoms of some true vipers (i.e. snakes of the family Viperidae). No such toxins have been identified in the venoms of elapids or their close relatives, the sea-snakes.

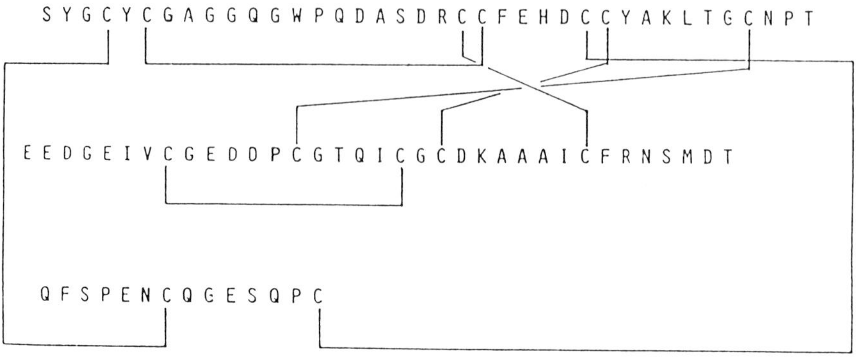

Fig. 2. The three chains of crotoxin A. Chain A (top) is homologous with residues 23–61 of typical crotalid phospholipases A_2; Chain B (middle) is homologous with residues 71–104; Chain C (bottom) is homologous with residues 112–125. Representative sequence data of crotalid phospholipases A_2 can be found in the compilation by Mebs & Claus (1991). Data presented in this figure were prepared by the author but derived from Aird *et al.* (1985) and Bouchier *et al.* (1988).

The β-bungarotoxins are unique phospholipases A_2 found only in the venoms of species of *Bungarus*, the kraits. The β-bungarotoxins consist of two covalently linked subunits, Chains A and B. Chain A is a polypeptide of 120 residues cross-linked by six disulphide bridges and is homologous with mammalian pancreatic phospholipase A_2. Three isoforms of Chain A have been identifed, A_1, A_2 and A_3. Chain B is a small polypeptide of only 60 residues. There are two variants of Chain B, B_1 and B_2. One Chain A and one Chain B are linked, to form the full molecule, by one disulphide bridge. Clearly there are six possible variants, A_1B_1, A_1B_2, A_2B_1, A_2B_2, A_3B_1 and A_3B_2.

The diversity of quaternary structure of venom phospholipases A_2 is striking and has a number of implications in terms of both the function of the toxic phospholipases A_2 and their evolution.

The functions of venom phospholipases A_2

Mebs & Claus (1991) tabulated the known sequences of phospholipases A_2 purified from snake venoms, including active enzymes and inactive homologues. The 67 listed polypeptides had been isolated from the venoms of 31 different species representing the four major families of venomous snakes — Elapidae, Hydrophiidae, Viperidae and Crotalidae. Clearly, phospholipase A_2 activity is a prominent feature of snake venom chemistry. Since the venom phospholipases A_2 share structural features in common with pancreatic phospholipases A_2, it is difficult to avoid the conclusion that the major function of the venom phospholipases A_2 is digestive. Thus, the inoculation of such enzymes into prey during a bite will ensure that the digestive process is initiated within the prey even before the prey has reached the stomach of the snake. This is particularly important when large prey items (relative to the body size of the predatory snake) are to be consumed.

Although some of the venom phospholipases A_2 are of low toxicity and presumably have a primary digestive role, many venom phospholipases A_2 are very toxic and are involved in the rapid killing of the prey. One of the most toxic of the venom phospholipases A_2 is notexin, the single chain phospholipase A_2 of 120 residues isolated from the venom of the Australian tiger snake, *Notechis scutatus scutatus*. The LD_{50} of notexin (mouse, i.v.) is 17 μg/kg. *Notechis* II-5 is a second phospholipase A_2 found in the venom of the tiger snake, and it has an LD_{50} of 49 μg/kg. The various forms of β-bungarotoxin have LD_{50}s ranging between 19 μg/kg and 130 μg/kg (Kondo *et al.* 1982a,b).

The toxic phospholipases A_2 of snake venoms exhibit two major properties related to their toxicity: neurotoxicity and myotoxicity. Neurotoxicity in man is manifest most commonly as neuromuscular weakness, ptosis, difficulties with swallowing and speaking and impaired respiration. Fatal neurotoxic poisoning results from respiratory paralysis. β-Bungarotoxin, isolated from the venom of the banded krait *Bungarus multicinctus*, was the first venom phospholipase A_2 to be characterized as a neurotoxin (Chang & Lee 1963). It was shown to cause neuromuscular paralysis by inhibiting the release of transmitter from motor nerve terminals — i.e. it was presynaptically active. A decade later, crotoxin and notexin were also shown to be presynaptically active neurotoxins (Vital-Brazil & Excell 1971; Harris, Karlsson & Thesleff 1973). It is now clear that many of the phospholipases A_2 of snake venoms are presynaptically active (for recent reviews see Harris 1991; Hawgood & Bon 1991).

The failure of neuromuscular transmission in isolated nerve muscle preparations exposed to these toxic phospholipases is usually multiphasic. The first phase is a partial inhibition of transmitter release. This is followed by a second phase of facilitation of release followed by a third phase of progressive failure of transmission until transmission fails completely (Fig. 3). The precise pattern of events depends on the toxic phospholipase A_2 used, the source of tissue and the conditions of the experiment (see Harris 1991). Phase 1 is often considered to be unrelated to hydrolytic activity. The second phase (facilitation) is Ca^{2+}-dependent and may be related to the transient depolarization of the nerve-terminal plasma membrane and the opening of voltage-sensitive Ca^{2+} channels, or to the block of a slowly inactivating K^+ channel at or close to the nerve terminal (Dreyer & Penner 1987; Rowan & Harvey 1988). This would lead to delayed repolarization of the nerve terminal membrane, prolonged opening of voltage activated Ca^{2+} channels and enhanced Ca^{2+} entry. The role of hydrolytic activity in phase II is not clear.

There is a similar lack of definitive information on the physiological basis of the longer-term decline of transmission that constitutes phase III. The suggestion that transmission failed because of the inhibition of high-affinity choline uptake into nerve terminals (Dowdall, Fohlmann & Eaker 1977) has not been accepted as an adequate explanation, although the specific high-affinity uptake of choline into nerve terminals would be expected to decline as the membrane potential fell. There is good morphological evidence that membrane recycling may be blocked at motor nerve terminals (Cull-Candy *et al.* 1976) but, in the absence of large increases in nerve-terminal plasma membrane, any inhibition of recycling is probably associated with a concurrent inhibition of vesicle docking during transmitter release and possibly the internal destruction of vesicles (Anderson & Parsons 1986; Noremberg & Parsons 1986). Hawgood, Smith & Strong (1988) suggested that 'the phospholipase A_2 subunit of crotoxin induces changes in areas of the axolemmal membrane involved in exocytosis and endocytotic retrieval possibly by localized alterations in membrane fluidity'. Theirs was only one of a number of similar suggestions that the presynaptically active neurotoxins hydrolysed specific populations of lipids in nerve terminal membranes, thereby inducing functional changes in major transmembrane proteins (see, for example, Strong, Heuser & Kelly 1977).

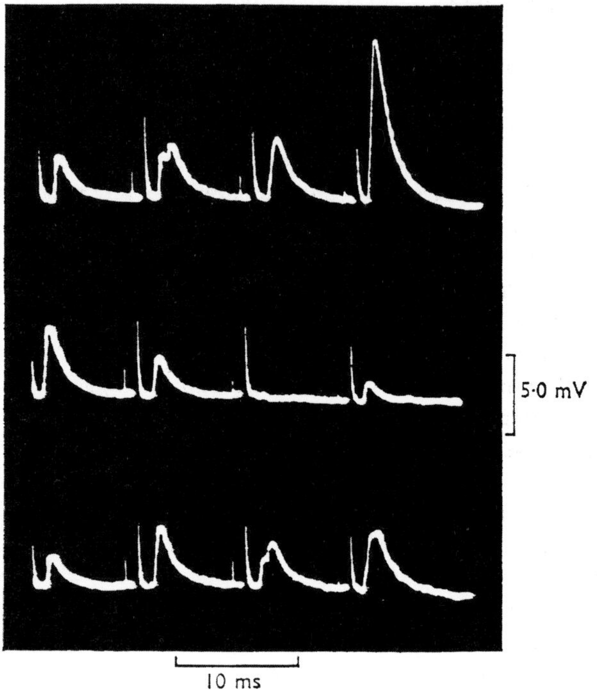

Fig. 3. Nerve stimulation was applied to a nerve/muscle preparation in late stages of poisoning by notexin. The sensitivity of the end-plate to the motor transmitter (ACh) was unimpaired, but transmitter release was so impaired that each pulse generated only a small end-plate potential rather than a muscle action potential. Data from Harris, Karlsson *et al.* (1973).

The possibility that neurotoxicity is expressed primarily as a result of the uptake of the neurotoxic phospholipases A_2 into the nerve terminal is particularly intriguing. Several authors have shown that during relatively early stages of neurotoxic activity the nerve terminals contain large numbers of clathrin-coated vesicles (Cull-Candy *et al.* 1976; Strong *et al.* 1977; Hamilton, Broad & Sutherland 1980). Such vesicles are commonly associated with the internalization of clostridial neurotoxins such as botulinum toxin (Morris & Saelinger 1989). Thus, the proposed mechanism of action of the presynaptically active phospholipases A_2 would be: binding to appropriate 'receptor' sites on the nerve terminal, uptake via clathrin-coated vesicles, release into the nerve terminal cytoplasm, disruption of terminal mitochondria and the synaptic vesicles and, ultimately, degeneration of the terminal. Strong *et al.* (1977) described the apparent uptake of peroxidase-β-bungarotoxin conjugates into murine motor nerve terminals *in vitro* but definitive evidence in favour of the internalization of the toxic phospholipases A_2 is still lacking.

Severe, systemic neuromuscular paralysis in man lasts for up to three weeks following envenomation by snakes whose venoms contain potent presynaptic phospholipases A_2, and the survival of the victim may require intensive care and assisted ventilation. Since the neurotoxic phospholipases A_2 are known to cause significant damage to neural tissue (Abe, Limbrick & Miledi 1976; Pittman, Oppenheim & Chu-Wang 1978; Landon *et al.* 1980; Rehm, Schaefer & Betz 1982; Gopalakrishnakone & Hawgood 1984), it is assumed that recovery is associated with the regeneration of the nerve terminal (and the muscle — see below) and the reconstruction of the neuromuscular junction (Grubb, Harris & Schoffield 1991).

The pronounced neurotoxicity exhibited by the presynaptically active phospholipases A_2 at the skeletal neuromuscular junction does not appear to be matched in the autonomic nervous system (see Harris 1991 for details) but it should be noted that there have been few *systematic* investigations of autonomic function in the presence of the presynaptically active phospholipases A_2.

The venom phospholipases A_2 are active in some parts of the CNS (see Harris 1991; Hawgood & Bon 1991) but since it is unlikely that the phospholipases A_2 cross the blood/brain barrier in intact animals or in man, it is not a topic discussed in this review.

As well as blocking transmitter release at the skeletal neuromuscular junction, many of the phospholipases A_2 of snake venoms are potent myotoxins (Fig. 4). Notexin was the first of the myotoxic phospholipases A_2 to be described (Harris, Johnson & Karlsson 1975) and many more have since been identified. The most potent myotoxic phospholipases A_2 have been isolated from the venoms of the Australian elapids *Notechis scutatus scutatus*, *Oxyuranus scutellatus* and *Pseudonaja textilis*, and the sea snake *Enhydrina schistosa*, but the crotoxin-related phospholipases A_2 from the venoms of several species of rattlesnake have potent myotoxic activity. Myotoxic activity usually coexists with presynaptic toxicity, but this is not always the case. β-Bungarotoxin is a potent presynaptically active compound but totally devoid of measurable myotoxic activity (J.B. Harris & R. Dixon unpublished).

The mechanism of action of the myotoxic phospholipases. A_2 is not known. Phospholipase A_2 activity is essential but not sufficient. Most phospholipases A_2, including mammalian pancreatic phospholipases A_2, are non-myotoxic. The inactivation of the hydrolytic activity of the potent myotoxic phospholipase A_2, notexin, abolishes neurotoxic and myotoxic activity (Harris & Johnson 1978; Harris & Zar 1978). The site of action of the myotoxic phospholipases A_2 is certainly the plasma membrane and early signs of damage include the formation of very small (10 μm) lesions in the muscle fibre membrane. These lesions are associated with regions of local hypercontraction of the fibre. With time the lesions become larger until the muscle fibre loses its typical architecture (Dixon & Harris in press). It is reasonable to suppose that hydrolysis of the lipids of the plasma membrane renders the membrane fluid leaky and incapable of maintaining normal ionic gradients. This results in the influx of Ca^{2+} and the initiation of the local contracture. The primary cause of an increase in fluidity may be the release of fatty acids and the formation of lysophosphatides, both of which are products of lipid hydrolysis and powerful detergents.

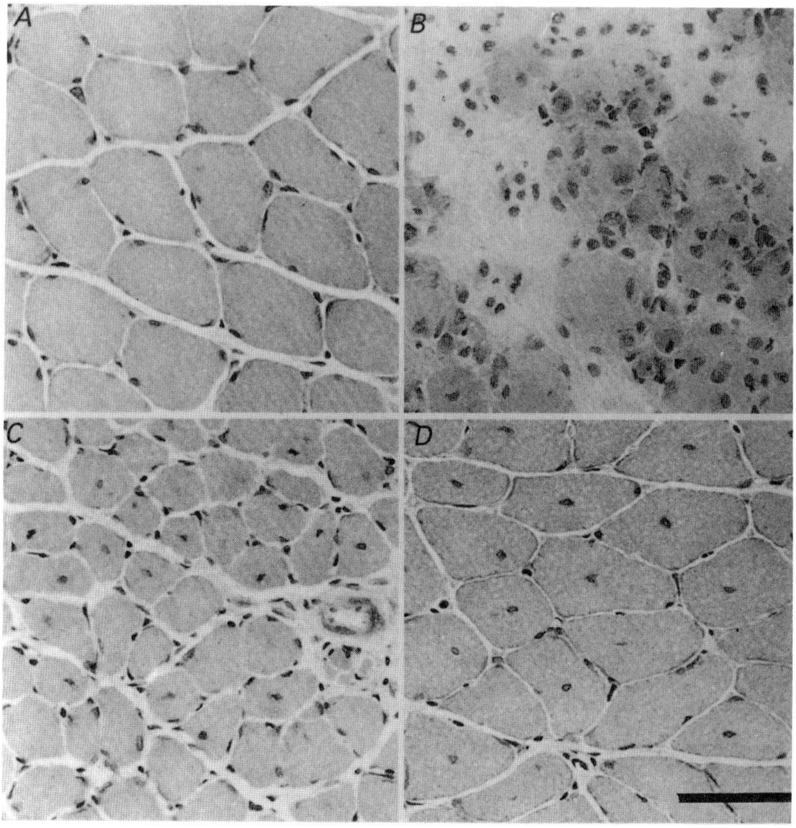

Fig. 4. Frozen sections of soleus muscles from the rat stained with H&E. (A) Control section; (B) 24 h after inoculation of notexin (2.0 μg s.c.); (C) 4 days after notexin; (D) 28 days after notexin. Note the total destruction of the muscle in B and the rapid regeneration of the tissue. From Grubb *et al.* (1991).

The degeneration of muscle fibres begins with the rapid degradation of the key cytoskeletal proteins desmin and titin (Vater, Cullen & Harris 1992) liberating the A bands (comprising, in the main, myosin) which are ingested by invading phagocytic cells (Harris & Maltin 1982). The mechanism whereby proteolytic activity within the muscle fibres is stimulated is not known.

Determining why some phospholipases A_2 are toxic when homologous enzymes are non-toxic is the subject of great activity. There is no structural feature of the venom phospholipases A_2 that correlates with either presynaptic neurotoxicity or myotoxicity. The toxic phospholipases A_2 may be single chain polypeptides — notexin for example, multi-subunit compounds — crotoxin and taipoxin for example, or covalently linked assemblies of two or more subunits — β-bungarotoxin for example. They may be Class I phospholipases A_2, like notexin, or Class II phospholipases A_2 such as crotoxin. The most consistent feature (and there are exceptions to this general rule) is that, if the toxic phospholipase A_2 is a single chain polypeptide, it is basic. If it is a complex of several subunits it contains a basic single chain. Almost invariably, the basic chain of a multi-subunit complex is the chain that is both hydrolytic and toxic. A second general rule is that not only is the basic chain in multi-subunit toxic phospholipases A_2 toxic on its own, but in its natural complex toxicity is enhanced by the presence of an inactive homologous subunit. Conversely, hydrolytic activity may be higher in the isolated subunit and suppressed when the complex is formed. This type of behaviour is best documented for crotoxin. The CA subunit of crotoxin is not hydrolytic and is devoid of both neurotoxic and myotoxic activity. CB is hydrolytic, neurotoxic and myotoxic. The reconstitution of the complex (achieved simply by mixing) results in a reduction in the hydrolytic activity of CB and a potentiation of its neurotoxicity and myotoxicity (Rübsamen, Breithaupt & Habermann 1971; Hawgood & Santana da Sa 1979; Bon *et al.* 1979; Marlas & Bon 1982; Kouyoumdjian, Harris & Johnson 1986).

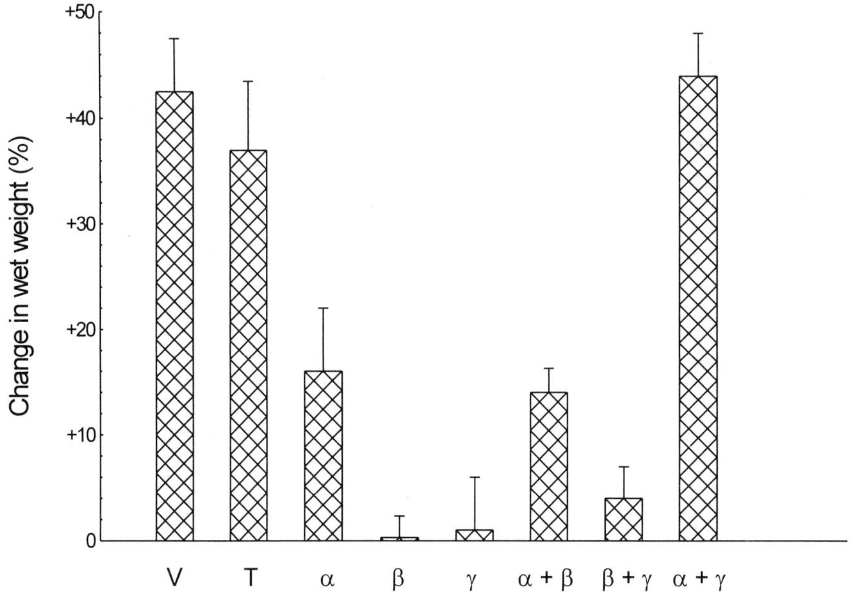

Fig. 5. Mean (± SEM) wet weight changes in soleus muscle (n = 5) of rat following the inoculation of venom (10 μg s.c.) of the Australian taipan (V) and equivalent amount of taipoxin (T) and the α, β and γ sub-units of taipoxin. Note the synergistic activity of α + γ sub-units.

Similar experiments have shown that α-taipoxin, the basic subunit of taipoxin, is the only toxic subunit and that its toxicity is enhanced by the addition of γ-taipoxin (Harris & Maltin 1982, and see Fig. 5). The non-toxic subunits of complex toxins (e.g. CA- and γ-taipoxin) do not enhance the toxicity of notexin (J.B. Harris unpubl.), which suggests that the 'enhancing' effect is highly specific.

The neurotoxic and myotoxic activity of many venom phospholipases A_2 are life-threatening and in clinical terms are by far the most significant toxic effects of the compounds. A number of phospholipases A_2 are haemolytic and some have anti-coagulant activity. Phospholipase A_2 activity can also activate the production of prostaglandins, thromboxanes and leukotrienes, all of which are physiologically active and could complicate an already complex syndrome. These activities of the venom phospholipases A_2 have not been systematically studied, probably because it is generally thought that they do not cause significant added risk to patients and because haemolytic, haemorrhagic and haematological activities tend to be the result of the presence in venoms of other groups of highly specialized toxins (see Rosenberg 1986, for a brief discussion of phospholipases A_2 and their involvement in haemolysis, and Kornalik 1991, for a discussion of anti-coagulant and pro-coagulant toxins).

Active domains of phospholipases A_2

There is general agreement that the full expression of the toxicity of the toxic phospholipases A_2 requires hydrolytic activity. It is, however, well established that many phospholipases A_2 of snake venoms are not toxic. This indicates not only that hydrolytic activity *per se* is an insufficient explanation of toxicity; it also suggests that the toxic phospholipases A_2 have, in addition to a hydrolytic domain, binding domains that allow binding to nerve terminals and muscle fibre membranes as appropriate and, possibly, other domains that enable the hydrolytic potential to be expressed. Numerous studies involving chemical derivatization (reviewed in detail by Rosenberg 1986), molecular and immunological techniques and physiological/biophysical techniques have confirmed that separate domains do exist, but every attempt to determine the location of those domains has failed. It is a topic of great significance to understanding the biology of this important group of toxins, and it generates a great deal of work (see Hawgood & Bon 1991; Harris 1991; Tzeng 1993, for extensive discussions of this problem). The only reasonable certainty is the identity of the hydrolytic centre involving the residues His 48, Asp 49, and Asp 99 (Verheij *et al.* 1980; Van Den Bergh *et al.* 1988, for example).

Binding of toxic phospholipases A_2 to target sites

The toxicity of the neurotoxic and myotoxic phospholipases A_2 is so specific that it is impossible to believe that the binding of the toxins to sensitive membrane sites is purely opportunistic. It seems far more likely that specific domains in the toxins bind with high affinity to specific 'receptive sites' or 'acceptors' at the nerve terminal or muscle fibre membrane.

The investigation of binding sites has been characterized by slow progress on a seemingly simple problem. The major difficulty is that most phospholipases A_2 cannot easily be labelled in such a way as to retain biological activity. Single chain toxins have proven virtually impossible to label and the only phospholipases A_2 successfully labelled so far have been β-bungarotoxin, crotoxin (but only in its native complex of CA + CB) and a rather poorly characterized group of phospholipases A_2 OS_1 OS_2 and OS_3, from the venom of the Australian taipan. The separated sub-units of crotoxin (CA and CB) cannot be successfully labelled. It would appear that radiolabelling tends to inactivate the toxic phospholipases A_2 because a crucial lysine is attacked. Colloidal gold, another frequently used label in such circumstances, probably disrupts activity by becoming attached to sulphur in the disulphide bonds, thereby distorting the tertiary structure of the proteins. It would seem that the molecular structure of crotoxin and β-bungarotoxin allows labelling whilst protecting vulnerable lysines.

Crotoxin and β-bungarotoxin have been used to obtain good evidence that the toxic phospholipases A_2 bind with high affinity to specific sites on excitable membranes. β-Bungarotoxin binds with an equilibrium dissociation constant (K_D) of 0.025–0.5 nmol/l to chick-brain synaptosomes, but the density of binding sites is only 50 fmol/mg protein. In rodent brain two classes of binding sites have been identified, one with a K_D of 0.2–0.6 nmol/l and the other a low-affinity binding site with a K_D of 6–10 nmol/l. Binding does not appear to be Ca^{2+}-dependent, but in the absence of Ca^{2+} (or any other divalent cation) binding is reduced. This may suggest that phospholipase activity is required to expose more binding sites to the toxin.

Crotoxin also binds to synaptosomal membranes. There appear to be two binding sites, one of relatively high affinity ($K_D = 4$ nmol/l) and one of very low-affinity ($K_D = 87$ nmol/l). The density of binding sites was low (2–4 pmol/mg protein).

Studies of crotoxin binding to post-synaptic membranes of the electric organ of *Torpedo* suggest that the CB subunit alone binds with low affinity and low specificity, but that in the presence of CA the low-affinity, non-specific binding is blocked and a high-affinity, saturable binding is revealed.

Photoaffinity labelling of radio-labelled crotoxin and β-bungarotoxin has been used in an attempt to identify binding proteins on the synaptosomal surface. Two polypeptides, of 45 kDa and 85 kDa, have been identified for crotoxin (with the possibility that two others of 50 kDa and 65 kDa exist in some preparations) and two have been identified for β-bungarotoxin — 25 kDa and 70–75 kDa. In neither case have the polypeptides been characterized. Crotoxin, in addition, may require negatively charged lipids as a binding site or to facilitate binding.

Data on binding are exceptionally difficult to analyse and interpret. Labelled toxins and chemically derivatized toxins are rarely fully characterized in terms of tertiary structure, hydrolytic activity and toxicity; binding data are highly variable and are inconsistent between investigations; little note is taken of the complications caused by concomitant hydrolysis of the cell membranes being used in binding studies; synaptosomes derived from the brains of immature animals cannot be compared with the motor nerve terminals in the peripheral somatic nervous system.

Very comprehensive reviews of this field have been recently prepared by Hawgood & Bon (1991) and Tzeng (1993) and the interested reader should refer to those reviews for original sources and critical surveys of the interactions between presynaptically active toxic phospholipases and membrane binding sites.

Evolution of toxic phospholipases A_2

The evolution of toxic phospholipases and their presence in venoms of snakes of all venomous families suggests that they are of significant advantage to the snake — few other classes of toxin are so universally distributed between families.

Sequence analysis of the toxic phospholipases A_2 of Class I has shown that in structural terms the phospholipases A_2 of the venoms of the Australian elapids and the sea-snakes are very similar and probably share a common ancestor. (This would be consistent with the view that the Australian elapids evolved from the sea-snakes.) A similar sequence analysis of the Class I phospholipases A_2 of African and Asian elapids suggests that they form a second group of polypeptides with a probable common ancestor. The phospholipases A_2 of the kraits are structurally distinct and form a third group of Class I phospholipases A_2. The phospholipases A_2 of the crotalid and viperid species form a fourth group of closely related Class II toxins (Dufton & Hider 1983; Tamiya 1985).

The structural differences between Class I and Class II phospholipases A_2 are so large that it suggests that two completely separate but convergent evolutionary processes were involved in their emergence as major components of snake venoms.

The common observation that many venoms contain more than one phospholipase A_2 homologue and that, in the main, these proteins have to have been produced from different genes, suggests that extremely strong evolutionary pressure must have been exerted to allow the gene duplication because of the high metabolic cost of maintaining multiple copies of a gene, most copies of which seem to be fully translated. It is interesting to speculate that the large number of toxic phospholipases present in many samples of venom allow the snake to respond to changes in the availability of prey species by utilizing venom constituents that are themselves selectively toxic.

Clinical aspects of toxic phospholipases A_2 in venom

Severe neurotoxic poisoning is a common feature of poisoning bites by elapid snakes and sea-snakes. The venoms of these animals contain post-synaptically active toxins that bind to acetylcholine receptors at the neuromuscular junction as well as (in many cases) presynaptically active phospholipases A_2. Acute neuromuscular poisoning is more likely, therefore, to be caused by the former. Prolonged paralysis, often requiring intensive care, is seen most frequently following bites by the kraits and the taipan of Papua New Guinea (Pearn 1971; A. Trevitt pers. comm.). In these cases, victims may be paralysed for up to three weeks. The paralysis is unresponsive to antivenom and anticholinesterase and is not necessarily associated with muscle damage. It is probable that such prolonged paralysis is presynaptic in origin and results from the activity of presynaptically active phospholipases A_2. This interpretation is based on experimental studies with the presynaptically active toxins, and refers to the time needed for the nerve terminal to regenerate and begin transmitting normally (Grubb *et al.* 1991).

Muscle damage is a feature of bites by a number of elapids including the Australian tiger snake (*Notechis scutatus*), crotalids, including the South American rattlesnake (*Crotalus durissus terrificus*), hydrophids, including *Enhydrina schistosa*, and viperids, including the Sri Lankan Russell's viper. It is difficult to treat acute degenerative muscle disease in such a way as to prevent the progression of damage and conservative management is essential. Regeneration will occur provided that (1) the microcirculation remains intact; (2) the innervation is undamaged; (3) no deep foci of infection are formed. Surgical intervention is inadvisable unless the release of excessive compartmental pressure makes it essential. Renal failure following the release of myoglobin, from the damaged muscle, is a common complication and renal function must be maintained.

Concluding comments

Despite the very large body of literature concerning the phospholipases A_2 of snake venoms, the areas of ignorance are huge. The evolutionary pressures that resulted in two major classes of phospholipases, several subclasses and multiple copies of genes in large numbers of snakes are not understood. The fact that such phenomena are common suggests that the pressure to maintain the evolutionary process is ongoing. It is an area of work not yet fully exploited by ecologists or evolutionary biologists.

Of the molecules themselves, we can neither define those regions that confer toxicity nor identify how the toxins interact with their targets in cell membranes. Their mode(s) of action is (are) subject to extensive speculation but little real fact.

The management of patients poisoned by the toxic phospholipases A_2 is still based on the use of antivenoms and conservative management of the patient. Where appropriate, intensive care, ventilatory support and the preservation of kidney function are essential.

Acknowledgements

The author thanks the graduate students and post-doctoral fellows who played a major role in the author's own work described herein. Financial support from MRC, Wellcome Trust, MDGB and Action Research is gratefully acknowledged.

References

Abe, T., Limbrick, A. R. & Miledi, R. (1976). Acute muscle denervation induced by β-bungarotoxin. *Proc. R. Soc. (B)* **194**: 545–553.

Aird, S. D., Kaiser, I. I., Lewis, R. V. & Kruggel, W. G. (1985). Rattlesnake presynaptic neurotoxins: primary structure and evolutionary origin of the acidic subunit. *Biochemistry* **24**: 7054–7058.

Anderson, D. C. & Parsons, S. M. (1986). Uncoupling of cholinergic synaptic vesicles by the presynaptic toxin β-bungarotoxin. *J. Neurochem.* **47**: 1305–1311.

Bon, C., Changeux, J-P., Jeng, T. W. & Fraenkel-Conrat, H. (1979). Postsynaptic effects of crotoxin and its isolated subunits. *Eur. J. Biochem.* **99**: 471–481.

Bouchier, C., Ducancel, F., Guignery-Frelat, G., Bon, C., Boulain, J.-C. & Ménez, A. (1988). Cloning and sequencing of cDNA's encoding the two subunits of crotoxin. *Nucleic Acid Res.* **16**: 9050.

Chang, C. C. & Lee, C. Y. (1963). Isolation of neurotoxins from the venom of *Bungarus multicinctus* and their modes of neuromuscular blocking action. *Archs int. Pharmacodyn. Thér.* **144**: 241–257.

Cull-Candy, S. G., Fohlman, J., Gustavsson, D., Lüllman-Rauch, R. & Theskeff, S. (1976). The effects of taipoxin and notexin on the function and fine structure of the murine neuromuscular junction. *Neuroscience* **1**: 175–180.

Dixon, R. & Harris, J. B. (In press). The early expression of myotoxicity and localization of binding sites of notexin in the soleus muscles of the rat. *Proc. Am. chem. Soc.*

Dowdall, M. J., Fohlman, J. P. & Eaker, D. (1977). Inhibition of high affinity choline transport in peripheral cholinergic endings by pre-synaptic snake venom toxins. *Nature, Lond.* **269**: 700–702.

Dreyer, F. & Penner, R. (1987). The actions of presynaptic snake toxins on membrane currents of mouse motor nerve terminals. *J. Physiol., Lond.* **386**: 455–464.

Dufton, M. J. & Hider, R. C. (1983). Classification of phospholipases A_2 according to sequence. Evolutionary and pharmacological considerations. *Eur. J. Biochem.* **137**: 545–551.

Faure, G. & Bon, C. (1987). Several isoforms of crotoxin are present in individual venoms from the South American rattlesnake *Crotalus durissus terrificus*. *Toxicon* **25**: 229–234.

Gopalakrishnakone, P. & Hawgood, B. J. (1984). Morphological changes induced by crotoxin in murine nerve and neuromuscular junction. *Toxicon* **22**: 791–804.

Grubb, B. D., Harris, J. B. & Schoffield, I. S. (1991). Neuromuscular transmission at newly formed neuromuscular junctions in the regenerating soleus muscle of the rat. *J. Physiol., Lond.* **441**: 405–421.

Hamilton, R. C., Broad, A. J. & Sutherland, S. K. (1980). Effects of Australian eastern brown snake (*Pseudonaja textilis*) venom on the ultrastructure of nerve terminals on the rat diaphragm. *Neurosci. Lett.* **19**: 45–50.

Harris, J. B. (1991). Phospholipases in snake venoms and their effects on nerve and muscle. In *Snake toxins*: 91–129. (Ed. Harvey, A. L.). Pergamon Press, New York, Oxford etc.

Harris, J. B. & Johnson, M. A. (1978). Further observations on the pathological responses of rat skeletal muscle to toxins isolated from the venom of the Australian tiger snake, *Notechis scutatus scutatus*. *Clin. exp. Pharmac. Physiol.* **5**: 587–600.

Harris, J. B., Johnson, M. A. & Karlsson, E. (1975). Pathological responses of rat skeletal muscle to a single subcutaneous injection of a toxin isolated from the venom of the Australian tiger snake, *Notechis scutatus scutatus*. *Clin. exp. Pharmac. Physiol.* **2**: 383–404.

Harris, J. B., Karlsson, E. & Thesleff, S. (1973). Effects of an isolated toxin from Australian tiger snake (*Notechis scutatus scutatus*) venom at the mammalian neuromuscular junction. *Br. J. Pharmac.* **47**: 141–146.

Harris, J. B. & MacDonell, C. A. (1981). Phospholipase A_2 activity of notexin and its role in muscle damage. *Toxicon* **19**: 419–430.

Harris, J. B & Maltin, C. A. (1982). Myotoxic activity of the crude venom and the principal neurotoxin, taipoxin, of the Australian taipan, *Oxyuranus scutellatus*. *Br. J. Pharmac.* **76**: 61–75.

Harris, J. B. & Zar, M. A. (1978). The effects of a toxin isolated from Australian tiger snake (*Notechis scutatus scutatus*) venom on autonomic neuromuscular transmission. *Br. J. Pharmac.* **62**: 349–358.

Hawgood, B. & Bon, C. (1991). Snake venom presynaptic toxins. In *Handbook of natural toxins. 5. Reptile venoms and toxins*: 3–52. (Ed. Tu, A. T.). Decker, New York.

Hawgood, B. J. & Santana da Sa, S. (1979). Changes in spontaneous and evoked release of transmitter induced by the crotoxin complex and its component phospholipase A_2 at the frog neuromuscular junction. *Neuroscience* **4**: 293–203.

Hawgood, B. J., Smith, I. C. H. & Strong, P. N. (1988). Early induction by crotoxin of biphasic frequency changes and giant miniature end-plate potentials at the frog neuromuscular junction. *Br. J. Pharmac.* **94**: 765–772.

Kondo, K., Toda, H., Narita, K. & Lee, C.-Y. (1982a). Amino acid sequences of β_2-bungarotoxin from *Bungarus multicinctus* venom. The amino acid substitutions in the B-chains. *J. Biochem., Tokyo* **91**: 1519–1530.

Kondo, K., Toda, H., Narita, K. & Lee, C.-Y. (1982b). Amino acid sequences of three β-bungarotoxins (β_3-, β_4-, β_5-bungarotoxins) from *Bungarus multicinctus* venom. Amino acid substitutions in the A-chains. *J. Biochem., Tokyo* **91**: 1531–1548.

Kornalik, F. (1991). The influence of snake venom proteins on blood coagulation. In *Snake toxins*: 323–383. (Ed. Harvey, A. L.). Pergamon Press, New York, Oxford etc.

Kouyoumdjian, J. A., Harris, J. B. & Johnson, M. A. (1986). Muscle necrosis caused by the sub-units of crotoxin. *Toxicon* **24**: 575–593.

Landon, D. N., Westgaard, R. H., MacDermot, T. & Thompson, E. J. (1980). The morphology of rat soleus neuromuscular junctions treated *in vitro* with purified β-bungarotoxin. *Brain Res.* **202**: 1–20.

Marlas, G. & Bon, C. (1982). Relationship between the pharmacological action of crotoxin and its phospholipase activity. *Eur. J. Biochem.* **125**: 157–165.

Mebs, D. & Claus, I. (1991). Amino acid sequences and toxicities of snake venom components. In *Snake toxins*: 425–447. (Ed. Harvey, A. L.). Pergamon Press, New York, Oxford etc.

Morris, R. E. & Saelinger, C. B. (1989). Entry of bacterial toxins into mammalian cells. In *Botulinum neurotoxin and tetanus toxin*: 121–152. (Ed. Simpson, L. L.). Academic Press, San Diego.

Noremberg, K. & Parsons, S. M. (1986). Selectivity and regulation in the phospholipase A_2-mediated attack on cholinergic synaptic vesicles by β-bungarotoxin. *J. Neurochem.* **47**: 1312–1317.

Pearn, J. (1971). Survival after snake bite with prolonged neurotoxic envenomation. *Med. J. Aust.* **2**: 259–261.

Pittman, R., Oppenheim, R. W. & Chu-Wang, I. (1978). Beta-bungarotoxin induced neuronal degeneration in the chick embryo spinal cord. *Brain Res.* **153**: 199–204.

Rehm, H., Schaefer, T. & Betz, H. (1982). β-bungarotoxin induced cell death of neurons in chick retina. *Brain Res.* **250**: 309–381.

Rosenberg, P. (1986). The relationship between enzymatic activity and pharmacological properties of phospholipases in natural poisons. In *Natural toxins*: 129–174. (Ed. Harris, J. B.). Oxford University Press, Oxford.

Rowan, E. G. & Harvey, A. L. (1988). Potassium channel blocking actions of β-bungarotoxin and related toxins on mouse and frog nerve terminals. *Br. J. Pharmac.* **94**: 839–847.

Rübsamen, J., Breithaupt, H. & Habermann, E. (1971). Biochemistry and pharmacology of the crotoxin complex: 1. Subfractionation and recombination of the toxin complex. *Naunyn-Schmierderberg's Arch. Pharmac.* **270**: 274–280.

Strong, P. N., Heuser, J. E. & Kelly, R. B. (1977). Selective enzymatic hydrolysis of nerve terminal phospholipids by β-bungarotoxin: biochemical and morphological studies. In *Cellular neurobiology*: 227–249. (Eds Hall, Z., Kelly, R. B. & Fox, C. F.). Alan Liss, New York.

Tamiya, N. (1985). A comparison of amino acid sequences of neurotoxins and of phospholipases of some Australian elapid snakes with those of other proteroglyphous snakes. In *Biology of Australasian frogs and reptiles*: 209–219. (Eds Grigg, G., Shine, R. & Ehmann, H.). Surrey Beatty & the Royal Zoological Society of New South Wales, Chipping Norton, N.S.W.

Tzeng, M.-C. (1993). Interaction of presynaptically active toxic phospholipases A_2 with membrane receptors and other binding sites. *Toxicol. Toxin Rev.* **12**: 1–62.

Van den Bergh, C. J., Slotboom, A. J., Verheij, H. M. & De Haas, G. H. (1988). The role of aspartic acid-49 in the active site of PLA_2. *Eur. J. Biochem.* **176**: 353–357.

Vater, R., Cullen, M. J. & Harris, J. B. (1992). The fate of desmin and titin during the degeneration and regeneration of the soleus muscle of the rat. *Acta neuropath.* **84**: 278–288.

Verheij, H. M., Volwerk, J. J., Jansen, E. H. J. M., Puyk, W. C., Dykstra, B. W., Drenth, J. & De Haas, G. H. (1980). Methylation of histidine-48 in pancreatic phospholipase A_2. Role of histidine and calcium ions in the mechanism. *Biochemistry* **19**: 743–750.

Vital-Brazil, O. & Excell, B. J. (1971). Action of crotoxin and crotactin from the venom of *Crotalus durissus terrificus* (South American rattlesnake) on the frog neuromuscular junction. *J. Physiol., Lond.* **212**: 348P.

18 The kinetics of snakebite envenoming and therapy

R. D. G. THEAKSTON

Synopsis

Treatment of systemic envenoming is related to the neutralizing efficacy of the antivenom used, the route by which it is administered and the dose. In the past, treatment of patients has depended almost entirely on the individual clinician's experience in assessing the severity of envenoming. The development of enzyme immunoassay has enabled a more scientific appraisal of the situation, allowing the accurate estimation of circulating specific venom and antivenom levels at any time in the patient's blood and other body fluids. It is therefore possible to measure accurately the efficacy of antivenom in the neutralization and clearance of venom antigen. In Brazil it has been shown that clinicians are treating patients with excessive amounts of highly efficient *Bothrops* polyspecific antivenom, with a resulting unacceptably high incidence of early anaphylactic reactions and, in Sri Lanka, the use of imported Indian antivenom is relatively ineffective in neutralizing the venoms of Sri Lankan snakes, demonstrating the real problem of venom variability within individual species. For example, Indian Russell's viper venom does not possess the neurotoxin present in the venom of the Sri Lankan subspecies. In West Africa the improved clearance of venom when a monospecific *Echis* antivenom is used in the treatment of *Echis ocellatus* envenoming, compared with that of the polyspecific antivenoms currently in use, has been demonstrated. Such clinically-based immunological studies provide an objective assessment of antivenom therapy which has not in the past been feasible. This should result in more efficient and controlled use of expensive antivenoms for treatment of systemic envenoming. Such studies also emphasize the importance of individual countries producing their own antivenoms for treatment of snakebite.

Introduction

The only scientifically and medically approved treatment of envenoming by snakes is the use of intravenously administered antivenom. This is usually prepared by hyperimmunization of large animals, usually horses, and subsequent salt precipitation and enzyme (pepsin) digestion of the immunoglobulin G to produce an active neutralizing F $(ab')_2$ fragment (Theakston & Warrell 1991). The dose of antivenom, which may be either monospecific (active primarily against a single venom species) or

Symp. zool. Soc. Lond. (1997) No. 70: 251–259
ISBN 0 19 854986 5

polyspecific (active against a range of 2–5 different venom species), depends on the neutralizing efficacy and also the route (usually intravenous) by which it is administered.

In the past, treatment of envenomed victims with antivenom has been carried out without any real scientific criteria as to the optimal dose. The amount of antivenom given to a patient currently depends on the individual clinician's experience in assessing the severity of envenoming on the basis of clinical signs (e.g. local and systemic haemorrhage, incoagulable blood, neurotoxicity, myotoxicity, local swelling and necrosis and a range of other effects) (Cardoso *et al.* 1993).

This review considers the application of more objective methods of assessment involving the use of enzyme immunoassay for estimating the course of envenoming by medically important snakes in human victims, and assessing the effect of antivenom therapy, first aid and traditional methods in eliminating venom antigenaemia.

Some areas of the world where snakebite is of medical importance

In Nigeria, bites by the carpet viper (*Echis ocellatus*) constitute a major problem in farming communities (Warrell & Arnett 1976; Pugh & Theakston 1980). Envenoming by this species results in local and systemic haemorrhage, incoagulable blood and local necrosis. On the basis of these signs, two ampoules (20 ml) of a monospecific *Echis* antivenom (South African Institute for Medical Research) or four ampoules (40 ml) of either a French (Institut Pasteur) or German (Behringwerke) polyspecific antivenom have been recommended as a starting dose in the past (Warrell, Davidson *et al.* 1974; Warrell, Warrell *et al.* 1980).

In Sri Lanka, the incidence of snakebite is currently one of the highest in the world with a mortality of 6/100 000 per year and an incidence exceeding 400/100 000 per year (Phillips *et al.* 1988). The most important species is Russell's viper, *Daboia russelli pulchella*, and systemic envenoming, involving haemorrhage, coagulopathy, neurotoxicity and myotoxicity, is normally treated with a starting dose of 5–15 ampoules (50–150 ml) of imported Indian (Haffkine or Serum Institute of India) polyspecific antivenom (Theakston & Warrell 1991).

In Brazil, patients bitten by the jararaca (*Bothrops jararaca*) are treated with a starting dose of four (40 ml), eight (80 ml) or more ampoules of *Bothrops* polyspecific antivenom (Theakston & Warrell 1991; Jorge *et al.* 1995). The dose depends on the severity of envenoming as assessed by the individual clinician treating the patient (Cardoso *et al.* 1993).

Methodology used for investigating the problem

The development of enzyme immunoassay (EIA) for detection of specific venom antigen (Theakston, Lloyd-Jones & Reid 1977; Ho, Warrell, Warrell *et al.* 1986) and for detection of therapeutic antivenom (Theakston, Fan *et al.* 1992) has enabled a more effective assessment of antivenom dosage and efficacy to be made. It is now possible, using this method, to detect and quantify specific venom in the blood or body fluids at any time after the bite, as well as to calculate the amount of therapeutic antivenom circulating at any time after antivenom administration. Figure 1 shows how the kinetics of envenoming and therapy can be assessed both when an effective and when a relatively ineffective antivenom is used.

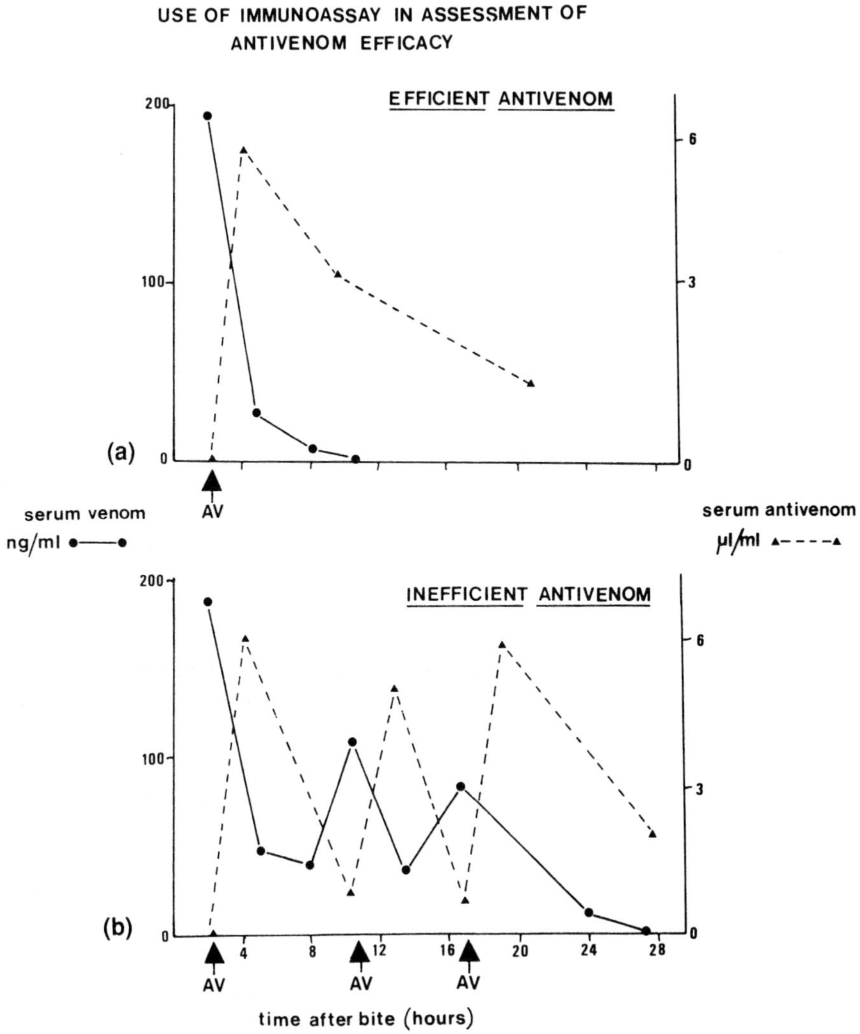

Fig. 1. Typical curves demonstrating serum venom and antivenom levels in envenomed patients treated with an effective (a) and a relatively ineffective (b) antivenom.

Summary of results obtained

In northern Nigeria the best antivenom used in the past has proved to be the monospecific South African Institute for Medical Research (SAIMR) antivenom raised against the venom of Kenyan *Echis* (Theakston & Warrell 1991). This, in most cases, cleared venom from the circulation within 6 h of intravenous administration of antivenom, with simultaneous permanent resolution of the clinical signs of envenoming (Pugh & Theakston 1987a; Rugman, Cottam & Theakston 1990) (Fig. 1a). The polyspecific German and French antivenoms tested were less effective and, although their use resulted in an initial decrease in venom antigenaemia and temporary resolution of the clinical signs after admission, this was often followed by an increase in levels of circulating venom and recurrence of clinical symptoms such as incoagulable blood, due presumably to further influx of venom from a depot area at the bite site (Warrell, Davidson *et al.* 1974; Warrell, Warrell *et al.* 1980). Owing to an unsustained and inadequate level of antivenom in the circulation, this additional venom was not neutralized and a further dose (or doses) of antivenom was required (Fig. 1b). In a case of envenoming by a Tunisian snake of the *Echis pyramidum* complex, a total dose of over 300 ml of three different antivenoms (one monospecific and two polyspecific) with activity against *Echis* species failed to neutralize circulating venom in the patient, demonstrating the lack of specificity of these antivenoms against the venom of this individual snake (Gillissen *et al.* 1994).

A similar situation occurred in Sri Lanka where imported Indian (Haffkine or Serum Institute of India) polyspecific antivenom is used. These antivenoms are not very effective against the venom of Sri Lankan *D. russelli pulchella* because they are prepared against the venom of Indian *D. russelli russelli* (Phillips *et al.* 1988; Theakston & Warrell 1991). Sri Lankan Russell's viper contains some different components to the Indian venom such as a presynaptically-acting phospholipase A_2 (PLA_2) which is also myotoxic (Phillips *et al.* 1988). These PLA_2 activities are not neutralized by the imported antivenom (Phillips *et al.* 1988).

In Brazil, the three major polyspecific *Bothrops* antivenoms have been compared clinically in a randomized comparative trial (Theakston, Fan *et al.* 1992; Cardoso *et al.* 1993). All the antivenoms produced a high degree of protection and a rapid rate of venom clearance after a single dose of the lowest amount of antivenom (four ampoules; 40 ml) recommended for the treatment of moderately envenomed patients (Cardoso *et al.* 1993). A more recent study demonstrated that even two ampoules (20 ml) was adequate in similarly envenomed patients (Jorge *et al.* 1995). The levels of circulating antivenom were maintained at a high level even after the venom antigenaemia had been abolished, and only finally cleared from the circulation after 37 days (Theakston, Fan *et al.* 1992); this permitted immediate neutralization of any additional venom entering the circulation from the depot at the bite site. The results of this particular study indicated both clinically and by immunoassay that the patients had received more antivenom than was actually necessary for neutralizing the circulating venom, especially in the higher dose ranges (four, eight and more ampoules) in cases of moderate envenoming.

All these antivenoms are $F(ab')_2$ fragment preparations (Theakston & Warrell 1991; Theakston, Fan *et al.* 1992; Laing, Theakston *et al.* 1992). More recently we have attempted to assess a novel ovine Fab antivenom, purified by papain digestion, raised against Nigerian *E. ocellatus* venom (Smith *et al.* 1992; Laing, Lee *et al.* 1995). Preliminary studies have yielded promising results. Such an antivenom, being a smaller fragment of the IgG molecule than $F(ab')_2$, has a larger volume of distribution and should theoretically also be less immunoreactive. Although it may have the disadvantage of being cleared more rapidly via the kidney, it may also have the advantage of being more potent than other available products and may also be more effective against local venom effects. Further studies, designed to fully assess this type of antivenom, are currently in progress.

Efficacy of antivenom in treating local effects of venom

The rationale for giving exceptionally high doses of antivenom in Brazil is that some clinicians consider that there may be some beneficial effect in reducing the extent of the local necrosis caused by the cytolytic enzymes present in *Bothrops* (and other) venoms. Both experimental (Iddon, Theakston & Ownby 1987) and clinical (R. D. G. Theakston & D. A. Warrell unpubl. obs.) studies, however, have shown that although a small amount of $F(ab')_2$ antivenom does gain eventual access to the area of local necrosis, this is well after the initial rapid and irreversible changes have occurred. In human victims, who usually arrive at hospital hours or even days after the bite, there appears to be no real effect on local lesions (Ribeiro & Jorge 1988).

First aid

Investigations in Burmese and Thai patients have attempted to assess the effectiveness of tourniquets as a first-aid measure by estimating venom levels before and after tourniquet release (Ho, Warrell, Looareesuwan *et al.* 1986), venom levels distal and proximal to the tourniquet (Tun-Pe *et al.* 1987) and admission venom levels in patients admitted with and without tourniquets (Khin Ohm Lwin *et al.* 1984). In these cases tourniquets did not show any enhanced inhibition of spread of venom into the general circulation, but it should be borne in mind that, in the field, tourniquets are often not applied or managed correctly. In the Philippines, tourniquets properly applied have been shown to delay the spread of the major neurotoxin present in the venom of the Philippine cobra, *Naja philippinensis*, into the circulation until release in hospital (Watt *et al.* 1988). Likewise in Nigeria one patient, admitted to hospital with *E. ocellatus* bite, had no clinical signs and no detectable venom on admission but developed significant clinical signs (incoagulable blood) with associated venom antigenaemia immediately after release of the tourniquet (Pugh & Theakston 1987a). Another died, not because of envenoming, but owing to pulmonary thromboembolus, preceded by thrombophlebitis, local necrosis and gas gangrene caused by the late removal of a tight tourniquet which had been in place for 48 h after the bite (Pugh & Theakston 1987b).

The possibility that a smaller Fab IgG fragment, given by the intramuscular route, may be of use in prehospital treatment is currently under investigation. If this system works it could represent a major advance in early therapy. Although preliminary results have indicated that absorption even of an Fab fragment into the circulation by this route is still too slow to be useful in early neutralization of venom, further, more detailed studies are necessary.

Other proposed future studies will involve the objective assessment, using EIA, of so-called traditional and other untested remedies, such as the 'black snake stone', high voltage/low current electric shock (Guderian, Mackenzie & Williams 1986) and other somewhat suspect and unproven treatments.

Conclusions

The first important stage in the evaluation of either a new or an existing antivenom is its testing in an animal (usually rodent) model using assessment of protection against the lethal and other venom effects such as haemorrhagic, local necrotizing, defibrinogenating and coagulant activities. WHO-recommended tests are available for this purpose (WHO 1981; Theakston & Reid 1983; Warrell, Looareesuwan *et al.* 1986; Theakston, Fan *et al.* 1992; Laing, Theakston *et al.* 1992). However, it should be stressed that it is dangerous to extrapolate from animal results to the situation in man for a wide range of reasons — for instance, unrealistic routes of administration in animals of a preincubated mixture of venom and antivenom (Theakston 1991; Laing, Theakston *et al.* 1992; Laing, Lee *et al.* 1995). The second stage, which is therefore the only truly meaningful method, is testing the experimentally approved antivenom in envenomed humans, using both clinical observations and objective measurements of both venom and antivenom levels.

An effective antivenom is one in which a high level of active circulating antibody is maintained in the circulation after initial venom clearance (e.g. Brazilian *Bothrops* antivenoms, SAIMR *Echis* antivenom in Nigeria) in human victims. This will be capable of neutralizing any venom subsequently entering the circulation from a depot. Such conditions are usually obtained when the venom or venoms used for producing the antivenom is obtained from snakes present in the country in which it is proposed to use the antivenom. Venoms, even within the same species, vary dramatically in composition from region to region (Táborská 1971; Williams *et al.* 1988).

The study carried out in Brazil strongly suggests that patients were receiving more antivenom than was necessary, and consequently experiencing a high incidence (37–87%) of early anaphylactic reactions requiring urgent treatment with adrenalin and antihistamines (Cardoso *et al.* 1993). A lower but effective dose should decrease the extent of this problem and also result in a reduction in cost: the latter is of major importance in developing countries. There is no real evidence that a high antivenom dose is effective in decreasing or eliminating local venom effects such as necrosis (Iddon *et al.* 1987; Jorge *et al.* 1995). Both clinical observations and EIA results support these observations.

There is also no convincing general evidence that tourniquets are effective. As there are so many obvious variables, the results of studies using EIA are, not surprisingly, inconclusive (Ho, Warrell, Looareesuwan *et al.* 1986; Tun-Pe *et al.* 1987; Khin Ohm Lwin *et al.* 1984; Watt *et al.* 1988; Pugh & Theakston 1987a,b). Likewise, further more detailed studies need to be performed on the possibilities of using lower molecular weight Fab fragments or even smaller antibody components, for early treatment following administration via the intramuscular route.

By using a combination of experimental, clinical and immunological assay systems, it is therefore possible to obtain a highly accurate assessment of the efficacy of antivenom and of current first-aid procedures.

References

Cardoso, J. L. C., Fan, H. W., França, F. O. S., Jorge, M. T., Leite, R. P., Nishioka, S. A., Avila, A., Sano-Martins, I. S., Tomy, S. C., Santoro, M. L., Chudzinski, A. M., Castro, S. C. B., Kamiguti, A. S., Kelen, E. M. A., Hirata, M. H., Mirandola, R. M. S., Theakston, R. D. G. & Warrell, D. A. (1993). Randomized comparative trial of three antivenoms in the treatment of envenoming by lance-headed vipers (*Bothrops jararaca*) in São Paulo, Brazil. *Q. Jl Med.* **86**: 315–325.

Gillissen, A., Theakston, R. D. G., Barth, J., May, B., Krieg, M. & Warrell, D. A. (1994). Neurotoxicity, haemostatic disturbances and haemolytic anaemia after a bite by a Tunisian saw-scaled or carpet viper (*Echis 'pyramidum'* complex): failure of antivenom treatment. *Toxicon* **32**: 937–944.

Guderian, R., Mackenzie, L. D. & Williams, J. F. (1986). High voltage shock treatment for snake bite. *Lancet* **331** (ii): 229.

Ho, M., Warrell, D. A., Looareesuwan, S., Phillips, R. E., Chanthavanich, P., Karbwang, J., Supanaranond, W., Hutton, R. A. & Vejcho, S. (1986). Clinical significance of venom antigen levels in patients envenomed by the Malayan pit viper (*Calloselasma rhodostoma*). *Am. J. trop. Med. Hyg.* **35**: 579–587.

Ho, M., Warrell, M. J., Warrell, D. A., Bidwell, D. & Voller, A. (1986). A critical reappraisal of the use of enzyme-linked immunosorbent assays in the study of snake bite. *Toxicon* **24**: 211–221.

Iddon, D., Theakston, R. D. G. & Ownby, C. (1987). A study of the pathogenesis of local skin necrosis induced by *Naja nigricollis* (spitting cobra) venom using simple histological staining techniques. *Toxicon* **25**: 665–672.

Jorge, M. T., Cardoso, J. L. C., Castro, S. C. B., Ribeiro, L., França, F. O. S., Sbroglio de Almeida, M. E., Kamiguti, A. S., Sano-Martins, I. S., Santoro, M. L., Moncau, J. E. C., Warrell, D. A. & Theakston R. D. G. (1995). A randomised blinded comparison of two doses of antivenom in the treatment of *Bothrops* envenoming in São Paulo, Brazil. *Trans. R. Soc. trop. Med. Hyg.* **88**: 111–114.

Khin Ohm Lwin, Aye Aye Myint, Tun-Pe, Thaingie Nwe & Min Naing (1984). Russell's viper venom levels in serum of snake bite victims in Burma. *Trans. R. Soc. trop. Med. Hyg.* **78**: 165–168.

Laing, G. D., Lee, L., Smith, D. C., Landon, J. & Theakston, R. D. G. (1995). Experimental assessment of a new low-cost antivenom for treatment of carpet viper (*Echis ocellatus*) envenoming. *Toxicon* **32**: 307–313.

Laing, G. D., Theakston, R. D. G., Leite, R. P., Dias da Silva, W., Warrell, D. A. & BIASG (1992). Comparison of the potency of three Brazilian *Bothrops* antivenoms using *in vivo* rodent and *in vitro* assays. *Toxicon* **30**: 1219–1225.

Phillips, R. E., Theakston, R. D. G., Warrell, D. A., Galigedara, Y., Abeysekera, D. T. D. J., Dissanayaka, P., Hutton, R. A. & Aloysius, D. J. (1988). Paralysis, rhabdomyolysis and haemolysis caused by bites of Russell's viper (*Vipera russelli pulchella*) in Sri Lanka: failure of Indian (Haffkine) antivenom. *Q. Jl Med.* **68**: 691–716.

Pugh, R. N. H. & Theakston, R. D. G. (1980). The incidence and mortality of snake bite in savanna Nigeria. *Lancet* **319** (ii): 1181–1183.

Pugh, R. N. H. & Theakston, R. D. G. (1987a). A clinical study of viper bite poisoning. *Ann. trop. Med. Parasit.* **81**: 135–149.

Pugh, R. N. H. & Theakston, R. D. G. (1987b). Fatality following use of a tourniquet after viper bite envenoming. *Ann. trop. Med. Parasit.* **81**: 77–78.

Ribeiro, L. A. & Jorge, M. T. (1988). Evaluación de efecto de la dosis del suero antibotropico. In *Resumenes del congreso do aniversario del Instituto de Medicina Tropical 'Pedro Kouri'*: 190. Instituto de Medicina Tropical 'Pedro Kouri', Havana.

Rugman, F. P., Cottam, S. N. & Theakston, R. D. G. (1990). A bite by a carpet viper (*Echis carinatus*) in a research institute. *Clin. Lab. Haematol.* **12**: 221–228.

Smith, D. C., Reddi, K. R., Laing, G., Theakston, R. D. G. & Landon, J. (1992). An affinity purified ovine antivenom for the treatment of *Vipera berus* envenoming. *Toxicon* **30**: 865–871.

Táborská, E. (1971). Intraspecies variability of the venom of *Echis carinatus*. *Physiologia bohemoslov.* **20**: 307–318.

Theakston, R. D. G. (1991). Immunological aspects of snake venom research. In *Reptile venoms and toxins*: 495–527. (Ed. Tu, A. T.). Marcel Dekker Inc., New York & Basel. (*Handb. nat. Toxins* 5.)

Theakston, R. D. G., Fan, H. W., Warrell, D. A., Dias da Silva, W., Higashi, H. G. & BIASG (1992). Use of immunoassay in assessing the efficacy of three Brazilian *Bothrops* antivenoms. *Am. J. trop. Med. Hyg.* **47**: 593–604.

Theakston, R. D. G., Lloyd-Jones, M. J. & Reid, H. A. (1977). Micro-ELISA for detecting and assaying snake venom and venom antibody. *Lancet* **313** (ii): 639–641.

Theakston, R. D. G. & Reid, H. A. (1983). Development of simple standard assay procedures for the characterization of snake venoms. *Bull. Wld Hlth Org.* **61**: 949–956.

Theakston, R. D. G. & Warrell, D. A. (1991). Antivenoms: a list of hyperimmune sera currently available for the treatment of envenoming by bites and stings. *Toxicon* **29**: 1419–1470.

Tun-Pe, Tin-Nu-Swe, Myint-Lwin, Warrell, D. A. & Than-Win (1987). The efficacy of tourniquets as a first-aid measure for Russell's viper bites in Burma. *Trans. R. Soc. trop. Med. Hyg.* **81**: 403–405.

Warrell, D. A. & Arnett, C. (1976). The importance of bites by the saw-scaled or carpet viper (*Echis carinatus*). Epidemiological studies in Nigeria and a review of the world literature. *Acta trop.* **33**: 307–341.

Warrell, D. A., Davidson, N. McD., Ormerod, L. D., Pope, H. M., Watkins, B. J., Greenwood, B. M. & Reid, H. A. (1974). Bites by the saw-scaled or carpet viper (*Echis carinatus*): trial of two specific antivenoms. *Br. med. J.* **251**: 437–440.

Warrell, D. A., Looareesuwan, S., Theakston, R. D. G., Phillips, R. E., Chanthavanich, P., Virivan, C., Supanaranond, W., Karbwang, J., Ho, M., Hutton, R. A. & Vejcho, S. (1986). Randomized comparative trial of three monospecific antivenoms for bites by the Malayan pit viper (*Calloselasma rhodostoma*) in southern Thailand: clinical and laboratory correlations. *Am. J. trop. Med. Hyg.* **35**: 1235–1247.

Warrell, D. A., Warrell, M. J., Edgar, W., Prentice, C. R. M., Mathison, J. & Mathison, J. (1980). Comparison of Pasteur and Behringwerke antivenoms in envenoming by the carpet viper (*Echis carinatus*). *Br. med. J.* **280**: 607–609.

Watt, G., Padre, L., Tuazon, M. L., Theakston, R. D. G. & Laughlin, L. W. (1988). Tourniquet application after cobra bite: delay in the systemic absorption of neurotoxins and dangers of sudden release. *Am. J. trop. Med. Hyg.* **38**: 618–622.

WHO (1981). Progress in the characterization of venom and the standardization of antivenoms. *Wld Hlth Org. Offset Publ.* No. 58: 1–44.

Williams, V., White, J., Schwaner, T. D. & Sparrow, A. (1988). Variation in venom proteins from isolated populations of tiger snakes (*Notechis ater niger*, *N. scutatus*) in South Australia. *Toxicon* **26**: 1067–1075.

Species index

Note: page numbers in *italics* refer to figures and tables.

Subject index

Note: page numbers in *italics* refer to figures and tables.